Artificial Intelligence Techniques in Smart Agriculture

Siddharth Singh Chouhan • Akash Saxena •
Uday Pratap Singh • Sanjeev Jain
Editors

Artificial Intelligence Techniques in Smart Agriculture

Springer

Editors
Siddharth Singh Chouhan
School of Computing Science
and Engineering
VIT Bhopal University
Sehore, Madhya Pradesh, India

Uday Pratap Singh
Department of Mathematics
Central University of Jammu
Jammu, Jammu and Kashmir, India

Akash Saxena
Department of Electrical Engineering
Central University of Haryana
Mahendragarh, Haryana, India

Sanjeev Jain
Central University of Jammu
Jammu, Jammu and Kashmir, India

ISBN 978-981-97-5877-7 ISBN 978-981-97-5878-4 (eBook)
https://doi.org/10.1007/978-981-97-5878-4

This Springer imprint is published by the registered company Springer Nature Singapore Pte Ltd.
The registered company address is: 152 Beach Road, #21-01/04 Gateway East, Singapore 189721, Singapore

To our family, friends, and colleagues.

Foreword

The applications of Artificial Intelligence (AI) techniques are spreading its wings almost in every walk of human life in the present times. AI is turning out to be a huge helping hand for improved productivity in manufacturing sector of world economy.

It has given a boost to the expertise of the workers by acquainting them with the unknown skills and also facilitating the work automation. A pinch of creativity and innovation in artificial intelligence is a boon for the humanity and the environment.

The existence of living beings in this macrocosm majorly hinges on the ever burgeoning productivity of the agricultural yields due to incessant humongous population rise and ecological balance.

We, as global citizens, are united in the conviction that the environmental challenges will be countered through automation promotion in the agricultural sector.

This would inevitably upswing the quantity and quality of farming in the near future.

Keeping these concerns in mind, the editors of this book have put their focus on this interesting topic which will be helpful for the society. The presented work will be interestingly insightful to various agricultural experts, scientists, researchers, and academicians for their better understanding and adopting the intelligence techniques. I also have this conviction that this book will be a milestone for myriad farmers in standardizing their life style as well.

I also congratulate the authors to choose such an interesting topic to dwell on and wish them all the very best for their future.

Central University of Haryana
Mahendergarh, Haryana, India

Tankeshwar Kumar

Preface

With the rapid development in the intelligent devices and high-end computing deployment of Artificial Intelligence techniques can be experienced nearly in all applications in day-to-day life. This methodology has paved off a way to utilize resources to an extent and enhance the outcome or productivity of a task. The human computer interaction is convenient in solving number of complex problems in less time and payload.

In the recent times, the concept of artificial intelligence has been widely deployed in the field of agriculture. This increases the chance of getting fruitful outcome in terms of crops yields by attain higher quality and quantity. Automation in agriculture has limited the traditional practices of farming. The farmers are also being convenient in using them. This methodology requires appropriate knowledge, understanding, and of course training. It is required to familiarize with the questions like why, when, and how to use the artificial intelligence. Therefore, in this book we have tried to incorporate basic terminologies of AI for better understanding and also some of the applications to get the better insight if it.

To start with in Chap. 1 some of the basic definitions, types, components, advantages, disadvantages, and applications of AI have been discussed. Authors in Chap. 2 present the various challenges that hurdles in achieving artificial intelligence in agriculture and also offer some solutions to overcome them. Authors in Chap. 3 discuss about the fundamental concepts of AI, its major components, and applications within the agricultural sector. Furthermore, future aspects, opportunities, and challenges in adoption of AI in agriculture have been discussed in detail.

AI has the ability to significantly change agriculture and create a future that is more resilient, productive, and sustainable if these challenges are successfully addressed. Certain challenges and limitations that restrict the farmers and various organization bodies to adapt the AI techniques implementation in agricultural field are briefly discussed in Chap. 4 along with some generic solutions that may give hands to the individuals reading the context. Chapter 5 presents a comprehensive view of how the integration of advanced technologies in smart farming management systems enhances agricultural sustainability, productivity, and food security in the face of evolving global challenges. Chapter 6 delves into the technical underpinnings of these innovations, exploring diverse devices, their functionalities, and the platforms that orchestrate their seamless integration. We embark on a journey, dissecting the hardware and software components of intelligent devices, including sensors, actuators, processors, and embedded operating systems. We then delve into the realm of AI and ML frameworks, elucidating their role in enabling intelligent behavior such as decision-making, pattern recognition, and adaptive control.

Chapter 7 explores how smart technology, like artificial intelligence (AI), can help farmers count and understand fruits better. It discusses why counting fruits accurately is crucial for farmers and how AI can make this process much easier. It also explains different ways AI, such as computer vision and machine learning, can be used for counting fruits. In Chap. 8, authors used an optimized convolutional neural network (CNN), for identifying and reliably detecting plant stress phenotyping. By using the capabilities of these sophisticated neural networks, we were able to successfully raise the stress detection process beyond the boundaries of traditional approaches. Plant stress phenotypic data from important crop species, including wheat, rice, cotton, and maize, are utilized to systematically assess the model's efficacy.

In Chap. 9, authors explore the profound impact of AI-driven computer vision, highlighting its crucial role in changing how we manage crops, detect diseases, and optimize resources in agriculture. This integration goes beyond just improving precision farming; it empowers effective crop monitoring and accurate yield prediction. The study emphasizes how this combination has enhanced operational efficiency, sustainability, and decision-making in farming. Chapter 10 highlights the comprehensive irrigation system that automates the process of irrigation will reduce human and resource consumption, as well as water consumption. AI can help us come up with smart irrigation strategies that apply water at the right time and with the right amount. Also, this chapter aims to outline the drawbacks of chemical pesticides, traditional pest management methods, and the evolving role of artificial intelligence in pest management.

Data integration platforms are crucial for combining data from diverse sources, while robust data management strategies address challenges like scalability, interoperability, and security. Looking ahead, emerging technologies like blockchain and internet-of-things (IoT) hold immense promise for further revolutionizing smart agriculture. Continuous innovation, coupled with a focus on sustainability and ethical considerations, will ensure that AI paves the way for a secure, efficient, and resource-conscious agricultural future are being discussed by authors in Chap. 11. Chapter 12 highlights promising AI approaches that can help farmers overcome water supply and pest control challenges. This chapter discusses the deployment of AI-powered pest detection, identification, classification, prediction, decision-making, and pesticide spraying in real time.

Artificial intelligence (AI) integration has ushered in a transformative era in agriculture known as Smart Agriculture. Chapter 13 comprehensively explores the AI concept within the context of smart agriculture. It conducts a systematic literature review, revealing a growing trend in AI-related publications and highlighting its applications and advantages. The evolution of agriculture from traditional methods to the current Agriculture 4.0 era is outlined, showcasing the pivotal role of advanced technologies such as Wireless Sensor Networks (WSN), IoT, and AI. Smart, sustainable farming is a new way of doing things that uses cutting-edge technologies like the IoT, cloud computing, and big data analytics to improve farming practices and care for the environment are discussed in Chap. 14.

Chapter 15 delves into the advantages of integrating drone technology into India's agricultural landscape. Over the past decade, advancements in information technology have propelled the widespread adoption of IoT devices, exemplified by the utilization of Unmanned Aerial Vehicles (UAVs) in environmental monitoring. Chapter 16 discusses the concept of smart agriculture, major sensors, and technologies used in IoT-based agricultural practices. Furthermore, data communication, storage, and application of IoT and AI in agriculture have been discussed in detail.

Artificial intelligence has helped bring about a transformation in farming. The use of this technology has protected agricultural productivity from a number of threats, such as rising populations, changing climates, employment difficulties, and problems with food safety. Architectures based on machine learning have been demonstrated to be among the best in regard to precision and resilience. By using systems based on AI, crop yields can be increased while using less water, fertilizer, and pesticides. In addition to it systems based on AI are used to improving the productivity of crops, reducing the cost and safety of workers etc. also intelligent technology can help reduce its effect on natural ecosystems, which will help keep food costs low and ensure more crops are grown to feed the world's expanding population.

Finally, we think that this book will be a great resource for academics, scientists, post-graduate students, field experts, agricultural specialists, and graduate students as they focus their work in this area and create models that benefit society and the environment. Plant pathologists, biotechnologists, experts in seed production, breeders, market managers, and other public and private sector stakeholders involved in underlying technology or market development can also find great value in this book.

Sehore, Madhya Pradesh, India — Siddharth Singh Chouhan
Mahendragarh, Haryana, India — Akash Saxena
Jammu, Jammu and Kashmir, India — Uday Pratap Singh
Jammu, Jammu and Kashmir, India — Sanjeev Jain

Acknowledgment

We would want to take this opportunity to thank everyone who helped us along the way and encouraged us to keep going. For us, the process of editing a book has been genuinely transformative. We are thankful to *God* for providing us with the chance to work with individuals who exhibit remarkable expertise and energy in their specialized domains.

We express our gratefulness to the School of Computing Science and Engineering, VIT Bhopal University, Department of Electrical Engineering Central University of Haryana, Mahendragarh, Department of Mathematics and Department of Computer Science and Engineering, Central University of Jammu for the contented environment, resources, inspiration, and positive mindset to achieve our goal.

We reverently recognize our parents' boundless benefits and inspiration. Mother and Father, who have been a continual source of inspiration and optimism throughout this journey. We owe a debt of gratitude to all of our friends and coworkers who helped us so much throughout this course and who continually encouraged and uplifted us.

We express our sincere gratitude to *Springer Nature Singapore Pvt. Ltd.,* the publisher, for providing us with the chance to author this book.

Finally, we wrap up by expressing our gratitude to everyone we forgot to include but who helped out through conversations or by being a silent supporter.

Contents

Editors and Contributors

About the Editors

Siddharth Singh Chouhan currently working as an Assistant Professor in School of Computing Science and Engineering, VIT Bhopal University, Kothri-Kalan, Sehore, 466114, Madhya Pradesh, India. He has completed Ph.D. in Computer Science and Engineering from Shri Mata Vaishno Devi University, Katra, 182320, UT of Jammu and Kashmir, India. He is a Post Doctorate from the University of Malta, Europe. He has received a B.Tech. degree in 2010 and an M.Tech. degree in 2013 in Computer Science and Engineering from RGPV University, Bhopal, India. His area of interest includes developing Soft Computing, Computer Vision, Drone Technology, Deep Learning, and Precision Agriculture. He had authored several research papers published in many reputed journals including Elsevier, Springer, etc.

School of Computing Science and Engineering, VIT Bhopal University, Kothri Kalan, Sehore, Madhya Pradesh, India

Akash Saxena received his Ph.D. and Master of Technology Degree (with Honors) in Electrical Engineering from Malaviya National Institute of Technology, Jaipur, in 2015 and 2008, respectively, and a Bachelor of Engineering Degree (with Honors) in Electrical Engineering from Engineering College Kota in 2001. Dr. Akash Saxena is working as Professor in the Department of Electrical Engineering and Director of Training and Placement and Professor in the Department of Electrical Engineering, Central University of Haryana, India, with almost two decades of teaching. As an academician, he has been associated with numerous institutions as Member, Doctoral

Research Committee and Member, Board of Studies. He has presented his research work at various platforms to showcase his findings to the fellow researchers and providing a pathway to the future researchers. His intellectual work has been published in leading journals in the form of short communication/letters / articles/research papers.

Department of Electrical Engineering, Central University of Haryana, Mahendragarh, Haryana, India

Uday Pratap Singh presently working as an Associate Professor in the Department of Mathematics, Central University of Jammu, 181143, Jammu, India and also an Associate Professor (on EoL) in the School of Mathematics, Shri Mata Vaishno Devi University, Katra. He has obtained B.Sc. and M.Sc. (Mathematics and Statistics) Degree from Dr. R.M.L. Awadh University, Faizabad, (U.P.), India, M.Sc. (Mathematics and Computing) Degree from the Indian Institute of Technology, Guwahati, India, and received Ph.D. in Computer Science from Barkatullah University, Bhopal. His area of interest includes Soft Computing, Nonlinear Systems, and Image Processing. He had authored several research papers in the many reputed journals and conferences. He is a life member of Soft Computing Research Society (SCRC), Bharata Ganita Parishad, Computer Society of India, and Member of IEEE and AMS.

Department of Mathematics, Central University of Jammu, Bagla Suchani, Jammu, Jammu and Kashmir, India

Sanjeev Jain currently working as a Vice Chancellor of the Central University of Jammu. He received a Master's degree in Computer Science and Engineering from IIT Delhi, New Delhi, India, in 1992, and a Ph.D. degree in Computer Science and Engineering. He has over 35 years of experience in teaching, research, and administration. He has the credit of making a significant contribution to research and development in the area of image processing, soft computing, IoT, and mobile ad hoc network. He had authored several research papers in the many reputed journals and conferences. He has undertaken a number of major Research and Development projects sponsored by the Government and Private

Agencies. He is a member of various societies like the Computer Society of India, IEEE.

Central University of Jammu, Bagla Suchani, Jammu, Jammu and Kashmir, India

Contributors

Akanksha Adhikari Department of Zoology, Radhey Hari Government Post Graduate College, Kashipur, Uttarakhand, India

Habib Ali Department of Agricultural Engineering, Khwaja Fareed University of Engineering and Information Technology, Rahim Yar Khan, Punjab, Pakistan

Rohit Anand ICAR-Indian Agricultural Research Institute, New Delhi, India

Dangi Pooja Arun Department of Agricultural Economics and Extension, Lovely Professional University, Phagwara, Punjab, India

K. Aswitha Department of Soil Science and Agricultural Chemistry, JKK Munirajah College of Agricultural Science, Erode, Tamil Nadu, India

Anjana J. Atapattu Agronomy Division, Coconut Research Institute, Lunuwila, Sri Lanka

Dattatray G. Bhalekar Center for Precision and Automated Agricultural Systems, Department of Biological Systems Engineering, Washington State University, Prosser, WA, USA

Tanmay Kumar Bhoi ICFRE-Arid Forest Research Institute (ICFRE-AFRI), Jodhpur, Rajasthan, India

Subir Kumar Chakraborty ICAR-Central Institute of Agricultural Engineering, Bhopal, India

Narendra Singh Chandel ICAR-Central Institute of Agricultural Engineering, Bhopal, India

Shubham Chaudhary ICAR-Indian Agricultural Research Institute, New Delhi, India

Pooja Chouhan ICAR-Central Institute of Agricultural Engineering Bhopal, Bhopal, India

Siddharth Singh Chouhan School of Computing Science and Engineering, VIT Bhopal University, Sehore, Madhya Pradesh, India

Manojit Chowdhury ICAR-Indian Agricultural Research Institute, New Delhi, India

Sanhita Chowdhury Department of Entomology, College of Agriculture, Punjab Agricultural University, Ludhiana, Punjab, India

Department of Entomology, Post Graduate College of Agriculture, Dr. Rajendra Prasad Central Agricultural University, Pusa, Samastipur, Bihar, India

Nuwandhya S. Dissanayaka Agronomy Division, Coconut Research Institute, Lunuwila, Sri Lanka

Vinod Kumar Dubey School of Agriculture and Veterinary Science, Shridhar University, Pilani, Rajasthan, India

Maheshwar Durgam Krishi Vigyan Kendra Ankushpur Ghazipur, Acharya Narendra Deva University of Agriculture and Technology, Kumarganj, Ayodhya, Uttar Pradesh, India

Rohit Bhojyareddy Gaddamwar ICAR-Indian Agricultural Research Institute, New Delhi, India

Somashekhar Gaddanakeri Pesticide Formulation Testing Laboratory, University of Agricultural Sciences, Raichur, Karnataka, India

Sanjeev Jain Central University of Jammu, Bagla Suchani, Jammu, Jammu and Kashmir, India

Dilip Jat ICAR-Central Institute of Agricultural Engineering, Bhopal, India

J. Jayanthi School of Computing Science and Engineering (SCSE), VIT Bhopal University, Kothrikalan, Sehore, Madhya Pradesh, India

Murugasridevi Kalirajan Department of Agricultural Entomology, Amrita School of Agricultural Sciences, Amrita Vishwa Vidyapeetham, Coimbatore, Tamil Nadu, India

Payel Kar Department of Entomology, Bidhan Chandra Krishi Viswavidyalaya, Mohanpur, Nadia, West Bengal, India

Manjit Kaur Department of Computer Science and Engineering, Akal University, Bathinda, Punjab, India

Upinder Kaur Department of Computer Science and Engineering, Akal University, Bathinda, Punjab, India

Suyog Balasaheb Khose Agricultural and Food Engineering Department, Indian Institute of Technology Kharagpur, Kharagpur, West Bengal, India

Uday Kiran M ICAR-Indian Agricultural Research Institute, New Delhi, India

K. Arun Kumar School of Bioscience, Engineering and Technology, VIT Bhopal University, Kothrikalan, Sehore, Madhya Pradesh, India

Madhusudan B. S ICAR-Indian Agricultural Research Institute, New Delhi, India

V. R. Mageshen Department of Soil Science and Agricultural Chemistry, Amrita School of Agricultural Sciences, Amrita Vishwa Vidyapeetham, Coimbatore, Tamil Nadu, India

Deepak Kumar Mahanta ICFRE-Forest Research Institute (ICFRE-FRI), Dehradun, Uttarakhand, India

Jyotirmay Mahapatra ICAR-Central Institute of Agricultural Engineering, Bhopal, Madhya Pradesh, India

College of Agricultural Engineering and Technology, Odisha University of Agriculture and Technology, Bhubaneswar, Odisha, India

Ayush Mishra Department of Agricultural Extension Education, CCS Haryana Agricultural University, Hisar, Haryana, India

Divyanshi Mishra Department of Agricultural Economics, College of Agriculture, Acharya Narendra Deva University of Agriculture and Technology, Kumarganj, Ayodhya, Uttar Pradesh, India

Harshit Mishra Department of Agricultural Economics, College of Agriculture, Acharya Narendra Deva University of Agriculture and Technology, Kumarganj, Ayodhya, Uttar Pradesh, India

Nikita Mishra Department of Food Technology, Dr. RPCAU Pusa, Samastipur, Bihar, India

Sripriyanka S. Nalla ICAR-Indian Agricultural Research Institute, New Delhi, India

Shaik Nasreen ICAR-Indian Agricultural Research Institute, New Delhi, India

Tharindu D. Nuwarapaksha Agronomy Division, Coconut Research Institute, Lunuwila, Sri Lanka

Abhishek Pandey ICAR-Indian Agricultural Research Institute, New Delhi, India

Lalith K. Perera Coconut Research Institute, Lunuwila, Sri Lanka

Chwadaka Pohshna Agricultural and Food Engineering Department, Indian Institute of Technology Kharagpur, Kharagpur, West Bengal, India

Ramesh K. Sahni ICAR-Central Institute of Agricultural Engineering, Bhopal, Madhya Pradesh, India

Center for Precision and Automated Agricultural Systems, Washington State University, Pullman, WA, USA

Pradeepta Kumar Sahoo College of Agricultural Engineering and Technology, Odisha University of Agriculture and Technology, Bhubaneswar, Odisha, India

M. Salomi Department of Computational Intelligence, Faculty of Engineering and Technology, SRM Institute of Science and Technology, Kattankulathur, Chennai, Tamil Nadu, India

M. SalomiSamsudeen Department of Computational Intelligence, Faculty of Engineering and Technology, SRM Institute of Science and Technology, Kattankulathur, Chennai, Tamil Nadu, India

Ipsita Samal ICAR-National Research Centre on Litchi, Mushahari, Ramna, Muzaffarpur, Bihar, India

S. Samsudeen Department of Computer Science and Engineering, GITAM University, Hyderabad, Telangana, India

Computer Science and Engineering, GITAM School of Technology, Hyderabad, Telangana, India

M. Saranya Department of Agricultural Entomology, Tamil Nadu Agricultural University, Coimbatore, Tamil Nadu, India

Akash Saxena Department of Electrical Engineering, Central University of Haryana, Mahendragarh, Haryana, India

Shashank Shekhar Krishi Vigyan Kendra Ankushpur Ghazipur, Acharya Narendra Deva University of Agriculture and Technology, Kumarganj, Ayodhya, Uttar Pradesh, India

D. J. Shrinivasa Department of Agricultural Engineering, Institute of Agricultural Sciences, Banaras Hindu University, Varanasi, Uttar Pradesh, India

Shreya Singh Ramlalit Singh Mahavidyalaya, Kailhat, Chunar, Uttar Pradesh, India

Uday Pratap Singh Department of Mathematics, Central University of Jammu, Bagla Suchani, Jammu, Jammu and Kashmir, India

Vyshnavi Sunil Department of Entomology, College of Agriculture, Kerala Agricultural University, Vellayani, Thiruvananthapuram, Kerala, India

Shashi S. Udumann Agronomy Division, Coconut Research Institute, Lunuwila, Sri Lanka

Assessing the Importance and Need of Artificial Intelligence for Precision Agriculture

1

Siddharth Singh Chouhan, Uday Pratap Singh, Akash Saxena, and Sanjeev Jain

Abstract

Innovations in Artificial Intelligence (AI) technology have made a significant impact on the living standard of human beings. Novelties in this technology have directed the future of almost every industry towards this domain. This technology has helped to automate a number of process and applications. Adopting intelligence upsurges the quality and quantity of the products. It enhances the workers skill and simulates the problem-solving mechanism in an easier way. In the recent times, the applications of AI can be found in almost all applications including military, agriculture, industries, surveillance, safety, among all. In this chapter, we will be discussing about the basic mechanism of artificial intelligence and its importance in the agricultural sector.

Keywords

Artificial intelligence (AI) · Precision agriculture (PA) · Machine learning (ML) · Deep learning (DL) · Robotics · Data science (DS)

S. S. Chouhan
School of Computing Science and Engineering, VIT Bhopal University, Sehore, Madhya Pradesh, India

U. P. Singh (✉)
Department of Mathematics, Central University of Jammu, Jammu and Kashmir, India

A. Saxena
Department of Electrical Engineering, Central University of Haryana, Mahendragarh, Haryana, India

S. Jain
Central University of Jammu, Jammu and Kashmir, India

S. S. Chouhan et al. (eds.), *Artificial Intelligence Techniques in Smart Agriculture*, https://doi.org/10.1007/978-981-97-5878-4_1

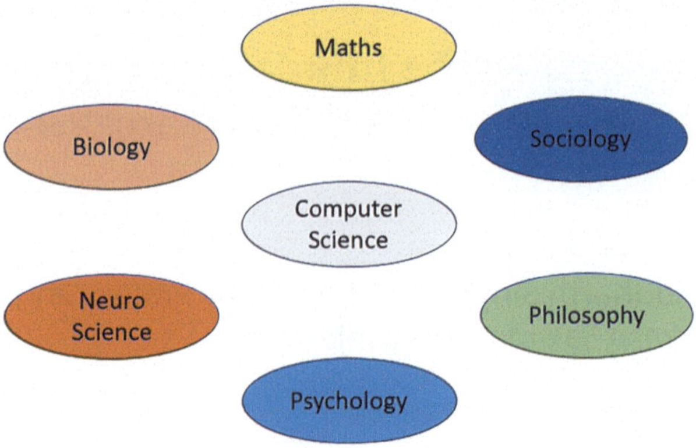

Fig. 1.1 Disciplines essential for achieving AI [https://www.javatpoint.com/artificial-intelligence-ai]

1.1 Introduction

Artificial Intelligence is the mechanism in which the human intelligence process is being simulated with the help of computer systems or machines or simply we can say that it is a science and engineering by which the machines are being trained to be intelligent. In the past decade the applications of AI has been found in healthcare system, defense, gaming, finance, transport, education, e-commerce, networking, robotics, data security, automation, entertainment, agriculture, etc. (Berger et al. 2024; Nihalani et al. 2024; Jain and Chouhan 2014; Chouhan and Sharma 2012).

In this chapter we will discuss the following points:

(a) Disciplines essential to achieve AI
(b) Types of AI
(c) Components of AI
(d) Advantages of AI
(e) Disadvantages of AI
(f) Assessment and need of AI in agriculture

1.1.1 Disciplines Essential to Achieve AI

There are lots of subjects required to achieve intelligence. Simulating the problem-solving skills of a human being is not an easy task to accomplish. Some of the disciplines required to do so are given by Fig. 1.1.

1.1.2 Types of AI

More frequently, based on capability and functionality, Artificial Intelligence theory is classified into two categories:

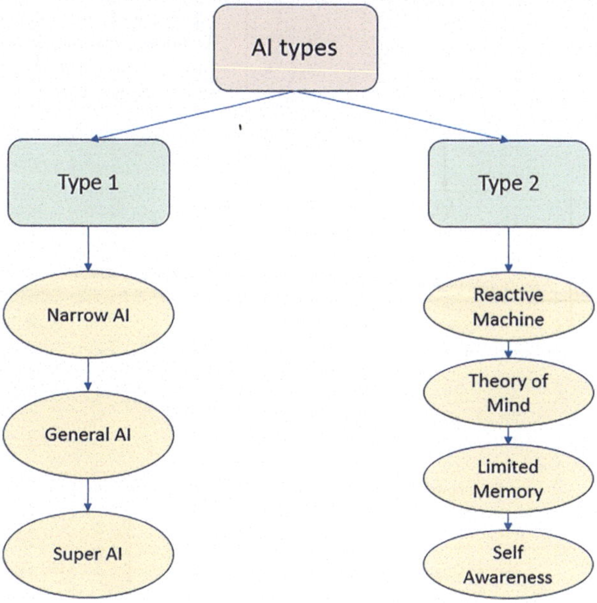

Fig. 1.2 Types of AI [https://www.geeksforgeeks.org/machine-learning-types-of-artificial-intelligence/]

1. Type 1 (based on capability)
2. Type 2 (based on functionality)

Further, Type 1 and Type 2 AI are categorized as given by Fig. 1.2.

1.1.3 Components of AI

Figure 1.3 shows the basic intelligent algorithm's hierarchy.

1.1.4 Advantages of AI

Some of the advantages of AI are as follows:

1. High efficiency
2. Higher accuracy
3. Minimal risk
4. Low chance of error
5. Digital support
6. Availability
7. Unbiased decision making capability
8. Usability in risky situations
9. Repetitive jobs
10. Cost effectiveness

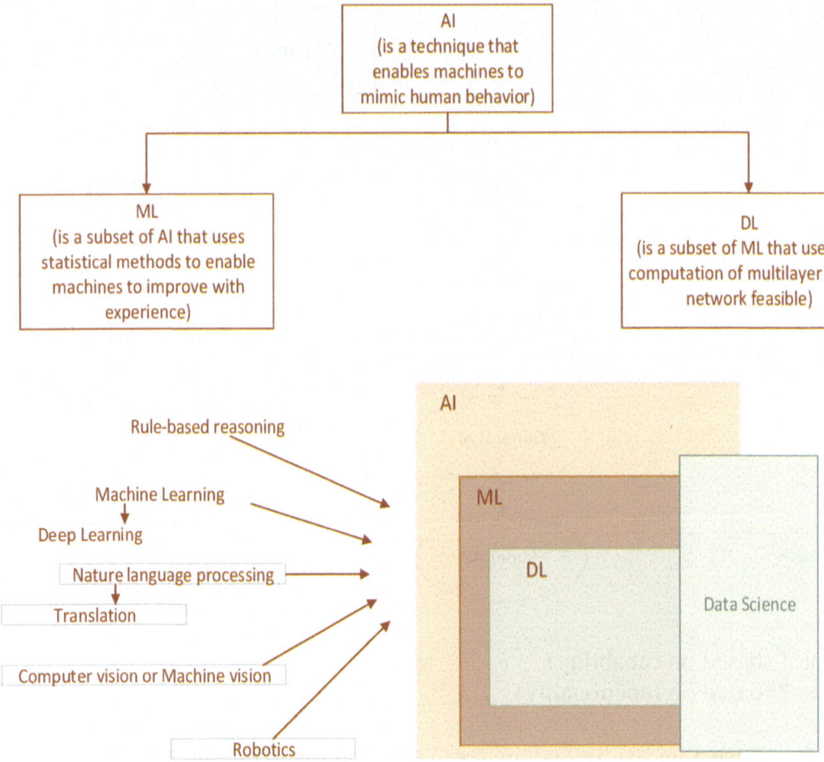

Fig. 1.3 Components of AI

1.1.5 Disadvantages of AI

Some of the disadvantages of AI are as follows:

1. High in cost
2. Less in creativity
3. Reduces employment
4. Ethical dilemmas
5. Data security and privacy concerns
6. Lack of explainability and transparency
7. Dependency
8. Absence of emotions
9. Ambiguity
10. Plagiarism

1.1.6 Assessment and Need of AI in Agriculture

To meet the necessary supply chain, it has become necessary to deploy intelligence in agricultural sector for overcoming the flaws of traditional practices. AI augments the farmer's working capability in an efficient and productive way. The farmers can be benefited in almost all application of agriculture sector (Islam et al. 2024; Al Mashhadany et al. 2024). AI can be used for the following purposes:

1. Monitor the soil moisture
2. Monitor crop growth
3. Protect crops from hazards
4. Spraying
5. Nutrient deficiency estimation
6. Supply chain
7. Surveying
8. Yield analysis
9. Quality assessment
10. Soil analysis

Adopting the intelligence in the agriculture fields can benefit a farmer by:

1. High productivity
2. Better product cost
3. Higher crop quality
4. Less laborer works
5. Assessing better health

These are however some of the advantages of deploying artificial intelligence in agriculture, which can improve the life of the farmer. Apart from this, a major concern that should always be highlighted before adopting any innovations is skill and knowledge about the underlying technology. So, appropriate training and information are imperative to make things fruitful (Chouhan et al. 2020; Ray 2024; Panchal et al. 2024). This book covers the basis of such phenomena and adoptability of artificial intelligence in different applications in the field of agriculture.

References

Al Mashhadany Y, Alsanad HR, Al-Askari MA et al (2024) Irrigation intelligence—enabling a cloud-based internet of things approach for enhanced water management in agriculture. Environ Monit Assess 196:438. https://doi.org/10.1007/s10661-024-12606-1

Berger T, Gimpel H, Stein A et al (2024) Hybrid intelligence for reconciling biodiversity and productivity in agriculture. Nat Food 5:270. https://doi.org/10.1038/s43016-024-00963-6

Chouhan SS, Sharma S (2012) Identification of current attacks and their counter measures in optical burst switched (OBS) network. Int J Adv Comput Res (IJACR) 2(3):61–64

Chouhan SS, Singh UP, Jain S (2020) Applications of computer vision in plant pathology: a survey. Arch Comput Methods Eng 27:611–632. https://doi.org/10.1007/s11831-019-09324-0

https://www.geeksforgeeks.org/machine-learning-types-of-artificial-intelligence/

https://www.javatpoint.com/artificial-intelligence-ai

Islam M, Bijjahalli S, Fahey T et al (2024) Destructive and non-destructive measurement approaches and the application of AI models in precision agriculture: a review. Precis Agric 25:1127. https://doi.org/10.1007/s11119-024-10112-5

Jain S, Chouhan SS (2014) Cyclic redundancy codes: study and implementation. Int J Emerg Technol Adv Eng 4:213–217

Nihalani R et al (2024) Long short-term memory (LSTM) model for Indian sign language recognition. J Intell Fuzzy Syst 46:11185–11203. https://doi.org/10.3233/JIFS-233250

Panchal B, Pranjal P, Patel RK, Sharma A, Chouhan SS (2024) Assessing the quantity of a crop field using aerial images. In: Chouhan SS, Singh UP, Jain S (eds) Applications of computer vision and drone technology in agriculture 4.0. Springer, Singapore. https://doi.org/10.1007/978-981-99-8684-2_11

Ray PP (2024) Generative AI and its impact on sugarcane industry: an insight into modern agricultural practices. Sugar Tech 26:325–332. https://doi.org/10.1007/s12355-023-01358-w

Challenges in Achieving Artificial Intelligence in Agriculture

2

Anjana J. Atapattu, Lalith K. Perera, Tharindu D. Nuwarapaksha, Shashi S. Udumann, and Nuwandhya S. Dissanayaka

Abstract

Artificial Intelligence (AI) holds significant potential for advancing global agriculture by increasing sustainability, efficiency, and productivity. Collaboratively addressing the numerous sophisticated technical and social issues is necessary for fully utilizing the potential of AI for small-scale farmers who are at risk. The primary issue that has been highlighted is the lack of high-quality data that represent different farm types, marginalized situations, and diverse areas. AI analytics skills are limited by biased and fragmented datasets centered on large industrial farms. Improved metadata standards, crowdsourcing, and open data platforms are essential for integrating AI in agriculture. As universal solutions are lacking in practical situations, the chapter also emphasizes the importance of developing AI technology improvements for farmer needs and geographical variances. Flexible, decentralized, and participative methods considered local socio-economic and policy constraints are necessary for the implementation of AI. Concerns about hidden biases and responsibility are raised by AI systems. Building trust requires redress channels, audits, and transparency systems. To create open, flexible data models, coordinated multi-stakeholder initiatives are also required to address interoperability challenges such as data compatibility, privacy, and policy variances. Public–private financing thus becomes necessary in this sector due to the lack of research funding and policies centered on sustainable AI concepts to help smallholders. To overcome the obstacles in achieving AI in agriculture, it is suggested that participatory training, flexible platforms, decentralized knowledge sharing, independent algorithm audits, and responsible

A. J. Atapattu (✉) · T. D. Nuwarapaksha · S. S. Udumann · N. S. Dissanayaka
Agronomy Division, Coconut Research Institute, Lunuwila, Sri Lanka

L. K. Perera
Coconut Research Institute, Lunuwila, Sri Lanka

S. S. Chouhan et al. (eds.), *Artificial Intelligence Techniques in Smart Agriculture*, https://doi.org/10.1007/978-981-97-5878-4_2

open data protocols be used to keep farmers at the center through cooperation, viability from a financial standpoint, empathy, and thoughtful solutions across the global diversity of agriculture.

Keywords

Artificial intelligence (AI) · Artificial general intelligence (AGI) · Computer vision (CV) · Fertilizer · Livestock monitoring · Yields

2.1 Introduction

The ability of a machine to carry out cognitive tasks including learning, thinking, problem-solving, perceiving, and even certain aspects of creativity is referred to as AI (Wang 2019). Since scientists and academics began investigating whether robots could be programmed to "think" in the 1950s, AI has existed (Natale and Ballatore 2020). The early years of AI research were filled with optimism about the possibility of creating a truly intelligent machine (Natale and Ballatore 2020). There were bold predictions that human-level AI would be achieved within a couple of decades.

Figure 2.1 illustrates the primary methods utilized in AI, encompassing machine learning algorithms like deep neural networks, techniques for knowledge representation and reasoning, natural language processing, computer vision models for image and video analysis, fuzzy logic, robotic technology, automation planning, and optimization (Sarker 2022). Simple tasks that humans perform unconsciously like recognizing objects, navigating a room, or understanding language proved extremely tough to replicate in machines (Vinciarelli et al. 2015). After the initial

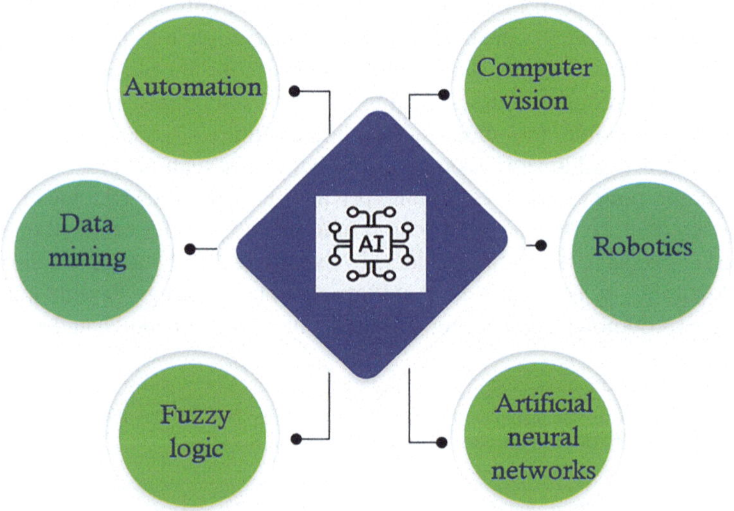

Fig. 2.1 Techniques involved in AI (Source: Figures by authors)

euphoria, there were periods where funding and interest in AI dried up as it failed to live up to its hype. AI research continued making slow by steady progress through the second half of the twentieth century (Patel et al. 2009). Much of this was driven by the realization that rather than aiming for human-like general intelligence, it made more sense to focus on narrow AI applications designed to perform specific functional tasks.

The current surge in enthusiasm for AI stems from the prominence of machine learning techniques and the accessibility of vast data sets and computational resources. Machine learning entails teaching algorithms using extensive data sets to identify patterns and provide predictions and suggestions (Saxena et al. 2024; Solanki et al. 2023a). When integrated with near unlimited processing capacity unlocked by cloud computing, machine learning has delivered impressive improvements in various domains (Solanki et al. 2023b). From Siri and Alexa-like virtual assistants to product recommendations, autonomous vehicles, diagnostic systems, and financial trading platforms, machine learning is driving a revolution in how humans and machines interact (Walker 2020). Current AI systems are often characterized as narrow or weak AI, as they are primarily focused on specific, predetermined tasks such as playing chess or identifying images (Eling et al. 2022). The next frontier for the field is to achieve Artificial General Intelligence (AGI), the ability to perform intellectual tasks across many domains just as humans can (Korteling et al. 2021). AGI could potentially match and even exceed human intelligence but leading experts are divided on how close we are to this milestone. As AI capabilities expand, it raises pressing questions around ethics, security, privacy, economics, and social impacts too (Carter et al. 2020). There are concerns that the ever-capable AI could take away jobs and be vulnerable to misuse or abuse. Governments will have a pivotal role to play in funding research and regulating certain sensitive use cases of AI technology.

In agriculture, AI technologies are being employed in numerous innovative ways to assist farmers in enhancing efficiency, cutting costs, and increasing production. One major application of AI in agriculture involves analyzing massive amounts of data to uncover hidden insights (Ayoub Shaikh et al. 2022). By collecting data from sensors in fields, drones, and satellites capturing imagery, equipment giving readings on metrics like soil moisture levels, and more, farmers are accumulating vast troves of data about their operations. Using machine learning algorithms, the systems can discover patterns and connections in the data that enable better decision-making around issues like when to plant, water, fertilize, or harvest crops (Elbasi et al. 2023). For example, some AI systems combine historical climate data, weather forecasts, and crop growth models to provide extremely precise recommendations to farmers on daily irrigation and fertilization amounts down to the square meter (Alahmad et al. 2023). This helps optimize crop yields while saving water, fertilizer costs, and labor. AI is also being used to monitor soil quality over time and suggest locations in fields that need nutrient amendments to maintain productivity. Satellite imagery analyzed with AI can quantify crop health metrics down to the plant level and spot signs of disease/pest damage so farmers know exactly which areas of fields need attention rather than having to scout whole fields manually (Zhang et al. 2019).

AI and advanced sensors are enabling completely autonomous agriculture equipment (Tiozzo Fasiolo et al. 2023). Such autonomous technologies will provide cost and time savings while also offsetting labor shortages on farms. At a broader level, AI systems are helping farmers optimize complex decisions in ways never before possible (Farmer et al. 2012). Choosing which crops to grow, varieties to plant, rotations to implement, and field assignments annually can have major economic implications that are extremely difficult to calculate. AI agricultural decision support systems can crunch thousands of data points on hyperlocal weather patterns, soil types, commodity market price projections, and more to provide individualized recommendations at every field zone on a farm (Javaid et al. 2023). This enables farmers to maximize planning efficiency and returns on limited resources. At research institutions and agriculture technology companies around the world, AI and machine learning models continue to be developed that increase the intelligence, adaptability, and capabilities of systems designed to aid farmers. From seed establishment, to harvest yield prediction, to price forecasting for commodity crops, the applications are vast and growing (Srivastava et al. 2022). Many solutions are becoming available to smallholder farmers through government programs and private sector business models, even though some advanced AI systems require significant technological investments (Fabregas et al. 2019). Although there are multiple advantages to utilizing AI in agriculture, there are also several obstacles that influence the widespread adoption of the technology. This chapter looks into the potential applications of AI in agriculture as well as its problems, solutions, and future possibilities.

2.2 Current Applications of AI in Agriculture

The current uses of AI in agriculture are addressed in this section. Through data collection, analysis, decision support systems, and implementation, AI transforms precision agriculture, optimizing resource use, raising agricultural yields, and minimizing environmental consequences (Linaza et al. 2021). AI-enabled advanced agricultural, soil, and livestock monitoring facilitates early problem detection and focused solutions. Farmers can benefit from optimal planting dates, pest control, and climate resilience strategies through weather forecasting based on AI climate modeling (Grigorieva et al. 2023). Accuracy can be achieved in tasks like planting, crop upkeep, and harvesting by using AI-powered agricultural robots and autonomous equipment (Mavridou et al. 2019). AI analyzes big data from many agricultural sources to find patterns that allow data-driven decision-making (Linaza et al. 2021). This process optimizes linked logistics, supply chain management, inventory forecasting, and agricultural product transportation routing. Agricultural yields, pest outbreaks, resource optimization, and the identification of desirable genetic features for agricultural breeding programs are all anticipated through predictive analytics using AI (Ayoub Shaikh et al. 2022).

2.2.1 Precision Agriculture

Traditional farming methods can be changed with precision agriculture combined with AI technologies to optimize resource use, increase crop yields, and reduce environmental impacts (Karunathilake et al. 2023). Through a systematic cycle of data sensing and collection, analysis, decision support, implementation, and evaluation (Fig. 2.2), the integration of AI and precision agriculture improves farming practices. Combining AI algorithms, data analytics, and sensor technologies in agricultural systems provides farmers with information on their fields, enabling data-based decision-making and sustainable management (Mandapuram et al. 2019). The foundation of precision agriculture is the decision made utilizing datasets from diverse sources like satellite imagery, drones, weather data, soil sensors, and crop monitoring (Cravero et al. 2022; Chouhan et al. 2024). To make the farmers aware of crop health, soil conditions, pests, and factors impacting productivity, these data are then analyzed and processed in actual instances. With the use of machine learning and predictive analytics, AI can recognize patterns, spot anomalies, and predict trends with accuracy, enabling farmers to maximize yields and improve operations (Strielkowski et al. 2023).

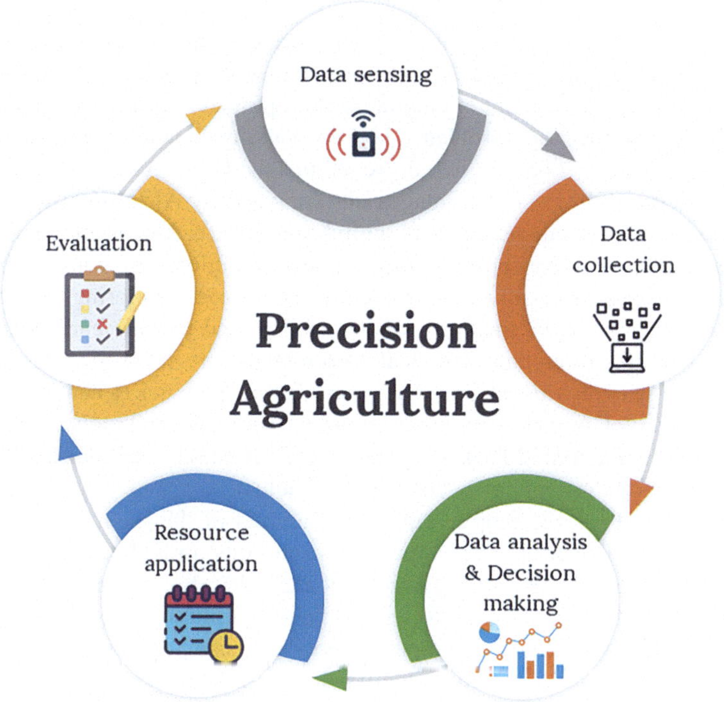

Fig. 2.2 Key steps in precision agriculture (Source: Figures by authors)

Furthermore, AI enables site-specific management strategies based on the unique characteristics of each field. Using GPS-guided machinery and variable rate technology, farmers can apply water, fertilizers, pesticides, and seeds based on localized soil and crop needs accurately (Oliveira et al. 2021). This approach optimizes resource utilization efficiency, and minimizes waste and environmental impact, contributing to sustainable farming. Predictive analytics based on AI mitigate risks and manage crops for precision agriculture. AI algorithms can predict potential challenges like droughts, diseases, or pest outbreaks by analyzing historical data and current environmental factors, allowing farmers to take protective measures for their crops (Balaska et al. 2023). Additionally, AI-based models improve planting schedules, predict ideal harvest times, and forecast market demand, enabling farmers to make strategic decisions that maximize profitability. Individual farms, the agricultural sector, and society get the benefits of AI in precision agriculture (Klerkx et al. 2019). By improving efficiency and productivity, precision agriculture ensures food security (Balasundram et al. 2023). Moreover, by optimizing the application of inputs, AI-based precision agriculture promotes environmental sustainability reducing the ecological impact of farming practices.

2.2.2 Crop, Soil, and Livestock Monitoring

Advanced real-time monitoring of crops, soil, and livestock with AI technologies, allows farmers to make rapid decisions (Subeesh and Mehta 2021). With continuous gathering and analyzing of data, AI systems can identify problems in advance. This allows farmers to protect plant or animal health, optimize yields, and reduce losses. Signs of diseases, irrigation problems, and nutrient deficiencies can be detected by frequent field surveys done by aerial drones installed with AI software (Neupane and Baysal-Gurel 2021). These images are analyzed by AI to identify emerging crop stresses and generate heat maps highlighting disturbed areas allowing farmers to take targeted action like applying fungicide only on infected regions (Abbas et al. 2023). Satellite systems with AI provide wider-scale crop monitoring by generating images to forecast yields, track growth anomalies, and provide early warning of adverse events like droughts. Statistical crop models integrated with AI platforms using weather data provide advanced yield prediction analytics (Chen et al. 2023a). At the ground level, AI-enabled sensors monitor factors like soil moisture and nutrient levels (Sishodia et al. 2020). They can guide irrigation schedules and fertilizer amounts tailored to microclimate zones within fields.

According to Fig. 2.3, deep learning and internet of things (IoT) technologies offer farmers and irrigation project managers solutions for addressing varying soil moisture levels in irrigated areas (Abioye et al. 2020). Employing IoT-controlled pump operations by the microcontroller unit helps to optimize operations at precise sites and precise time. High throughput phenotyping using drones and wireless soil moisture sensors can efficiently identify locations requiring pump operation. Thermal imagery collected by drones presents a promising method for assessing moisture content across fields, with thermal image processing alongside

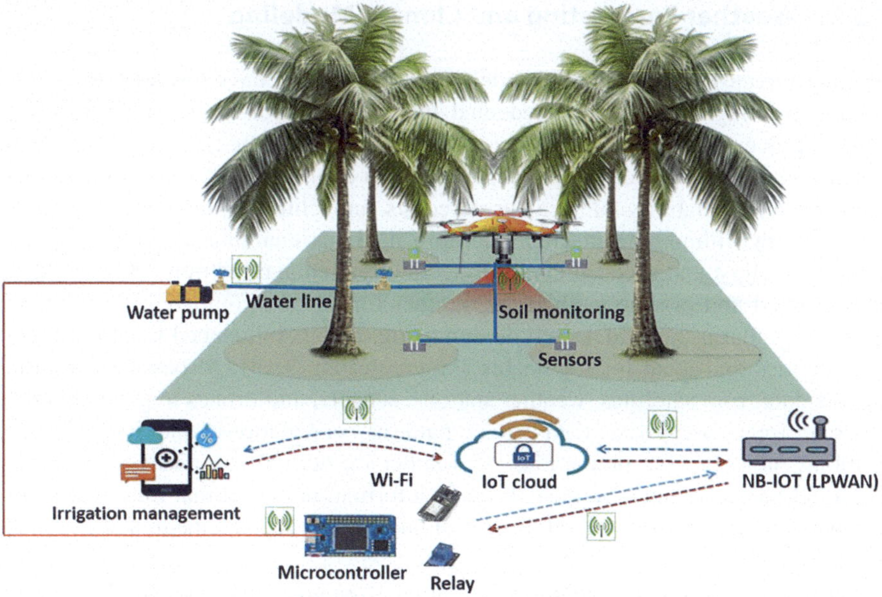

Fig. 2.3 IoT-based irrigation management system in coconut cultivation (Source: Figures by authors)

orthomosaic datasets enabling the creation of water-stress maps, aiding in identifying areas where crops lack sufficient water (Jin et al. 2021). Accordingly, based on the above pieces of information, this system releases the precise amount of water for the plants at the precise time. In greenhouse condition, AI systems can continually optimize climate conditions and resources to boost yields. Sensors also enable detection of diseases or pests early before widespread impacts. For livestock sector, AI facilitates monitoring of health and welfare (Neethirajan and Kemp 2021). Video cameras use computer vision techniques to track cattle movements, feeding behaviors, and activity levels, which reveal early signs of illness and injuries (Nasirahmadi et al. 2017). Integrated AI-enabled collars or ear tags can monitor temperature, movement, gastrointestinal health, breeding cycles, and other parameters (Aharwal et al. 2021). Some systems link to individual animal medical records for improved care. AI guides drones to perform wellness checks without disturbing animals (Lahoz-Monfort and Magrath 2021). For trouble spots like manure ponds, drones can provide continuous surveillance not possible manually. With global population rising, AI could prove vital in sustainably scaling up food production.

2.2.3 Weather Forecasting and Climate Modeling

Weather forecasting and climate modeling powered by AI have revolutionized agriculture, offering farmers unprecedented insights and tools to enhance decision-making processes, optimize resource allocation, and mitigate risks associated with weather variability and climate change (Hachimi et al. 2022). AI has improved the accuracy and reliability of weather forecasts and climate predictions by using advanced algorithms and machine learning methods enabling agricultural stakeholders to adjust and succeed in changing environmental conditions. AI provides a more refined and precise method for weather forecasting compared to traditional approaches that use past data and deterministic models influenced by atmospheric complexities and local microclimates (Mauree et al. 2019). AI-based forecasting utilizes data from satellites, weather stations, sensors, and drones to generate more accurate forecasts. These data detect patterns, correlations, and irregularities, enabling more precise predictions of weather events (Nayak and Ghosh 2013). These AI-based weather forecasts provide information on optimal times for planting, harvesting, irrigation scheduling, pest and disease management, and livestock management.

Customized farming techniques and input strategies suggested by AI systems assist farmers in maximizing crop productivity while minimizing input costs and environmental effects (Shams et al. 2024). AI is critical for long-term climate modeling for agriculture in addition to short-term weather forecasts (Chen et al. 2023b). Climate change raises significant threats to global food security due to changes in weather patterns. Relationships between the biosphere, land surface, oceans, and atmosphere are included in AI-generated climate models. According to Rawat et al. (2024), these models predict future climatic scenarios and assess possible impacts on crop viability, soil quality, water resources, and agricultural output. AI for climate-resilient agricultural systems helps to establish strong adaptation and mitigation plans by simulating climatic scenarios. These models are utilized by academics, policymakers, and farmers. Some examples include developing crop varieties that are sensitive to climate change, improving water management, agroforestry, soil preservation, and diverse cropping systems that are tailored to regional climates and socioeconomic situations. According to Tamasiga et al. (2023), enhancing agricultural resilience, sustainability, and productivity through AI-driven weather forecasting and climate modeling is critical as climate change becomes more unpredictable and unstable.

2.2.4 Agriculture Robots and Autonomous Equipment

New farming techniques facilitated by the use of AI-driven robots and autonomous machines have increased efficiency, precision, and sustainability (Rasool and Ammar 2023). These cutting-edge technologies program different processes along the agricultural value chain. They make farmers aware to maximize resource use efficiency, increase production, reduce labor costs, and minimize the impact on the

Fig. 2.4 Potential of agricultural robotics for AI-based farming (Source: Figures by authors)

environment by integrating robotic capabilities with data-driven decision-making (Wang et al. 2021). Robots and automated equipment integrated with AI can optimize field operations in precision agriculture. These robots equipped with advanced sensors, cameras, and GPS technology can plant, apply fertilizer, monitor crops, and harvest precisely in real time (Fig. 2.4). This data is analyzed by machine learning algorithms to enable farmers to implement site-specific management plans based on the unique needs of individual fields (Hegedus et al. 2023). Pesticides, herbicides, and fertilizers can be applied accurately by robotic systems, minimizing input waste and environmental contamination while maximizing crop yields. AI-powered robots do the planting and seeding tasks, improving accuracy, speed, and efficiency by identifying the best planting spots, adjusting seeding rates, and navigating uneven terrain (Oliveira et al. 2021). By ensuring uniform seed spacing and depth, these robots optimize plant growth and stand establishment increasing yields and crop quality. AI-driven robots carry out harvesting processes across various crops, including fruits, vegetables, and grains (Hanson et al. 2003). Harvesting robots equipped with advanced vision systems and robotic arms can identify, pick, and sort accurately.

2.2.5 Analyzing Agricultural Big Data

There are both benefits and drawbacks to the increasing development in agricultural information generation (Shaikh et al. 2022). AI has become an important tool for gathering values from vast and complex data sources by monitoring crop status, livestock health, microclimate, soil chemistry, equipment performance, and other aspects of the farm-generate data (Akhtar et al. 2021). Satellite photography and aerial drones provide more detailed information on field conditions and vegetation. Farm management decisions also depend on market data, such as consumer patterns and commodity pricing (Newton et al. 2013). Supply chain data offer traceability from origin to endpoint. Forecasting models are created using global climate and weather data. IoT sensor networks and cloud computing are needed to scale and link these diverse data streams (Villari et al. 2016). Real-time decision-making and improved operational efficiency have been enabled by the instant incorporation of data into AI analytics and farm automation systems by the integration of 5G connectivity and edge computing. Human analytical capacity is not enough to keep up with the volume, diversity, and velocity of organized and unstructured data. To find patterns, anomalies, correlations, and conclusions, AI capabilities are essential (Elshoush and Osman 2011).

Machine learning models can analyze and combine various datasets to enhance decision-making in areas such as irrigation management, fertilizer application, pest management, yield predictions, commodity pricing, and other related aspects (Kamilaris et al. 2017). As data grows, AI models become more accurate through continuous self-learning. Nevertheless, the foundation of dependable AI lies in the presence of quality data essential for its training. Investments in ground truthing, data hygiene, and metadata standards are critical (Leipzig et al. 2021). Simple dashboards that translate AI insights into digestible recommendations can make the technology more accessible to the average farm. Cloud computing enables running complex AI algorithms on behalf of farmers with basic hardware. But connectivity issues can hamper rural access. Developing AI edge computing solutions suited for remote farms is an active area of innovation (Sittón-Candanedo et al. 2019). While promising, AI in agriculture raises concerns about data privacy, security, quality, and ethics. Policies and best practices for agricultural big data management continue evolving (Javaid et al. 2023). As long as data quality and responsible use remain priorities, AI can transform data into an invaluable asset for the industry.

2.2.6 Agricultural Predictive Analytics

AI agricultural predictive analytics is changing the farmers' approach to crop management, providing information on environmental conditions, crop health, and optimal farming practices (Ayoub Shaikh et al. 2022). AI algorithms can predict various agricultural parameters by utilizing advanced algorithms and extensive datasets. These predictions help farmers make decisions to increase productivity, sustainability, and profitability. Crop yield forecasting is a primary application of AI agricultural predictive analytics (Naga Swetha and Balaji 2021). By examining historical yield data on weather patterns, soil quality, and pest infestations, AI models can forecast future yields accurately. This allows farmers to project market trends, optimize resource distribution, and lessen potential losses. Moreover, AI predictive analytics help in pest and disease management (Ayoub Shaikh et al. 2022). By monitoring environmental conditions and crop health indicators continuously in real time, AI algorithms can identify early signs of pest infestations or diseases. This early warning system assists farmers to implement preventive measures (Heeb et al. 2019).

AI agricultural predictive analytics also assists in optimizing resource usage (Redhu et al. 2022). By analyzing soil composition, moisture levels, and climate data, AI models can recommend precise irrigation schedules and fertilizer application rates for specific crop needs. This approach optimizes yield and reduces water and fertilizer waste ensuring sustainable farming practices. Another advantage of AI predictive analytics in agriculture is its contribution to crop breeding and genetic improvement (Ersoz et al. 2020). Through the analysis of genetic data and environmental factors, AI algorithms can identify favorable traits to desirable yield, resilience to diseases, and adaptation to climate change. This information supports breeders develop new crop varieties with improved characteristics contributing to global food security. AI predictive analytics facilitates the optimization of field-level management decisions (Miles 2019). By integrating data from satellite imagery, drones, and sensors established in the field, AI algorithms generate detailed maps of soil variability, plant health, and moisture levels. With this information, farmers can implement site-specific variable-rate seeding, targeted pesticide application, and precision irrigation, optimizing input usage and maximizing yield potential (Rayhan Shaheb et al. 2022).

2.2.7 Logistics and Supply Chain Management

The incorporation of AI into logistics and supply chain management in the agricultural sector can change how products are obtained, transported, and distributed to consumers (MacCarthy and Ivanov 2022). Advanced algorithms, real-time data analytics, and IoT technology in AI-connected logistics systems optimize the movement of goods throughout the supply chain, improving efficiency, visibility, and responsiveness to dynamic market demands. AI-connected logistics in agriculture improve inventory management and demand forecasting (Cannas et al. 2023). By analyzing historical sales data, market trends, weather patterns, and other relevant factors, AI algorithms can forecast future demand for agricultural products. This supports suppliers and retailers to uphold optimal inventory levels, mitigate stock-outs, and minimize surplus inventory, consequently enhancing customer satisfaction and profitability. Furthermore, AI-powered predictive analytics assist risk management in supply chain operations (Brintrup et al. 2023). By continuously monitoring weather conditions, traffic patterns, and supplier performance, AI systems can identify potential disruptions and bottlenecks before they occur. This allows supply chain managers to implement timely mitigation strategies (Sudan et al. 2023). Figure 2.5 demonstrates the steps of an AI-based supply chain from the farm gate to the consumer.

AI-connected logistics systems also enhance transportation efficiency and route optimization (Soori et al. 2023). By analyzing real-time traffic data, weather conditions, and vehicle performance metrics, AI algorithms can separate the most efficient routes for transporting agricultural products from farms to processing facilities, distribution centers, and retail outlets. This not only lowers transportation costs but also diminishes fuel consumption, greenhouse gas emissions, and overall

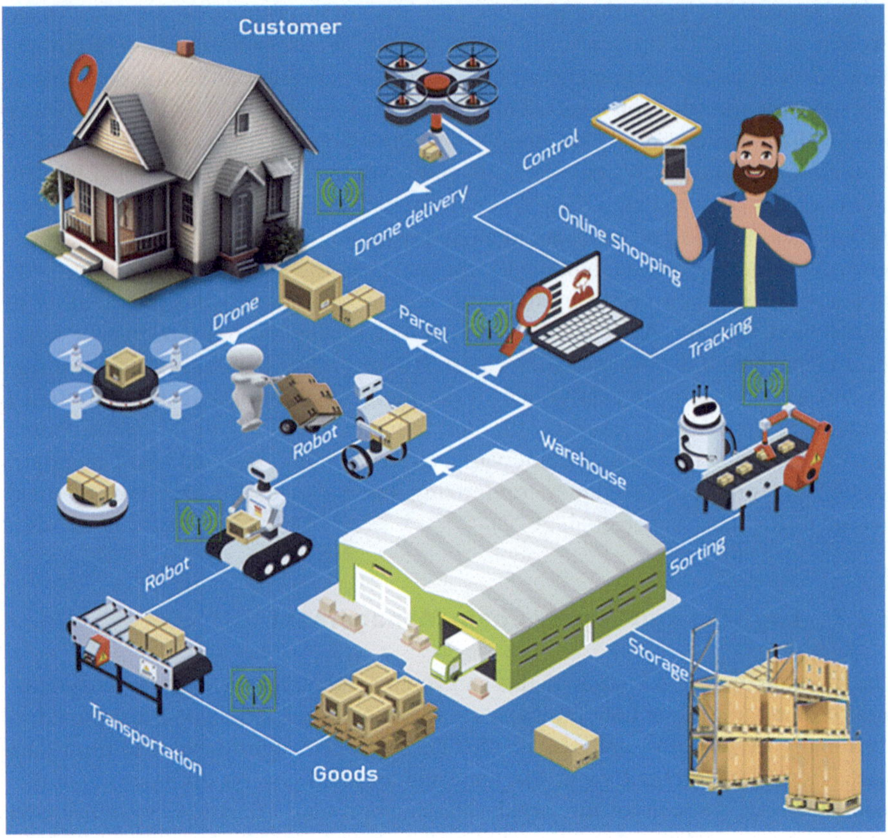

Fig. 2.5 Visual concept of AI-based supply chain steps from the farm gate to the consumer (Source: Figures by authors)

environmental impact (Lv and Shang 2023). Furthermore, AI-powered supply chain visibility platforms provide stakeholders with real-time insights into the status and location of shipments throughout the entire logistics network. By integrating data from IoT-enabled sensors, GPS tracking devices, and RFID tags, these platforms enable end-to-end visibility and traceability of agricultural products from farm to fork (Da Costa et al. 2022). This enhances transparency and accountability and enables faster customer responses. The integration of AI into connected logistics and supply chain management improves efficiency, visibility, and sustainability across the entire value chain.

2.3 Challenges for Achieving AI in Agriculture

This section explores key challenges and barriers to the adoption and utilization of AI in agriculture. The lack of availability and quality of agricultural data is a major challenge. Collecting large, consistent, and representative datasets is hard due to the inherent variability and complexity of agricultural systems. Farmers face difficulties in accessing and utilizing AI technologies due to high costs, lack of infrastructure, and skill gaps (Mhlanga 2021). The development and adoption of AI in agriculture are also challenging due to the high costs involved, including data collection, specialized hardware, talent acquisition, and ongoing maintenance (Barnes et al. 2019). Social and cultural barriers, such as concerns over job losses, privacy, and lack of trust in technology, can also reduce the adoption of AI in agriculture (Cubric 2020), while stakeholders hesitate to adopt AI because of the lack of clear regulations, guidelines, and standardized testing or certification processes for agricultural AI creates policy barriers.

2.3.1 Lack of Data Availability and Quality of Data

The lack of availability of quality data is one of the major challenges in the utilization of AI in agriculture (Jha et al. 2019). AI systems require large, high-quality datasets to program algorithms effectively. Collecting agricultural data over large geographic areas and seasons is difficult and expensive. These data are highly variable and inconsistent due to factors like weather, soil conditions, crop genetics, and management practices (Matthews et al. 2002) as agricultural production happens outdoors in complex natural systems. Even though precision agriculture techniques allow farmers to collect data at high resolution on different factors, this data is often inadequate (Khanal et al. 2020). Key information like weather, watering schedules, fertilizer applications, and pest pressure may not be recorded or linked with the sensor data. Without these data, it is hard to program AI models to make accurate predictions and recommendations (Zhang et al. 2021). The variability in agricultural practices also reduces data quality. If different growers use different preventive measures for the same disease, it reduces the consistency of data across farms. To align and integrate multiple datasets, shared common concepts are required (Kumar et al. 2021). Still, agricultural stakeholders are often hesitant to share data due to concerns over privacy, proprietary practices, and competitiveness.

2.3.2 Barriers to Accessing High Technology

Even though AI is vital for improving agricultural productivity, sustainability, and efficiency, many farmers face difficulties in accessing and utilizing these advanced technologies (Balaska et al. 2023). The costs of AI software, hardware, and infrastructure are significantly high for smallholder farmers in developing countries. Even for larger farms in developed countries, the application of AI solutions like robotic harvesters or computer vision weeding requires higher investment and expertise (Kakani et al. 2020). Basic technology gaps also reduce AI adoption in agricultural practices and research projects. Unreliable electricity, lack of internet connectivity, sensor faults, and absence of technical support constrain technology access in remote areas (Bali et al. 2023; Philip and Williams 2019). Most current AI applications rely on cloud computing and big data analytics that require always-on internet access. Without connectivity in fields and barns, real-time AI is infeasible. Lack of basic digital skills among farmers further hampers technology utilization even when infrastructure is present (Dhillon and Moncur 2023). AI algorithms also suffer from data and social bias if the training data does not represent diverse farms. As most agricultural AI today is developed using data from large industrial farms, it risks excluding or failing smaller farms with fewer resources (Sparrow et al. 2021). Developing AI tools tailored to smallholder farming systems requires collecting field data across a variety of contexts. But it has been problematic because of the challenges associated with lack of adequate accessibility to the technology.

2.3.3 High Level of Input Cost

The development and adoption of AI technologies in the agricultural sector faces major hurdles due to the high costs involved (Dwivedi et al. 2021). Collecting the large datasets required to train robust AI models is expensive for both public researchers and private companies. Deploying fleets of sensors in farms to capture millions of data points on variables like soil chemistry, crop growth, and weather patterns requires huge upfront investment (Saiz-Rubio and Rovira-Más 2020). The expenditures associated with data storage, processing, and labeling are also noteworthy. Specialized hardware like aerial drones, camera-enabled robots, and autonomous tractors needed to apply agricultural AI does not come cheap (Uzhinskiy 2023). Smaller farms especially lack the capital needed to purchase high-tech solutions. Maintenance costs further add to the burden. Scaling costs as operations expand is also a challenge. The talent required to develop and implement the complex algorithms, programming, and analytics is scarce and therefore pricey (Philip Chen and Zhang 2014). As agriculture competes with other industries for skilled technical professionals, salaries climb higher. Continual model updating and monitoring by qualified personnel also contributes to sustained high costs (Aldoseri et al. 2023). Moreover, intellectual property protections on proprietary AI technologies deter competition and allow monopolistic pricing. Startup companies that achieve breakthroughs get acquired by large tech firms who can then charge premium prices for their AI products (Makridakis 2017).

2.3.4 Social Cultural Barriers

The adoption of AI technologies in agriculture has significant social barriers (Cubric 2020). Many farmers and agricultural workers hesitate to accept AI due to concerns over job losses, privacy, and lack of familiarity or trust in the technology. One major barrier is the potential impact of AI automation on agricultural employment (Lassoued et al. 2021). Though AI can improve efficiency and productivity, this would cause unemployment in many roles including equipment operators, farm laborers, and food processing workers. This has the potential to worsen the existing financial challenges in rural areas. Establishing trust between technology companies and farmer communities is important for addressing unemployment concerns. Another barrier is the lack of accessibility and digital literacy among agricultural workers (Soma and Nuckchady 2021). Introducing advanced technologies requires basic computer and technology skills. Training programs can help make AI systems more accessible and user-friendly for relevant stakeholders (Shaw et al. 2006). There are also concerns over data privacy, transparency, and ethical AI practices. Farmers are reluctant to adopt AI systems without fully understanding how their data is used (Kerr 2004). Clear communication about data collection, analysis, and storage is needed, along with robust cybersecurity measures and fair, accountable AI algorithms. Cultural perceptions and attitudes are another challenge. The agriculture industry is usually traditional and opposed to rapidly changing technologies. Younger farmers are typically more approachable to innovations than older ones (Rose et al. 2016). Establishing user-friendly AI interfaces along with educational efforts can gradually reduce negative opinions over time.

2.3.5 Policy Barriers

The use of AI in agriculture has policy challenges that impact both farmers and technology companies (Dhillon and Moncur 2023). Unclear regulations and guidelines related to AI have made agricultural stakeholders hesitant to accept AI technologies due to privacy, security, and ethics. Currently, there are no specific laws governing how agricultural AI systems can collect, use, or share farmer data. This inhibits technology adoption due to concerns over responsible data handling (Mohammed and Tejay 2017). Another major issue is the lack of clear responsibility rules for errors, biases, or harms caused by agricultural AI. Without strong frameworks, farmers may be unwilling to rely on AI systems, and developers may avoid installing potentially risky technology (Marda 2018). Policymakers need to establish reasonable liability rules for responsible AI innovation. Additionally, the absence of standard testing or certification processes for agricultural AI is another barrier. Despite potential benefits, without formal validation of safety, security, and efficacy, farmers consider AI tools unreliable (Galaz et al. 2021). Developing standardized oversight mechanisms can help build trust in the technology. Moreover, connectivity limitations reduce farmers' access to AI systems depending on cloud computing and data transfer. Policies targeted at overcoming the rural digital gap through infrastructure investment and subsidies can help make AI more accessible to the agriculture sector (Rotz et al. 2019).

2.4 Strategies to Overcome Achieving AI Challenges in Agriculture

This section outlines strategies for overcoming challenges in implementing AI in agriculture. One major obstacle is the lack of high-quality agricultural data, which can be addressed using technologies such as drones, robots, smart sensors, and IoT networks for data collection. Combining these with AI-assisted data can improve data quality (Razzaq et al. 2021). Improving technology accessibility requires expanding infrastructure, implementing low-cost sensing solutions, adopting efficient edge computing, and utilizing open-source tools. Reducing the high costs associated with developing and installing AI systems can be achieved by declining hardware prices, utilizing open-source software, employing energy-harvesting materials, exploring innovative business models, and offering government incentives (Zaman et al. 2022). Addressing social barriers involves participatory design with end-users, implementing explainable AI for transparency, and designing appropriate user interfaces. Evidence-based policies that create a balance between innovation and supervision of data rights, privacy, and responsibility are important (Sharma et al. 2024; Panchal et al. 2024). These policies can be developed through inclusive governance that involves a diverse range of stakeholders (Li et al. 2023). By employing these supportive strategies, AI can achieve its potential for modifying agriculture in a sustainable manner.

2.4.1 Strategies to Increase Data Availability and Quality of Data

Lack of quality, well-labeled data that is highly relevant and representative of diverse agricultural contexts poses one of the biggest obstacles in effectively leveraging AI and machine learning to advance agriculture (Hu and Shu 2023). Several technologies show promise on this front—drones equipped with multispectral cameras and sensors can rapidly survey fields, gathering rich spatial data on crop growth patterns, soil moisture levels, drainage issues, etc. Ground-based robots and tractors outfitted similarly augment this data collection across temporal scales (Botta et al. 2022). Smart sensors installed directly on farm equipment can track metrics like water, fertilizer, and pesticide use as well as engine, device, and irrigation dynamics. IoT sensor networks powered by energy harvesting provide in situ monitoring of air, soil, and plant health indicators (Liu et al. 2021). Wearables facilitate precision tracking of livestock indicators, while satellites offer additional aerial measurements. Advancements in robotic phenotyping platforms also facilitate rapid capture of crop traits (Atefi et al. 2021). Synthesizing useful datasets from this heterogeneous collection of platforms, sensors, and measurement types is enabled by AI-assisted data integration pipelines, common data models, semantic frameworks, and analytics dashboards bolstering quality and actionable insights. Improved data storage, movement and computing infrastructures streamline efficient analysis (Dash et al. 2019). Cross-sector data sharing partnerships and community managed agricultural data lakes allow collective benefit. This comprehensive digital infrastructure unlocks the full potential of agricultural big data serving to ultimately accelerate and enhance AI innovation aimed at supporting the future food needs of our growing world.

2.4.2 Strategies to Increase High Technology Accessibility

This technology accessibility barrier creates a disproportionate disadvantage, negatively impacting food and income security for these farmers (Rotz et al. 2019). It also significantly restricts the real-world data available from these contexts to inform AI innovation in this domain. Multipronged approaches across infrastructural development, farmer training with technical guidance, innovative low-cost sensing, and efficient edge computing are needed to expand technological accessibility (Agrawal and Bangale 2023). Extending reliable, affordable high-speed internet via coordinated government and telecom efforts helps enable connectivity. Community mesh networks powered by solar microgrids and networks of cell tower enhance signals. Where internet remains unreliable, AI-enabled image recognition and analytics models can be packaged into apps on low-cost smartphones and tablets leveraging internal storage and processing capabilities with minimal connectivity needs just for model updates (Dixit et al. 2023). Low power wide area networks similarly connect networks of sensors across entire fields to collect and transmit key agricultural data. Smart feature phones app and basic SMS services allow access to field-specific AI advisories. Drones, automated ground rovers equipped with multispectral and thermal imaging capabilities, soil probes, and other agricultural data collection platforms are becoming extremely cost effective, allowing more farms to leverage bio-sensing (Mohsan et al. 2023). Open-source software, cloud credits, and subsidized access to advanced analytics dashboards also bolster accessibility and AI integration. These approaches prevent accessibility barriers from exacerbating gaps in agricultural prosperity across regions and communities. Democratized technological access powered by interfacing connectivity infrastructure expansion, efficient edge devices, and sensors with responsible data sharing allows AI to benefit agriculture universally.

2.4.3 Strategies to Reduce High Input Cost

The tremendous promise of AI and machine learning in accelerating agricultural advancement is severely hampered by the high costs associated with developing, deploying, and maintaining these advanced systems (Issa et al. 2022). Ongoing costs like cloud services for data storage, model training, and inference have also been obstacles. Several technology trends indicate progress on cost reduction prices of core components like graphics processing units (GPUs) and radio frequency (RF) sensors and other hardware building blocks are declining rapidly, boosted by mass manufacturing. Open-source modular software stacks reduce coding costs. Energy harvesting materials, battery chemistry advances, and low power AI chipsets lower power needs and therefore longer-term operational costs (Latif et al. 2021). Clever ultralow-cost innovation around repurposing old smartphones as edged devices and mesh networking leveraging used phones enhances affordability. Streamlined sensors and multifunctional robots performing numerous intelligent agricultural tasks optimally balance capabilities and costs (Kumar et al. 2023). Innovative business models around leasing or equipment as a service rather than outright purchasing also eases farm cash flow barriers. Government subsidies and public–private partnerships focused on subsidizing or community ownership of sensing, connectivity, and edge analytics infrastructure in rural areas can amortize costs across entire regions (Morano et al. 2021). Venture capital innovation funds, startup accelerators, innovation prizes, and contests incentivize cost-effective solutions. These complementary technological and institutional approaches on optimizing sensors, connectivity modules, analytics packages, and business models provide pathways to minimizing life cycle costs, thus allowing AI-enabled digital agriculture tools to scale affordability and maximize social impact.

2.4.4 Strategies to Overcome Social Barriers

Achieving widespread adoption of AI innovations among agricultural communities faces significant social obstacles beyond just access and economics. Multidisciplinary teams including social scientists and domain experts in participatory design of AI-driven interfaces, tools, and system workflows are key to overcoming these barriers (Subramonyam et al. 2022). Collaboratively designing solutions with diverse groups of end-user farmers, rather than prioritizing technology, improves cultural and contextual alignment. Iteratively incorporating user feedback during the build-deploy-improve development cycle is equally important. At the core of the system, explainable AI techniques provide transparency into model features, confidence bounds, and reasoning building trust. Predictions qualified by their uncertainty and what data they are most influenced by provides helpful nuance for users (Chatterjee and Dethlefs 2021). Providing flexibility in certain parameters for customization gives them more control. For instance, allowing adjustments of risk thresholds and balances between yield potential and resilience reflects a farmer's ability to take risks and personal preferences (Findlater et al. 2019). Creating communication channels using chatbots, interactive voice response calls, and SMS text alerts are suitable for communities with lower digital literacy while still improving access to AI advisories. Additionally, using highly visual interfaces with intuitive charts, graphs, and maps enhances usability (Alsabhan and Dudin 2023). Adopting this human-centered design approach prioritizes farmers' needs, making them the primary experts in ethically and responsibly promoting digital agriculture. Overcoming social adoption barriers through participatory methods improves cultural impact, trust in AI systems, and user independence resulting in a broader and more impactful implementation of agricultural intelligence.

2.4.5 Strategies to Overcome Policy Barriers

Poorly designed policies reduce the progress and adoption of technologies (Ghaffarianhoseini et al. 2017). Similarly, a lack of adequate regulation minimizes public trust and leaves individuals and surroundings vulnerable to exploitation. To achieve this balance, policymakers must be participatory and evidence-based in addressing data rights, privacy, accountability, and barriers to access. Policies should promote responsible data sharing and transparency while placing reasonable limits on third-party use (Naz et al. 2019). Granting farmers complete ownership over the data generated by their equipment and farms encourages their active involvement and participation. Furthermore, rules should stop leaving out minority groups from datasets, which harms marginalized communities (Kumar et al. 2022). Safety and accountability policies are vital, requiring thorough testing and clear liability. Policymakers need to proceed carefully to avoid suppressing innovation, especially for smaller entities lacking legal resources. Accessibility policies, including universal connectivity and fair subsidy structures, promote widespread access to maximize societal benefits (Hamraie 2018).

Inclusive policymaking involving various perspectives of agricultural scientists, legal experts, economists, ethicists, and farmers, promotes balanced, evidence-based regulations. Collaborative efforts between public–private sectors and academics facilitate responsible progress in agricultural AI (Robles and Mallinson 2023). Adaptive authority, which responds to technological changes while prioritizing ethics and experiential evidence, is important for achieving the potential of AI in agriculture.

2.5 Future Directions

AI has the power to change agriculture by providing innovative solutions to address industry challenges and improve productivity. The future of AI in agriculture includes a diverse range of applications. These range from AI-driven robotic systems that can independently plant, maintain, and harvest crops to intelligent drones equipped with AI algorithms that monitor crop health and soil conditions (Kabir et al. 2023). Predictive analytics powered by AI will empower farmers to make data-driven decisions regarding planting, irrigation, and fertilization. Meanwhile, machine learning algorithms will optimize resource allocation to reduce inputs and maximize yields (Ayoub Shaikh et al. 2022). AI-enabled sensors will provide real-time monitoring of environmental factors, guiding precise actions in response to changing conditions. Autonomous vehicles guided by AI will revolutionize field tasks such as spraying pesticides and applying fertilizers, while AI-driven predictive models forecast market demand, informing production decisions (Mandapuram et al. 2019). Moreover, AI will facilitate seamless communication between farmers and agricultural systems through natural language processing technologies, and AI-driven chatbots will offer instant support and advice. As AI continues to advance, it promises to revolutionize agriculture, ushering in an era of unprecedented efficiency, sustainability, and innovation.

2.6 Conclusion

AI can revolutionize agriculture and address the challenge of feeding the world's growing population sustainably. However, realizing this potential requires overcoming several key barriers related to data, technology access, costs, social perceptions, and policy gaps. Successful adoption of AI in agriculture depends on the availability of high-quality, well-labeled datasets that represent diverse agronomic contexts. Advanced sensing technologies, such as satellites, drones, robotics, and on-farm IoT devices, can facilitate the rapid collection and aggregation of agricultural big data to feed AI systems. Establishing common standards, shared data platforms, and hybrid analytics that support both on-farm and collective datasets will enhance usability and understanding over time. Improving connectivity infrastructure and broadband access policies allows for widespread dissemination of AI advisories, benefiting communities socially. Using architectural approaches that facilitate model training on centralized cloud computing resources while deploying inferences to affordable farm edge devices presents cost-saving opportunities, including reduced bandwidth requirements and longer hardware lifespan. Business model innovations, like equipment leasing services, mitigate financial constraints for farmers.

Achieving substantial cost reductions depends on collaborative efforts through public–private partnerships that prioritize shared infrastructure and resources. Developing advanced AI systems for agriculture requires involving farmers in the design process, ensuring that solutions meet their needs and preferences. By engaging farm operators throughout the development process, we can refine and validate solutions for real-world use. User-friendly interfaces and transparent AI models help build trust among farmers while preserving their independence in decision-making. Training programs that enhance digital skills allow users to be self-reliant. Communication channels like telephone and SMS can be utilized for communities with limited access to smartphones or technical knowledge. Ensuring responsible data privacy measures and testing for biases are essential for ethical development. To encourage ethical innovation, policies regarding data rights, accessibility, and accountability must evolve alongside advancements in AI technology. In conclusion, a collaborative, research-driven approach that balances technological advancement with social responsibility holds the key to sustainable progress in agricultural AI.

References

Abbas A, Zhang Z, Zheng H, Alami MM, Alrefaei AF, Abbas Q, Naqvi SAH, Rao MJ, Mosa WFA, Abbas Q, Hussain A, Hassan MZ, Zhou L (2023) Drones in plant disease assessment, efficient monitoring, and detection: a way forward to smart agriculture. Agronomy 13(6):1524. https://doi.org/10.3390/agronomy13061524

Abioye EA, Abidin MSZ, Mahmud MSA, Buyamin S, Ishak MHI, Rahman MKIA, Otuoze AO, Onotu P, Ramli MSA (2020) A review on monitoring and advanced control strategies for precision irrigation. Comput Electron Agric 173:105441. https://doi.org/10.1016/j.compag.2020.105441

Agrawal KN, Bangale RA (2023) Mechanizing Indian agriculture with precision farming technologies: challenges and perspective. RASSA J Sci Soc 5(2–3):129–138. https://doi.org/10.5958/2583-3715.2023.00020.5

Aharwal B, Roy B, Meshram S, Yadav A (2021) Worth of artificial intelligence in the epoch of modern livestock farming: a review. Agric Sci Digest Res J 43(1):1–9. https://doi.org/10.18805/ag.D-5355

Akhtar M, Shaikh AJ, Khan A, Awais H, Bakar EA, Othman AR (2021) Smart sensing with edge computing in precision agriculture for soil assessment and heavy metal monitoring: a review. Agriculture 11(6):475. https://doi.org/10.3390/agriculture11060475

Alahmad T, Neményi M, Nyéki A (2023) Applying IoT sensors and big data to improve precision crop production: a review. Agronomy 13(10):2603. https://doi.org/10.3390/agronomy13102603

Aldoseri A, Al-Khalifa KN, Hamouda AM (2023) Re-thinking data strategy and integration for artificial intelligence: concepts, opportunities, and challenges. Appl Sci 13(12):7082. https://doi.org/10.3390/app13127082

Alsabhan W, Dudin B (2023) Real-time flood forecasting and warning: a comprehensive approach toward HCI-centric Mobile app development. Multimodal Technol Interact 7(5):44. https://doi.org/10.3390/mti7050044

Atefi A, Ge Y, Pitla S, Schnable J (2021) Robotic technologies for high-throughput plant phenotyping: contemporary reviews and future perspectives. Front Plant Sci 12:611940. https://doi.org/10.3389/fpls.2021.611940

Ayoub Shaikh T, Rasool T, Rasheed Lone F (2022) Towards leveraging the role of machine learning and artificial intelligence in precision agriculture and smart farming. Comput Electron Agric 198:107119. https://doi.org/10.1016/j.compag.2022.107119

Balaska V, Adamidou Z, Vryzas Z, Gasteratos A (2023) Sustainable crop protection via robotics and artificial intelligence solutions. Mach Des 11(8):774. https://doi.org/10.3390/machines11080774

Balasundram SK, Shamshiri RR, Sridhara S, Rizan N (2023) The role of digital agriculture in mitigating climate change and ensuring food security: an overview. Sustain For 15(6):5325. https://doi.org/10.3390/su15065325

Bali A, Chouhan SS, Kumar G, Kumar R, Singh UP (2023) Adaptive fault-tolerant control for pure-feedback stochastic nonlinear systems with sensor and actuator faults. Circuits Syst Signal Process 42(10):5838–5867. https://doi.org/10.1007/s00034-023-02366-7

Barnes AP, Soto I, Eory V, Beck B, Balafoutis A, Sánchez B, Vangeyte J, Fountas S, van der Wal T, Gómez-Barbero M (2019) Exploring the adoption of precision agricultural technologies: a cross regional study of EU farmers. Land Use Policy 80:163–174. https://doi.org/10.1016/j.landusepol.2018.10.004

Botta A, Cavallone P, Baglieri L, Colucci G, Tagliavini L, Quaglia G (2022) A review of robots, perception, and tasks in precision agriculture. Appl Mech 3(3):830–854. https://doi.org/10.3390/applmech3030049

Brintrup A, Kosasih E, Schaffer P, Zheng G, Demirel G, MacCarthy BL (2023) Digital supply chain surveillance using artificial intelligence: definitions, opportunities and risks. Int J Prod Res 1:22. https://doi.org/10.1080/00207543.2023.2270719

Cannas VG, Ciano MP, Saltalamacchia M, Secchi R (2023) Artificial intelligence in supply chain and operations management: a multiple case study research. Int J Prod Res 1:28. https://doi.org/10.1080/00207543.2023.2232050
Carter SM, Rogers W, Win KT, Frazer H, Richards B, Houssami N (2020) The ethical, legal and social implications of using artificial intelligence systems in breast cancer care. Breast 49:25–32. https://doi.org/10.1016/j.breast.2019.10.001
Chatterjee J, Dethlefs N (2021) Scientometric review of artificial intelligence for operations and maintenance of wind turbines: the past, present and future. Renew Sustain Energy Rev 144:111051. https://doi.org/10.1016/j.rser.2021.111051
Chen L, Chen Z, Zhang Y, Liu Y, Osman AI, Farghali M, Hua J, Al-Fatesh A, Ihara I, Rooney DW, Yap PS (2023a) Artificial intelligence-based solutions for climate change: a review. Environ Chem Lett 21(5):2525–2557. https://doi.org/10.1007/s10311-023-01617-y
Chen L, Han B, Wang X, Zhao J, Yang W, Yang Z (2023b) Machine learning methods in weather and climate applications: a survey. Appl Sci 13(21):12019. https://doi.org/10.3390/app132112019
Chouhan SS, Singh UP, Jain S (2024) Introduction to computer vision and drone technology. In: Chouhan SS, Singh UP, Jain S (eds) Applications of computer vision and drone technology in agriculture 4.0. Springer, Singapore. https://doi.org/10.1007/978-981-99-8684-2_1
Cravero A, Pardo S, Galeas P, López Fenner J, Caniupán M (2022) Data type and data sources for agricultural big data and machine learning. Sustain For 14(23):16131. https://doi.org/10.3390/su142316131
Cubric M (2020) Drivers, barriers and social considerations for AI adoption in business and management: a tertiary study. Technol Soc 62:101257. https://doi.org/10.1016/j.techsoc.2020.101257
Da Costa TP, Gillespie J, Cama-Moncunill X, Ward S, Condell J, Ramanathan R, Murphy F (2022) A systematic review of real-time monitoring technologies and its potential application to reduce food loss and waste: key elements of food supply chains and IoT technologies. Sustain For 15(1):614. https://doi.org/10.3390/su15010614
Dash S, Shakyawar SK, Sharma M, Kaushik S (2019) Big data in healthcare: management, analysis and future prospects. J Big Data 6(1):54. https://doi.org/10.1186/s40537-019-0217-0
Dhillon R, Moncur Q (2023) Small-scale farming: a review of challenges and potential opportunities offered by technological advancements. Sustain For 15(21):15478. https://doi.org/10.3390/su152115478
Dixit S, Bohre K, Singh Y, Himeur Y, Mansoor W, Atalla S, Srinivasan K (2023) A comprehensive review on AI-enabled models for Parkinson's disease diagnosis. Electronics 12(4):783. https://doi.org/10.3390/electronics12040783
Dwivedi YK, Hughes L, Ismagilova E, Aarts G, Coombs C, Crick T, Duan Y, Dwivedi R, Edwards J, Eirug A, Galanos V, Ilavarasan PV, Janssen M, Jones P, Kar AK, Kizgin H, Kronemann B, Lal B, Lucini B, Williams MD (2021) Artificial intelligence (AI): multidisciplinary perspectives on emerging challenges, opportunities, and agenda for research, practice and policy. Int J Inf Manag 57:101994. https://doi.org/10.1016/j.ijinfomgt.2019.08.002
Elbasi E, Zaki C, Topcu AE, Abdelbaki W, Zreikat AI, Cina E, Shdefat A, Saker L (2023) Crop prediction model using machine learning algorithms. Appl Sci 13(16):9288. https://doi.org/10.3390/app13169288
Eling M, Nuessle D, Staubli J (2022) The impact of artificial intelligence along the insurance value chain and on the insurability of risks. Geneva Pap Risk Insurance Issues Pract 47(2):205–241. https://doi.org/10.1057/s41288-020-00201-7
Elshoush HT, Osman IM (2011) Alert correlation in collaborative intelligent intrusion detection systems—a survey. Appl Soft Comput 11(7):4349–4365. https://doi.org/10.1016/j.asoc.2010.12.004
Ersoz ES, Martin NF, Stapleton AE (2020) On to the next chapter for crop breeding: convergence with data science. Crop Sci 60(2):639–655. https://doi.org/10.1002/csc2.20054
Fabregas R, Kremer M, Schilbach F (2019) Realizing the potential of digital development: the case of agricultural advice. Science 366(6471):eaay3038. https://doi.org/10.1126/science.aay3038
Farmer JD, Gallegati M, Hommes C, Kirman A, Ormerod P, Cincotti S, Sanchez A, Helbing D (2012) A complex systems approach to constructing better models for managing financial mar-

kets and the economy. Eur Phys J Spec Top 214(1):295–324. https://doi.org/10.1140/epjst/e2012-01696-9

Findlater KM, Satterfield T, Kandlikar M (2019) Farmers' risk-based decision making under pervasive uncertainty: cognitive thresholds and hazy hedging. Risk Anal 39(8):1755–1770. https://doi.org/10.1111/risa.13290

Galaz V, Centeno MA, Callahan PW, Causevic A, Patterson T, Brass I, Baum S, Farber D, Fischer J, Garcia D, McPhearson T, Jimenez D, King B, Larcey P, Levy K (2021) Artificial intelligence, systemic risks, and sustainability. Technol Soc 67:101741. https://doi.org/10.1016/j.techsoc.2021.101741

Ghaffarianhoseini A, Tookey J, Ghaffarianhoseini A, Naismith N, Azhar S, Efimova O, Raahemifar K (2017) Building information modelling (BIM) uptake: clear benefits, understanding its implementation, risks and challenges. Renew Sustain Energy Rev 75:1046–1053. https://doi.org/10.1016/j.rser.2016.11.083

Grigorieva E, Livenets A, Stelmakh E (2023) Adaptation of agriculture to climate change: a scoping review. Climate 11(10):202. https://doi.org/10.3390/cli11100202

Hachimi CE, Belaqziz S, Khabba S, Sebbar B, Dhiba D, Chehbouni A (2022) Smart weather data management based on artificial intelligence and big data analytics for precision agriculture. Agriculture 13(1):95. https://doi.org/10.3390/agriculture13010095

Hamraie A (2018) Mapping access: digital humanities, disability justice, and sociospatial practice. Am Q 70(3):455–482. https://doi.org/10.1353/aq.2018.0031

Hanson K, Ranson MK, Oliveira-Cruz V, Mills A (2003) Expanding access to priority health interventions: a framework for understanding the constraints to scaling-up. J Int Dev 15(1):1–14. https://doi.org/10.1002/jid.963

Heeb L, Jenner E, Cock MJW (2019) Climate-smart pest management: building resilience of farms and landscapes to changing pest threats. J Pest Sci 92(3):951–969. https://doi.org/10.1007/s10340-019-01083-y

Hegedus PB, Maxwell B, Sheppard J, Loewen S, Duff H, Morales-Luna G, Peerlinck A (2023) Towards a low-cost comprehensive process for on-farm precision experimentation and analysis. Agriculture 13(3):524. https://doi.org/10.3390/agriculture13030524

Hu L, Shu Y (2023) Enhancing decision-making with data science in the internet of things environments. Int J Adv Comput Sci Appl 14(9):01409120. https://doi.org/10.14569/IJACSA.2023.01409120

Issa H, Jabbouri R, Palmer M (2022) An artificial intelligence (AI)-readiness and adoption framework for AgriTech firms. Technol Forecast Soc Chang 182:121874. https://doi.org/10.1016/j.techfore.2022.121874

Javaid M, Haleem A, Khan IH, Suman R (2023) Understanding the potential applications of artificial intelligence in agriculture sector. Adv Agrochem 2(1):15–30. https://doi.org/10.1016/j.aac.2022.10.001

Jha K, Doshi A, Patel P, Shah M (2019) A comprehensive review on automation in agriculture using artificial intelligence. Artif Intell Agric 2:1–12. https://doi.org/10.1016/j.aiia.2019.05.004

Jin H, Köppl CJ, Fischer BMC, Rojas-Conejo J, Johnson MS, Morillas L, Lyon SW, Durán-Quesada AM, Suárez-Serrano A, Manzoni S, Garcia M (2021) Drone-based hyperspectral and thermal imagery for quantifying upland Rice productivity and water use efficiency after biochar application. Remote Sens 13(10):1866. https://doi.org/10.3390/rs13101866

Kabir MSN, Reza MN, Chowdhury M, Ali M, Samsuzzaman AMR, Lee KY, Chung SO (2023) Technological trends and engineering issues on vertical farms: a review. Horticulturae 9(11):1229. https://doi.org/10.3390/horticulturae9111229

Kakani V, Nguyen VH, Kumar BP, Kim H, Pasupuleti VR (2020) A critical review on computer vision and artificial intelligence in food industry. J Agric Food Res 2:100033. https://doi.org/10.1016/j.jafr.2020.100033

Kamilaris A, Kartakoullis A, Prenafeta-Boldú FX (2017) A review on the practice of big data analysis in agriculture. Comput Electron Agric 143:23–37. https://doi.org/10.1016/j.compag.2017.09.037

Karunathilake EMBM, Le AT, Heo S, Chung YS, Mansoor S (2023) The path to smart farming: innovations and opportunities in precision agriculture. Agriculture 13(8):1593. https://doi.org/10.3390/agriculture13081593

Kerr D (2004) Factors influencing the development and adoption of knowledge based decision support systems for small, owner-operated rural business. Artif Intell Rev 22(2):127–147. https://doi.org/10.1007/s10462-004-4305-x

Khanal S, Kushal KC, Fulton JP, Shearer S, Ozkan E (2020) Remote sensing in agriculture—accomplishments, limitations, and opportunities. Remote Sens 12(22):3783. https://doi.org/10.3390/rs12223783

Klerkx L, Jakku E, Labarthe P (2019) A review of social science on digital agriculture, smart farming and agriculture 4.0: new contributions and a future research agenda. NJAS Wageningen J Life Sci 90(1):1–16. https://doi.org/10.1016/j.njas.2019.100315

Korteling JH, van de Boer-Visschedijk GC, Blankendaal RAM, Boonekamp RC, Eikelboom AR (2021) Human- versus artificial intelligence. Front Artif Intell 4:622364. https://doi.org/10.3389/frai.2021.622364

Kumar G, Basri S, Imam AA, Khowaja SA, Capretz LF, Balogun AO (2021) Data harmonization for heterogeneous datasets: a systematic literature review. Appl Sci 11(17):8275. https://doi.org/10.3390/app11178275

Kumar IE, Hines KE, Dickerson JP (2022) Equalizing credit opportunity in algorithms: aligning algorithmic fairness research with us fair lending regulation. In: Proceedings of the AAAI/ACM conference on AI, ethics, and society, Oxford, UK, pp 357–368. https://doi.org/10.1145/3514094.3534154

Kumar S, Mohan S, Skitova V (2023) Designing and implementing a versatile agricultural robot: a vehicle manipulator system for efficient multitasking in farming operations. Mach Des 11(8):776. https://doi.org/10.3390/machines11080776

Lahoz-Monfort JJ, Magrath MJL (2021) A comprehensive overview of technologies for species and habitat monitoring and conservation. Bioscience 71(10):1038–1062. https://doi.org/10.1093/biosci/biab073

Lassoued R, Macall DM, Smyth SJ, Phillips PWB, Hesseln H (2021) Expert insights on the impacts of, and potential for, agricultural big data. Sustainability 13(5):2521. https://doi.org/10.3390/su13052521

Latif S, Driss M, Boulila W, Ze H, Jamal SS, Idrees Z, Ahmad J (2021) Deep learning for the industrial internet of things (IIoT): a comprehensive survey of techniques, implementation frameworks, potential applications, and future directions. Sensors 21(22):7518. https://doi.org/10.3390/s21227518

Leipzig J, Nüst D, Hoyt CT, Ram K, Greenberg J (2021) The role of metadata in reproducible computational research. Patterns 2(9):100322. https://doi.org/10.1016/j.patter.2021.100322

Li W, Yigitcanlar T, Browne W, Nili A (2023) The making of responsible innovation and technology: an overview and framework. Smart Cities 6(4):1996–2034. https://doi.org/10.3390/smartcities6040093

Linaza MT, Posada J, Bund J, Eisert P, Quartulli M, Döllner J, Pagani AG, Olaizola I, Barriguinha A, Moysiadis T, Lucat L (2021) Data-driven artificial intelligence applications for sustainable precision agriculture. Agronomy 11(6):1227. https://doi.org/10.3390/agronomy11061227

Liu L, Guo X, Lee C (2021) Promoting smart cities into the 5G era with multi-field internet of things (IoT) applications powered with advanced mechanical energy harvesters. Nano Energy 88:106304. https://doi.org/10.1016/j.nanoen.2021.106304

Lv Z, Shang W (2023) Impacts of intelligent transportation systems on energy conservation and emission reduction of transport systems: a comprehensive review. Green Technol Sustain 1(1):100002. https://doi.org/10.1016/j.grets.2022.100002

MacCarthy BL, Ivanov D (2022) The digital supply chain-emergence, concepts, definitions, and technologies. In: MacCarthy BL, Ivanov D (eds) The digital supply chain. Elsevier, Amsterdam, pp 3–24. https://doi.org/10.1016/B978-0-323-91614-1.00001-0

Makridakis S (2017) The forthcoming artificial intelligence (AI) revolution: its impact on society and firms. Futures 9046:60. https://doi.org/10.1016/j.futures.2017.03.006

Mandapuram M, Mahadasa R, Surarapu P (2019) Evolution of smart farming: integrating IoT and AI in agricultural engineering. Glob Discl Econ Bus 8(2):165–178. https://doi.org/10.18034/gdeb.v8i2.714
Marda V (2018) Artificial intelligence policy in India: a framework for engaging the limits of data-driven decision-making. Philos Trans R Soc A Math Phys Eng Sci 376(2133):20180087. https://doi.org/10.1098/rsta.2018.0087
Matthews R, Stephens W, Hess T, Middleton T, Graves A (2002) Applications of crop/soil simulation models in tropical agricultural systems. Adv Agron 76:31–124. https://doi.org/10.1016/S0065-2113(02)76003-3
Mauree D, Naboni E, Coccolo S, Perera ATD, Nik VM, Scartezzini JL (2019) A review of assessment methods for the urban environment and its energy sustainability to guarantee climate adaptation of future cities. Renew Sustain Energy Rev 112:733–746. https://doi.org/10.1016/j.rser.2019.06.005
Mavridou E, Vrochidou E, Papakostas GA, Pachidis T, Kaburlasos VG (2019) Machine vision systems in precision agriculture for crop farming. J Imaging 5(12):89. https://doi.org/10.3390/jimaging5120089
Mhlanga D (2021) Artificial intelligence in the industry 4.0, and its impact on poverty, innovation, infrastructure development, and the sustainable development goals: lessons from emerging economies? Sustainability 13(11):5788. https://doi.org/10.3390/su13115788
Miles C (2019) The combine will tell the truth: on precision agriculture and algorithmic rationality. Big Data Soc 6(1):205395171984944. https://doi.org/10.1177/2053951719849444
Mohammed ZA, Tejay GP (2017) Examining privacy concerns and ecommerce adoption in developing countries: the impact of culture in shaping individuals' perceptions toward technology. Comput Secur 67:254–265. https://doi.org/10.1016/j.cose.2017.03.001
Mohsan SAH, Othman NQH, Li Y, Alsharif MH, Khan MA (2023) Unmanned aerial vehicles (UAVs): practical aspects, applications, open challenges, security issues, and future trends. Intell Serv Robot 16(1):109–137. https://doi.org/10.1007/s11370-022-00452-4
Morano P, Tajani F, Di Liddo F, Amoruso P (2021) The public role for the effectiveness of the territorial enhancement initiatives: a case study on the redevelopment of a building in disuse in an Italian small town. Buildings 11(3):87. https://doi.org/10.3390/buildings11030087
Nasirahmadi A, Edwards SA, Sturm B (2017) Implementation of machine vision for detecting behaviour of cattle and pigs. Livest Sci 202:25–38. https://doi.org/10.1016/j.livsci.2017.05.014
Natale S, Ballatore A (2020) Imagining the thinking machine: technological myths and the rise of artificial intelligence. Convergence Int J Res New Media Technol 26(1):3–18. https://doi.org/10.1177/1354856517715164
Nayak MA, Ghosh S (2013) Prediction of extreme rainfall event using weather pattern recognition and support vector machine classifier. Theor Appl Climatol 114(3–4):583–603. https://doi.org/10.1007/s00704-013-0867-3
Naz M, Al-zahrani FA, Khalid R, Javaid N, Qamar AM, Afzal MK, Shafiq M (2019) A secure data sharing platform using blockchain and interplanetary file system. Sustain For 11(24):7054. https://doi.org/10.3390/su11247054
Neethirajan S, Kemp B (2021) Digital livestock farming. Sens Bio-Sens Res 32:100408. https://doi.org/10.1016/j.sbsr.2021.100408
Neupane K, Baysal-Gurel F (2021) Automatic identification and monitoring of plant diseases using unmanned aerial vehicles: a review. Remote Sens 13(19):3841. https://doi.org/10.3390/rs13193841
Newton P, Agrawal A, Wollenberg L (2013) Enhancing the sustainability of commodity supply chains in tropical forest and agricultural landscapes. Glob Environ Change 23(6):1761–1772. https://doi.org/10.1016/j.gloenvcha.2013.08.004
Oliveira LFP, Moreira AP, Silva MF (2021) Advances in agriculture robotics: a state-of-the-art review and challenges ahead. Robotics 10(2):52. https://doi.org/10.3390/robotics10020052
Panchal B, Pranjal P, Patel RK, Sharma A, Chouhan SS (2024) Assessing the quantity of a crop field using aerial images. In: Chouhan SS, Singh UP, Jain S (eds) Applications of

computer vision and drone technology in agriculture 4.0. Springer, Singapore. https://doi.org/10.1007/978-981-99-8684-2_11

Patel VL, Shortliffe EH, Stefanelli M, Szolovits P, Berthold MR, Bellazzi R, Abu-Hanna A (2009) The coming of age of artificial intelligence in medicine. Artif Intell Med 46(1):5–17. https://doi.org/10.1016/j.artmed.2008.07.017

Philip Chen CL, Zhang CY (2014) Data-intensive applications, challenges, techniques and technologies: a survey on big data. Inf Sci 275:314–347. https://doi.org/10.1016/j.ins.2014.01.015

Philip L, Williams F (2019) Remote rural home-based businesses and digital inequalities: understanding needs and expectations in a digitally underserved community. J Rural Stud 68:306–318. https://doi.org/10.1016/j.jrurstud.2018.09.011

Rasool A, Ammar A (2023) The dawn of AI in agriculture: from predictive analysis to autonomous farming. Trends Anim Plant Sci 1:73–83. https://doi.org/10.62324/TAPS/2023.010

Rawat A, Kumar D, Khati BS (2024) A review on climate change impacts, models, and its consequences on different sectors: a systematic approach. J Water Clim Change 15(1):104–126. https://doi.org/10.2166/wcc.2023.536

Razzaq A, Kaur P, Akhter N, Wani SH, Saleem F (2021) Next-generation breeding strategies for climate-ready crops. Front Plant Sci 12:620420. https://doi.org/10.3389/fpls.2021.620420

Redhu NS, Thakur Z, Yashveer S, Mor P (2022) Artificial intelligence: a way forward for agricultural sciences. In: Bioinformatics in agriculture. Elsevier, Amsterdam, pp 641–668. https://doi.org/10.1016/B978-0-323-89778-5.00007-6

Robles P, Mallinson DJ (2023) Catching up with AI: pushing toward a cohesive governance framework. Policy Polit 51(3):355–372. https://doi.org/10.1111/polp.12529

Rose DC, Sutherland WJ, Parker C, Lobley M, Winter M, Morris C, Twining S, Ffoulkes C, Amano T, Dicks LV (2016) Decision support tools for agriculture: towards effective design and delivery. Agric Syst 149:165–174. https://doi.org/10.1016/j.agsy.2016.09.009

Rotz S, Gravely E, Mosby I, Duncan E, Finnis E, Horgan M, LeBlanc J, Martin R, Neufeld HT, Nixon A, Pant L, Shalla V, Fraser E (2019) Automated pastures and the digital divide: how agricultural technologies are shaping labour and rural communities. J Rural Stud 68:112–122. https://doi.org/10.1016/j.jrurstud.2019.01.023

Saiz-Rubio V, Rovira-Más F (2020) From smart farming towards agriculture 5.0: a review on crop data management. Agronomy 10(2):207. https://doi.org/10.3390/agronomy10020207

Sarker IH (2022) AI-based modeling: techniques, applications and research issues towards automation, intelligent and smart systems. SN Comput Sci 3(2):158. https://doi.org/10.1007/s42979-022-01043-x

Saxena A, Chouhan SS, Aziz RM, Agarwal V (2024) A comprehensive evaluation of marine predator chaotic algorithm for feature selection of COVID-19. Evol Syst 1:14. https://doi.org/10.1007/s12530-023-09557-2

Shaheb MR, Sarker A, Shearer S (2022) Precision agriculture for sustainable soil and crop management. In: Soil science—emerging technologies, global perspectives and applications, vol 49. IntechOpen. https://doi.org/10.5772/intechopen.101759

Shaikh TA, Mir WA, Rasool T, Sofi S (2022) Machine learning for smart agriculture and precision farming: towards making the fields talk. Arch Comput Methods Eng 29(7):4557–4597. https://doi.org/10.1007/s11831-022-09761-4

Shams MY, Gamel SA, Talaat FM (2024) Enhancing crop recommendation systems with explainable artificial intelligence: a study on agricultural decision-making. Neural Comput Appl 1:20. https://doi.org/10.1007/s00521-023-09391-2

Sharma A, Patel RK, Pranjal P, Panchal B, Chouhan SS (2024) Computer vision-based smart monitoring and control system for crop. In: Chouhan SS, Singh UP, Jain S (eds) Applications of computer vision and drone technology in agriculture 4.0. Springer, Singapore. https://doi.org/10.1007/978-981-99-8684-2_5

Shaw B, Gustafson DH, Hawkins R, McTavish F, McDowell H, Pingree S, Ballard D (2006) How underserved breast cancer patients use and benefit from eHealth programs. Am Behav Sci 49(6):823–834. https://doi.org/10.1177/0002764205283803

Sishodia RP, Ray RL, Singh SK (2020) Applications of remote sensing in precision agriculture: a review. Remote Sens 12(19):3136. https://doi.org/10.3390/rs12193136

Sittón-Candanedo I, Alonso RS, Corchado JM, Rodríguez-González S, Casado-Vara R (2019) A review of edge computing reference architectures and a new global edge proposal. Future Gener Comput Syst 99:278–294. https://doi.org/10.1016/j.future.2019.04.016

Solanki S, Singh UP, Chouhan SS (2023a) Brain tumor classification using ML and DL approaches. In: 2023 IEEE 5th International conference on cybernetics, cognition and machine learning applications (ICCCMLA), Hamburg, Germany, pp 204–208. https://doi.org/10.1109/ICCCMLA58983.2023.10346854

Solanki S, Singh UP, Chouhan SS, Jain S (2023b) A systematic analysis of magnetic resonance images and deep learning methods used for diagnosis of brain tumor. Multimed Tools Appl 1:38. https://doi.org/10.1007/s11042-023-16430-6

Soma T, Nuckchady B (2021) Communicating the benefits and risks of digital agriculture technologies: perspectives on the future of digital agricultural education and training. Front Commun 6:259. https://doi.org/10.3389/fcomm.2021.762201

Soori M, Arezoo B, Dastres R (2023) Artificial neural networks in supply chain management, a review. J Econ Technol 1:179–196. https://doi.org/10.1016/j.ject.2023.11.002

Sparrow R, Howard M, Degeling C (2021) Managing the risks of artificial intelligence in agriculture. NJAS Impact Agric Life Sci 93(1):172–196. https://doi.org/10.1080/27685241.2021.2008777

Srivastava A, Jain S, Maity R, Desai VR (2022) Demystifying artificial intelligence amidst sustainable agricultural water management. Curr Dir Water Scarcity Res 7:17–35. https://doi.org/10.1016/B978-0-323-91910-4.00002-9

Strielkowski W, Vlasov A, Selivanov K, Muraviev K, Shakhnov V (2023) Prospects and challenges of the machine learning and data-driven methods for the predictive analysis of power systems: a review. Energies 16(10):4025. https://doi.org/10.3390/en16104025

Subeesh A, Mehta CR (2021) Automation and digitization of agriculture using artificial intelligence and internet of things. Artif Intell Agric 5:278–291. https://doi.org/10.1016/j.aiia.2021.11.004

Subramonyam H, Im J, Seifert C, Adar E (2022) Solving separation-of-concerns problems in collaborative design of human-AI systems through leaky abstractions. In: CHI Conference on human factors in computing systems, vol 1, p 21. https://doi.org/10.1145/3491102.3517537

Sudan T, Taggar R, Jena PK, Sharma D (2023) Supply chain disruption mitigation strategies to advance future research agenda: a systematic literature review. J Clean Prod 425:138643. https://doi.org/10.1016/j.jclepro.2023.138643

Swetha ND, Balaji S (2021) Agriculture cloud system based emphatic data analysis and crop yield prediction using hybrid artificial intelligence. J Phys Conf Ser 2040(1):012010. https://doi.org/10.1088/1742-6596/2040/1/012010

Tamasiga P, Ouassou EH, Onyeaka H, Bakwena M, Happonen A, Molala M (2023) Forecasting disruptions in global food value chains to tackle food insecurity: the role of AI and big data analytics—a bibliometric and scientometric analysis. J Agric Food Res 14:100819. https://doi.org/10.1016/j.jafr.2023.100819

Tiozzo Fasiolo D, Scalera L, Maset E, Gasparetto A (2023) Towards autonomous mapping in agriculture: a review of supportive technologies for ground robotics. Robot Auton Syst 169:104514. https://doi.org/10.1016/j.robot.2023.104514

Uzhinskiy A (2023) Advanced technologies and artificial intelligence in agriculture. AppliedMath 3(4):799–813. https://doi.org/10.3390/appliedmath3040043

Villari M, Al-Anbuky A, Celesti A, Moessner K (2016) Leveraging the internet of things: integration of sensors and cloud computing systems. Int J Distrib Sens Netw 12(7):9764287. https://doi.org/10.1177/155014779764287

Vinciarelli A, Esposito A, André E, Bonin F, Chetouani M, Cohn JF, Cristani M, Fuhrmann F, Gilmartin E, Hammal Z, Heylen D, Kaiser R, Koutsombogera M, Potamianos A, Renals S, Riccardi G, Salah AA (2015) Open challenges in modelling, analysis and synthesis of human behaviour in human–human and human–machine interactions. Cogn Comput 7(4):397–413. https://doi.org/10.1007/s12559-015-9326-z

Walker T (2020) "Alexa, are you a feminist?" Virtual assistants doing gender and what that means for the world. IJournal Grad Stud J Fac Inf 6(1):1–16. https://doi.org/10.33137/ijournal.v6i1.35264

Wang P (2019) On defining artificial intelligence. J Artif Gen Intell 10(2):1–37. https://doi.org/10.2478/jagi-2019-0002

Wang T, Xu X, Wang C, Li Z, Li D (2021) From smart farming towards unmanned farms: a new mode of agricultural production. Agriculture 11(2):145. https://doi.org/10.3390/agriculture11020145

Zaman U, Imran Mehmood F, Iqbal N, Kim J, Ibrahim M (2022) Towards secure and intelligent internet of health things: a survey of enabling technologies and applications. Electronics 11(12):1893. https://doi.org/10.3390/electronics11121893

Zhang J, Huang Y, Pu R, Gonzalez-Moreno P, Yuan L, Wu K, Huang W (2019) Monitoring plant diseases and pests through remote sensing technology: a review. Comput Electron Agric 165:104943. https://doi.org/10.1016/j.compag.2019.104943

Zhang Q, Lu J, Jin Y (2021) Artificial intelligence in recommender systems. Complex Intell Syst 7(1):439–457. https://doi.org/10.1007/s40747-020-00212-w

Introduction to Artificial Intelligence Techniques in Agricultural Applications and Their Future Aspects

3

Nikita Mishra and Ramesh Kumar Sahni

Abstract

Agriculture plays an important role in providing food, fiber, and raw materials to sustain human progress. In recent years, technology has revolutionized agriculture, increasing efficiency, productivity, and sustainability. One of the main drivers of this change is the integration of artificial intelligence (AI) technologies encompassing machine learning (ML), computer vision and data analytics into agriculture. AI has become a revolutionary tool in agriculture, changing all aspects of agriculture and helping farmers make better decisions, use resources more efficiently and plan to increase profits, reduce waste, and improve management. However, for AI to be approved and used effectively in agriculture, issues such as initial costs, data management, education, rules and regulations, and public recognition need to be resolved. By delving deeper into research, fostering collaboration, and employing AI in a responsible manner, we can unleash the complete potential of this technology, revolutionizing farming practices for the betterment of both farmers and the agricultural industry. This chapter discusses the fundamental concepts of AI, its major components, and applications within the agricultural sector. Furthermore, future aspects, opportunities, and challenges in adoption of AI in agriculture have been discussed in detail.

Keywords

Machine learning (ML) · Gross domestic product (GDP) · Expert systems · Natural language processing (NLP) · Neural networks (NNs) · Crop · Spraying

N. Mishra
Department of Food Technology, Dr. RPCAU Pusa, Samastipur, Bihar, India

R. K. Sahni (✉)
ICAR-Central Institute of Agricultural Engineering, Bhopal, Madhya Pradesh, India
e-mail: ramesh.sahni@icar.gov.in

S. S. Chouhan et al. (eds.), *Artificial Intelligence Techniques in Smart Agriculture*, https://doi.org/10.1007/978-981-97-5878-4_3

3.1 Introduction

The challenge for global agriculture is to meet the demands of a rapidly rising population, which is predicted to reach around 10 billion by 2050, while agricultural productivity increased by over 50% in a context of limited financial development. Approximately 37.7% of the land is now used for crop production in the world (FAO 2017). Agriculture plays an important role in terms of job creation and contribution to national income. In India, the agricultural industry contributes for 18% of gross domestic product (GDP) and employs half of the country's workers (Singh and Sahni 2019). It significantly bolsters the economic prosperity of industrialized nations and holds a pivotal position in the economies of emerging nations (Sahni et al. 2018c; Tiwari et al. 2019b). Surprisingly, agriculture, while being the least computerized, has seen a surge in agricultural technology development and commercialization.

Artificial intelligence (AI) has increasingly become integral to our daily lives, stretching our horizons, and empowering us to reshape our surroundings (Talaviya et al. 2020). AI, a term derived from "Artificial" and "Intelligence," encompasses the creation of nonnatural entities through human ingenuity and the infusion of human-like intelligence into machines, enabling them to undertake complex tasks akin to humans (McCarthy 2007). This implies that intelligent machines have the capacity to autonomously make decisions based on specific circumstances. Moreover, they possess learning capabilities, enabling them to accumulate experience from mistakes and acquire problem-solving skills. Additionally, AI machines are equipped with perception abilities, enabling them to sense and comprehend the environment, thus facilitating informed decision-making.

AI has emerged as a critical technology in a wide range of industries, including robotics, banking, agriculture, education, and more. It is revolutionizing the agriculture industry and playing a critical role in the sector. AI serves as a crucial safeguard for the agriculture sector against various challenges, including climate change, insect outbreak, labor shortages, and ensuring food safety. The integration of AI has propelled today's agricultural systems to unprecedented levels of efficiency and effectiveness. The yield of crops has increased along with real-time monitoring, harvesting, processing, and marketing thanks to AI. Numerous advanced computer-based systems are intended to identify numerous crucial parameters, including crop quality, yield estimation, weed identification, and many others.

AI technologies are used in areas such as agricultural planning, livestock management, sustainable resource management, and disease control (Fuentes et al. 2022). Drones, robots, and intelligent surveillance systems have been successfully implemented by many industries and are now expected to revolutionize agriculture (Smith 2006). AI is a unique technology that is urgently needed in cattle production. AI helps farmers collect and analyze profiles to accurately predict performance based on purchasing and key criteria. To increase yield, reduce waste, and reduce environmental impact, AI is making current and future agricultural machinery more efficient. AI-based intelligent observation can create solutions that can help farmers get more from their products while using sustainable resources, creating health products, disease control, mobility monitoring, and labor reduction, thus constantly becoming part of the technological transformation of the economy (Rotaru et al. 2021).

Additionally, AI and machine learning (ML) play an important role in precision agriculture, allowing farmers to adapt their methods to specific areas of their farms. The technology uses sensors, drones, and satellite imagery to provide detailed information on soil quality, moisture content, and nutrient deficiencies, allowing farmers to reduce waste and costs from fertilizer and pesticide use. The aim is not only to grow healthy products, but also to reduce the environmental footprint of traditional agriculture. The combination of AI and ML also extends to agriculture and automation. Smart robots and autonomous ground and aerial vehicles equipped with smart technology can perform many tasks, from planting to sowing, from weeding to harvesting. These automation advances not only streamline labor-intensive processes but also increase efficiency and reduce dependence on manual labor. Farmers can increase productivity and sustainability by focusing on more productive activities such as data analysis, decision-making, and innovation. This chapter discusses the fundamental concept of AI, its key components, and applications in agricultural domain. Furthermore, future aspects, opportunities, and challenges in adoption of AI in agriculture have been discussed in detail.

3.2 Key Components and Techniques in AI

AI covers a variety of techniques and components, such as machine learning, neural networks, natural language processing, computer vision, expert systems, and more. These elements work together to create intelligent systems capable of performing tasks that have traditionally required human intelligence (Fig. 3.1). Below are some of the key components and techniques used in AI:

1. **Machine Learning (ML)**: ML includes the creation of algorithms, which permit devices to adapt by data and enhance their effectiveness through time requiring conscious coding. Supervised as well as autonomous learning and reinforcement programming are common types of ML approaches.
2. **Neural Networks**: Based on the design and operations of the human brain, neural networks are computational frameworks employed in ML for pattern recognition and predictive tasks (Solanki et al. 2023). Deep learning, which is a branch of neural networks, entails the training of expansive neural networks comprising multiple layers (known as deep neural networks) to tackle intricate tasks (Solanki et al. 2024).
3. **Natural Language Processing (NLP)**: NLP is concerned with how computers and human languages interact, allowing machines to comprehend, interpret, and produce human language. This makes tasks like speech recognition, language translation, and chatbot conversations possible.
4. **Computer Vision**: Computer vision involves training machines to understand and make choices using visual information (Nihalani et al. 2024). Computer

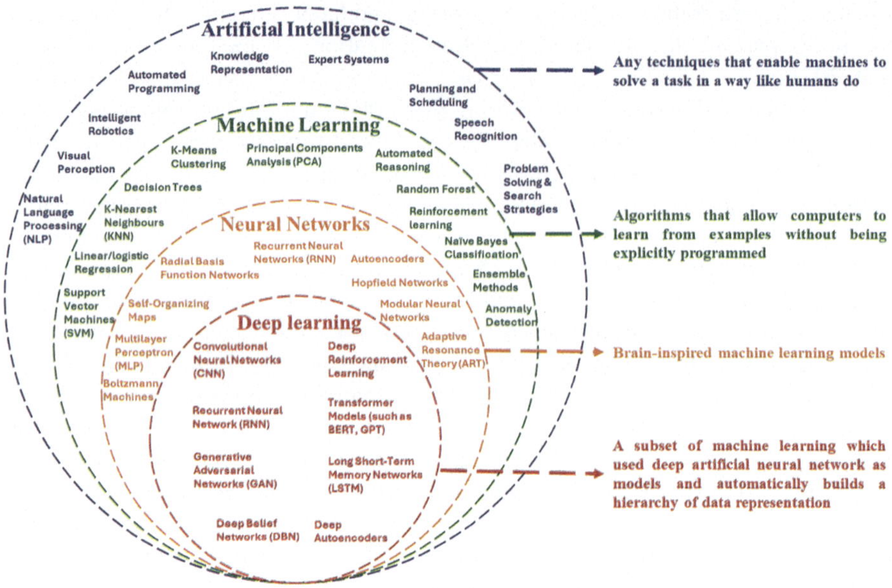

Fig. 3.1 Techniques used in artificial intelligence

vision is used in image and video recognition, object detection, and autonomous vehicles.

5. **Expert Systems**: Expert systems are software applications crafted to emulate the decision-making skills of a human expert within a particular field. It uses a knowledge base and inference engine to draw conclusions and make recommendations.

These components and techniques are continually evolving as researchers and practitioners explore new ways to advance the capabilities of AI systems.

3.3 Major Applications of AI in Agriculture

AI could transform agriculture by enhancing its efficiency, accuracy, and sustainability. AI in agriculture optimizes crop monitoring, disease detection, and precision farming. It employs autonomous vehicles for tasks like planting and harvesting, enhances decision-making with predictive analytics, and ensures sustainability through smart greenhouses. Livestock monitoring using AI aids health management. These applications improve efficiency, reduce costs, and promote sustainable practices in agriculture. Figure 3.2 shows some of the major applications of AI in the agriculture field. Below are some of the major applications of AI in agriculture.

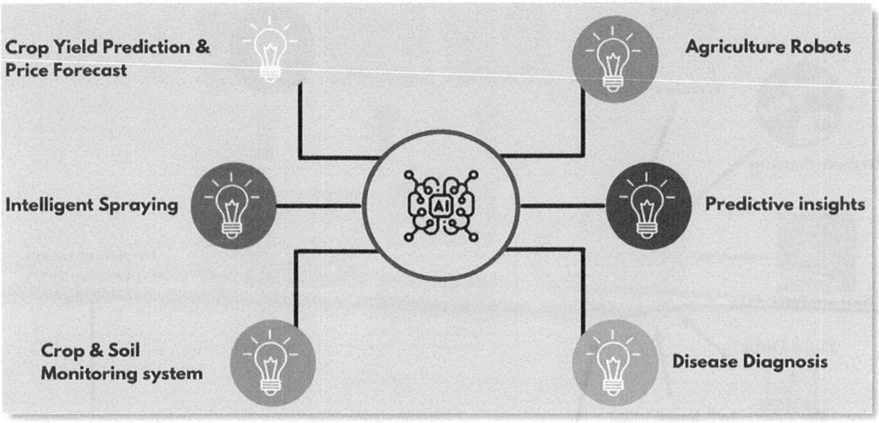

Fig. 3.2 Application of artificial intelligence in agriculture

3.3.1 Precision Farming Management

Precision farming is all about doing "Right things, Right products, Right place and Right Time." Precision farming is the process of optimizing many aspects of agriculture using technology and data-driven vision (Sahni et al. 2018b; Tiwari et al. 2019a). It uses data science and high-performance AI. It employs technology like GPS, sensors, weather monitoring, yield prediction, soil moisture and nutrient analysis, and data analytics to optimize agricultural practices and enhances efficiency by tailoring operations to specific field conditions, minimizing resource use, and maximizing crop yields (Fig. 3.3). Automated machinery, IoT, and smart data analysis contribute to informed decision-making, fostering sustainable and productive farming. Below are several ways AI contributes to precision farming:

1. **Satellite and Drone Imaging**: Algorithms analyze satellite and drone images (Fig. 3.4) to provide detailed information about crop health, growth patterns, and potential problems (Anand et al. 2023). This assists farmers in making well-informed choices regarding irrigation, fertilization, and pesticide use (Sahni et al. 2024).
2. **Crop Monitoring**: AI sensors and cameras placed in the fields monitor crops in real time. These systems detect early signs of disease, malnutrition, and pest infestation, allowing for timely intervention (Chowdhury et al. 2024) (Fig. 3.5).
3. **Automated Machinery and Robotics**: Promote the development of autonomous tractors, planters, sprayers (Fig. 3.6) and combine harvesters. These machines can work with high efficiency and perform planting, harvesting, and other agricultural activities.
4. **Variable Rate Technology (VRT)**: Analyze data on soil health, topography, and crop history to optimize the use of inputs such as water, fertilizer, and pesticides

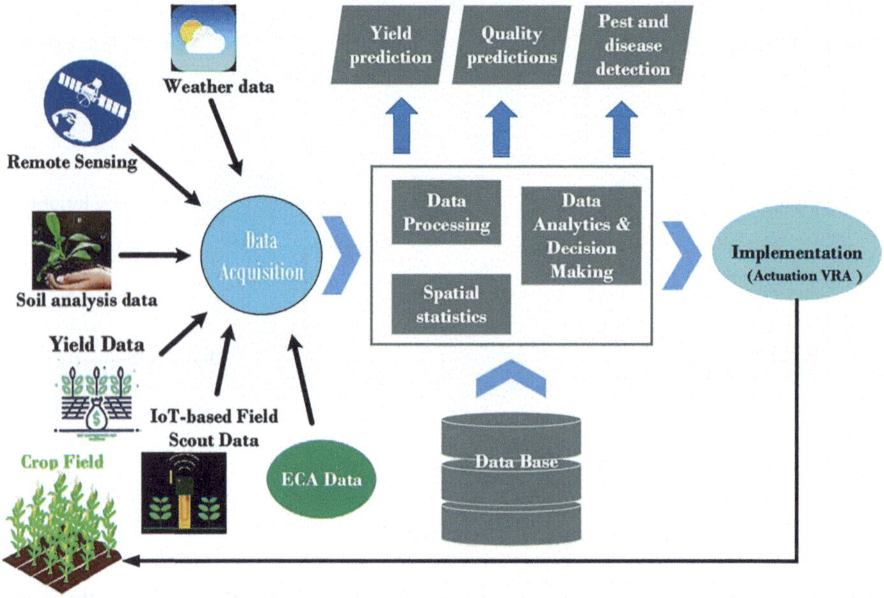

Fig. 3.3 Schematic representation of precision farming management

Fig. 3.4 Drone imaging (**a**) drone flying for image collection, (**b**) image collection using multi-spectral camera, and (**c**) imaging path visible in the digital remote

(Rathnayake et al. 2022). This allows farmers to use resources more efficiently under certain conditions (Fig. 3.7).

5. **Weather Predictions and Climate Modeling**: Analyze historical weather data and predict future weather patterns (Fig. 3.8). This information helps farmers plan their activities, optimize irrigation timing, and reduce the impact of extreme weather conditions.
6. **Precision Irrigation**: Irrigation systems use real-time data on soil moisture, water level in the field, weather conditions, and crop water needed to optimize irrigation (Fig. 3.9). This reduces water use, reduces runoff, and improves water quality (Kumar et al. 2018; Chouhan et al. 2024; Sharma et al. 2024).

Fig. 3.5 (**a**) Camera and sensors installed in the orchard, and (**b**) on the vehicle for real-time crop monitoring

Fig. 3.6 Automated sprayer for spraying pesticide in the orchards

Fig. 3.7 Variable rate spraying technology (**a**) intelligent sprayer for spraying agrochemical in the orchard based on canopy volume, and (**b**) spraying with intelligent sprayer in the vineyard

Fig. 3.8 One-in-all weather station installed in the vineyard to monitor and record real-time weather parameters

7. **Automated Weed and Pest Control**: Image recognition systems can differentiate between crops and weeds, allowing for targeted application of herbicides. Similarly, smart traps and sensors use AI to detect and manage pest infestations more precisely.
8. **Data Analytics for Yield Prediction**: Algorithms examine past data, such as crop yields, soil quality, and weather patterns, to forecast upcoming harvests. This data aids farmers in making well-informed choices regarding resource distribution and marketing strategies.
9. **IoT (Internet of Things) Integration**: It is used with IoT devices to collect and analyze data from various sensors and devices in agriculture. This includes data from soil sensors, weather stations, and monitoring equipment.
10. **Farm Management Software**: Farm management platforms provide farmers with centralized systems for monitoring and controlling various aspects of their operations. This includes crop planning, resource allocation, and equipment

(a) (b)

Fig. 3.9 Real-time precision irrigation system: (**a**) automated irrigation system installed in paddy field and (**b**) controller to operate pump in real time based on the feedback from the field installed system

management. By leveraging AI in precision farming, farmers can optimize resource use, improve crop yields, reduce environmental impact, and enhance overall farm efficiency. The integration of AI technologies allows for more precise, data-driven decision-making, contributing to sustainable and economically viable agricultural practices.

3.3.2 Soil and Irrigation Management

The key elements that determine the best crop yield are irrigation and soil. Appropriate knowledge of the soil resources is necessary to achieve improved crop yield and preserve the soil's qualities. Scheduling irrigations turns vital in times of limited water resources. Thus, to guarantee a potential crop yield, soil- and irrigation-related issues need to be handled carefully and appropriately. In this sense, methods based on AI have demonstrated the potential to address crop-related problems with irrigation and soil.

3.3.2.1 Soil Management

Soil management through AI involves the application of artificial intelligence techniques to analyze and optimize various aspects of soil health, fertility, and supervision observers. Below are some ways in which this technique is employed in soil management:

1. **Soil Health Assessment**: It is used to analyze soil samples and provide comprehensive assessments of soil health. ML models can interpret data on nutrient levels, organic matter, pH, and other soil properties to offer insights into the overall health of the soil.

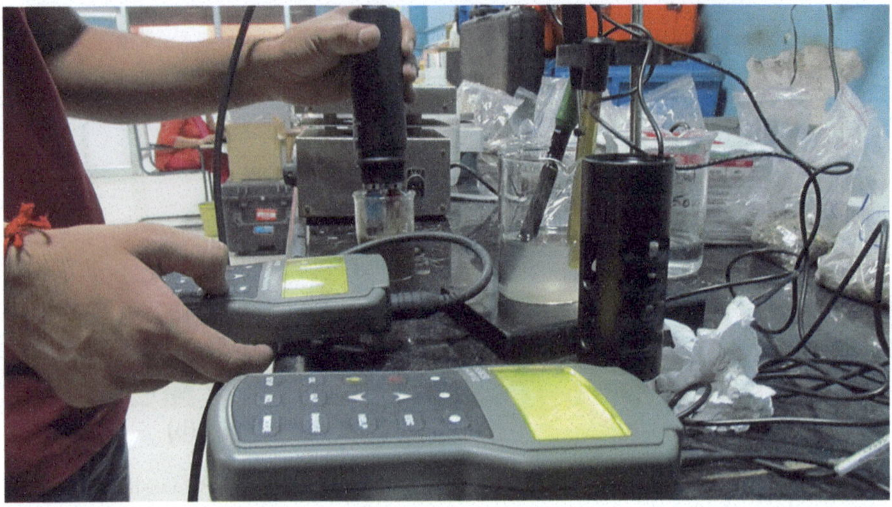

Fig. 3.10 Soil sample analysis in the lab to monitor the soil health of the agricultural field

2. **Nutrient Management**: To analyze data from soil tests, historical crop performance, and weather patterns to recommend optimal nutrient management plans (Fig. 3.10). This helps farmers adapt their fertilizer applications to meet the specific needs of their crops (Sahni et al. 2018a).
3. **Disease and Pest Prediction**: AI can predict the occurrence of soil-borne diseases and pests by analyzing historical data, weather conditions, and crop rotation patterns. This enables farmers to take preventive measures, such as adjusting planting schedules or using targeted treatments.
4. **Crop Rotation Planning**: It helps in planning optimal crop rotation schedules by considering factors such as soil nutrient depletion, disease resistance, and overall soil health. This promotes sustainable farming practices and minimizes soil degradation.
5. **Decision Support Systems**: AI driven decision support systems incorporate multiple data sources, such as crop performance records, weather data, and soil information, to help farmers make well-informed decisions regarding soil management techniques.
6. **Predictive Analytics for Soil Erosion**: To analyze the data, weather patterns, and land use information to predict areas at risk of soil erosion. This allows for the implementation of erosion control measures in vulnerable areas.
7. **Soil Mapping**: Soil mapping technologies create detailed maps of soil properties across a field. This information helps farmers understand spatial variations in soil characteristics, enabling them to tailor management practices to specific areas.

By incorporating AI into soil management practices, farmers can optimize resource use, enhance productivity, and contribute to sustainable and environmentally friendly agricultural practices. The data-driven insights provided by this innovation enable more precise and targeted interventions, leading to improved soil health and long-term sustainability.

3.3.2.2 Irrigation Management

AI-based irrigation management involves streamlining and mechanizing the agricultural watering process. It is essential for raising total agricultural productivity, cutting down on resource waste, and increasing water efficiency. Below are ways in which AI can be applied to irrigation management:

1. **Sensor Integration**: To provide real-time information about soil conditions and weather patterns, analyze data from a variety of sensors, including weather stations, drones, and soil moisture sensors.
2. **Data Fusion**: Setting up an in-depth understanding of the irrigation requirements through the integration of data from various sources enables more precise decision-making.
3. **Prophetic Analytics**: These algorithms can forecast future irrigation requirements by examining patterns and data from the past. This facilitates the process of creating irrigation schedules that are more precisely based on local conditions and crop requirements.
4. **Weather Forecasting**: It can utilize meteorological data to forecast temperature, humidity, and rainfall, enabling schedule modifications for irrigation.
5. **Automation and Control**: It control irrigation systems based on real-time data, ensuring that crops receive the right amount of water at the right time.
6. **Remote Monitoring**: Farmers can monitor and control irrigation systems remotely through mobile applications or web interfaces, providing flexibility and convenience.

3.3.3 Livestock Management

AI plays a crucial role in modern livestock management, offering innovative solutions to enhance animal welfare, efficiency, and overall farm productivity. AI plays an important role in animal identification, providing accurate and efficient solutions for tracking and managing livestock.

3.3.3.1 Animal Identification

1. **Facial Recognition**: AI-powered facial recognition can be used to identify and track an animal based on its unique facial features. This is especially useful for animals with distinctive facial markings or patterns. Facial recognition is nothing new, but it is now being used in cattle and humans. While "smart" cattle management is becoming more common, current systems still require the use of physical tracking equipment.
2. **Biometric Identification**: Biometric data such as iris scans, ear tags, or hoof prints can be used for unique identification. AI algorithms can analyze these

biometric features to create a database and match them for accurate identification. Simply scan the barcode to access full details about a particular animal and its history.
3. **RFID Technology**: Radio-Frequency Identification (RFID) tags are commonly used for animal identification. AI can enhance RFID systems by processing and analyzing data from these tags, providing real-time information on an animal's location, health status, and more.

3.3.3.2 Automated Weighing Systems

1. **Dynamic Weight Estimation**: AI algorithms can analyze data from sensors, such as load cells in feeding stations or water troughs, to estimate the weight of individual animals in real time. Highly responsive sensors remove the necessity of manual scans by swiftly and accurately measuring weight, then automatically logging the data in the database. This information can be analyzed by an AI system to gain valuable insights that can enhance agricultural techniques. The system can also establish connections between an animal's weight and its past records by recognizing individual animals. This dynamic weight estimation helps in tracking growth rates and identifying anomalies promptly.
2. **Behavior Analysis**: AI can analyze the behavior of livestock around weighing stations. Changes in behavior, such as reluctance to approach the weighing area, may indicate health issues or stress. Behavioral analysis can complement weight data to provide a more holistic view of animal well-being. Farmers can use data processed by AI systems to identify animals with unusual diets that may indicate behavioral or health problems. They can also use the data collected to see the relationship between certain foods and animal weight and health. Changes in movement, eating habits, or social interactions may indicate stress or illness.
3. **Feed Conversion Efficiency (FCE)**: AI can optimize feed management by analyzing the relationship between feed consumption and weight gain. This helps in identifying the most efficient feed conversion practices and adjusting diets for improved growth rates.

3.3.3.3 Automated Health Monitoring

AI algorithms can analyze data from various sources, such as sensors and cameras, to identify early signs of diseases or abnormalities in livestock. This allows for timely intervention and preventive measures. Automated health monitoring through AI in livestock management offers numerous benefits, enabling early detection of diseases, improving overall herd health, and enhancing farm productivity. Below are some of the ways AI can be applied to automate health monitoring in livestock:

1. **Vocalization Analysis**: AI can process and analyze the sounds and vocalizations of livestock. Changes in the frequency, duration, or pitch of vocalizations can be indicative of distress or health issues, aiding in early disease detection.
2. **Image and Video Analysis**: AI-powered cameras can capture and analyze visual data to detect signs of illness or abnormalities. For example, changes in body

posture, coat condition, or lameness can be identified through image recognition.
3. **Biometric Data Monitoring**: Sensors and wearable that collect biometric data, such as heart rate, temperature, and respiratory rate, can be integrated with AI algorithms. Abnormalities in these parameters can trigger alerts for veterinary attention.
4. **Predictive Analytics**: AI has the capability to examine past health records, environmental circumstances, and additional pertinent elements to anticipate disease outbreaks or health concerns. This allows farmers to take preventive measures and optimize health management practices.
5. **Smart Collars and Wearable**: AI-enabled smart collars or wearable can continuously monitor vital signs and activity levels. These devices can send real-time data to a centralized system, providing immediate alerts for potential health issues.

By leveraging AI in livestock management, farmers can improve animal health, optimize resource use, and enhance overall farm productivity, contributing to more sustainable and efficient agricultural practices.

3.3.4 Aquaculture System

AI plays a significant role in advancing and helping to enhance efficiency, sustainability, and overall productivity in the aquaculture system. Below are several ways in which AI is utilized in aquacultural engineering:

3.3.4.1 Automated Monitoring and Control

1. **Water Quality Management**: Sensors have the capability to continuously monitor various water parameters such as pH, dissolved oxygen, temperature, and nutrient levels. AI algorithms are then utilized to analyze this data in real time, allowing for prompt adjustments to be made in order to maintain optimal conditions for aquatic organisms.
2. **Feeding Systems**: It optimizes feeding processes by analyzing fish behavior, feeding patterns, and environmental factors. Automated feeding systems can adjust feed amounts and timings based on the specific needs of the fish, reducing waste and improving growth rates.

3.3.4.2 Precision Aquaculture

1. **Fish Health Monitoring**: Computer vision and AI algorithms can analyze images or videos of fish to detect early signs of diseases, parasites, or stress. This enables early intervention, preventing the spread of illnesses and reducing the need for antibiotics.
2. **Predictive Modeling**: It helps create predictive models that forecast environmental conditions, disease outbreaks, and growth rates. This helps farmers make informed decisions to optimize production.

The incorporation of AI in aquaculture enhances overall operational efficiency, reduces environmental impact, and ensures the sustainable growth of the aquaculture industry by addressing challenges related to disease management, resource optimization, and production planning.

3.3.5 Supply Chain Management

AI plays an important role in increasing productivity in many sectors, including agriculture. As consumers become more interested in the origins and production processes of the food they consume, there is a growing trend towards exploring methods to ensure food traceability and safety. One such method gaining attention is the use of blockchain technology. Blockchain, a distributed ledger system, has the capability to securely store and share information (Shehabi et al. 2021). By utilizing blockchain, various participants in the food supply chain, from producers to consumers, can be interconnected (Fig. 3.11). This increased connectivity has the potential to enhance transparency and build trust among stakeholders involved in the food supply chain. Below are some of the uses of AI techniques that can help ensure a good supply chain:

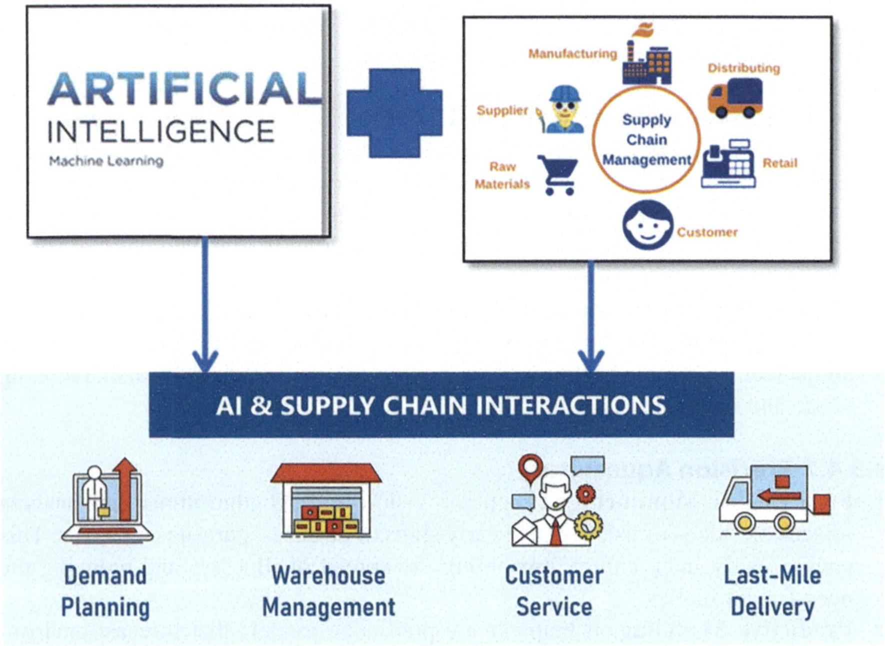

Fig. 3.11 Schematic representation of supply chain management using artificial intelligence

1. **Demand Estimating**: To forecast future demand, examine past data, current market trends, and outside influences more precisely. This lowers the possibility of overstocking or stock outs and optimizes inventory levels.
2. **Inventory Management**: Real-time inventory level observing gives you data about when and how much to reorder. By implementing this approach, waste is reduced, holding costs are minimized, and product availability is ensured.
3. **Warehouse Automation**: Self-driving and robots with AI support picking, packing, and sorting tasks in warehouses. This minimizes errors, lowers labor costs, and increases efficiency.
4. **Supplier Relationship Management**: Finds the best suppliers by analyzing data on lead times, prices, and supplier performance. This guarantees a consistent and affordable raw material supply.
5. **Risk Management**: Evaluates a range of risks, including supply chain interruptions brought on by natural disasters, international events, or other causes. Having this knowledge aids in creating backup plans to reduce possible hazards.
6. **Blockchain Integration**: AI and blockchain technologies can be integrated to enhance transparency and traceability in the supply chain. This is particularly important in industries like agriculture, where consumers are increasingly interested in the origin of their products.
7. **Collaborative Planning**: Facilitates collaboration among different stakeholders in the supply chain by providing a platform for real-time information sharing. This helps in synchronized planning and decision-making.

Implementing AI in supply chain management leads to more agile, responsive, and efficient operations. It enables companies to adapt to changing market conditions, reduce costs, and improve overall customer satisfaction. In agriculture, these optimizations can lead to better management of perishable goods, reduced waste, and a more sustainable supply chain.

3.4 Future Aspects of AI in Agriculture

The future of AI in agriculture holds immense potential for transformative advancements, addressing key challenges and optimizing various aspects of the industry. Some key future aspects include:

1. **Collaborative Robotics (Cobots)**: AI-driven collaborative robots can work alongside farmers, assisting with tasks such as planting, weeding, and harvesting. This can enhance productivity and reduce physical strain on farm workers.
2. **Hyper-Precision Farming**: AI algorithms will evolve to provide hyper-precise recommendations for crop management, leveraging real-time data from sensors, drones, and satellite imagery to optimize resource usage at the individual plant level.

3. **Automated Machinery**: AI can be integrated into farm machinery, such as tractors and harvesters, enabling them to operate autonomously. This can improve efficiency and reduce labor costs. Smart farming equipment can use computer vision and ML to identify and manage crops, weeds, and pests in real time.
4. **AI-Integrated Crop Breeding**: AI will play a significant role in accelerating crop breeding programs by analyzing genomic data, predicting desirable traits, and developing new, resilient, and high-yielding crop varieties.
5. **Crop Monitoring and Disease Detection**: AI-driven systems have the capability to observe crops for indications of diseases, pests, or nutrient deficiencies through the analysis of images and sensor data. The ability to detect these issues early enables prompt action, ultimately minimizing losses in crop yield. ML models can be trained on vast datasets of plant diseases to enhance accuracy in diagnosis.
6. **Climate Prediction and Adaptation**: AI has the capability to examine past climate data and forecast upcoming weather patterns more precisely. This information helps farmers plan for changing climate conditions and make informed decisions about planting and harvesting times. AI can contribute to the development of climate-resilient crops through genetic analysis and breeding techniques.
7. **AI in Aquaculture and Vertical Farming**: Expanded applications of AI in aquaculture for improved water management, disease prevention, and sustainable fish farming. Additionally, AI will play a significant role in optimizing vertical farming systems, ensuring efficient resource utilization in controlled environments.
8. **Agri-FinTech**: AI-powered financial technologies can help farmers manage risk, access credit, and optimize financial decision-making based on various data inputs, including crop performance and market trends.

The future of AI in agriculture is characterized by a convergence of technologies, data-driven insights, and collaborative efforts to create more resilient, sustainable, and efficient farming practices. As advancements continue, the agriculture industry is likely to experience a significant transformation, meeting the growing demand for food production while minimizing environmental impact.

3.5 Opportunity of AI Technology in Agriculture

AI offers numerous opportunities to revolutionize the agricultural sector, enhancing efficiency, productivity, and sustainability. Below are some of the areas where AI can make a significant impact in agriculture:

1. **Smart Agriculture**: AI gives farmers accurate information on crop health, nutrient content, and soil moisture levels, allowing them to maximize resource use. This data can be collected through various means such as drones, satellites, and sensors. AI algorithms analyze this data to offer insights for precise planting, irrigation, fertilization, and pest control, ultimately maximizing yields while

minimizing inputs. AI-powered systems can continuously monitor crops using image recognition techniques to identify diseases, plant stresses, insect pest, and nutrient deficiencies.
2. **Robotic and Drones' Application**: AI-driven autonomous machinery, such as autonomous machines, robotic harvesters, and drones, can be used to perform several unit operations like planting, chemical spraying, and harvesting with higher accuracy and efficiency. These machines can be operated around the clock, reducing labor costs and increasing productivity.
3. **Genetic Improvement**: AI and ML techniques can analyze genomic data to accelerate crop breeding programs. By identifying genetic markers associated with desirable traits such as yield, drought tolerance, and disease resistance, AI can help breed crops that are better suited to specific environmental conditions.
4. **Predictive Analytics**: To forecast crop yields, market trends, and possible hazards, artificial intelligence (AI) algorithms can examine historical data, weather patterns, and other pertinent variables. This aids farmers in choosing crops, scheduling plantings, and developing marketing plans. AI can streamline logistics, manage inventory, and estimate demand to optimize the agricultural supply chain. This ensures that crops reach markets and consumers efficiently, reducing waste and improving profitability.
5. **Resource Optimization**: AI can optimize resource allocation by dynamically adjusting irrigation schedules, fertilizer applications, and energy usage based on real-time inputs and predictive models. This reduces the impact on the environment while simultaneously increasing resource efficiency.
6. **Food Safety and Quality Assurance**: AI systems can enhance food safety by detecting contaminants, pathogens, and quality issues throughout the supply chain. Machine vision and spectroscopy techniques can inspect food products for defects and ensure compliance with regulatory standards.

Overall, the integration of AI with agriculture presents an array of opportunities and can help in satisfying the food demand of growing world population while promoting sustainability and resilience in the face of climate change.

3.6 Challenges in Adoption of AI Technology in Agriculture

Despite several benefits of AI in agriculture, there are a few challenges that hinder its widespread adoption. Some of these challenges are mentioned below:

1. **Cost of Implementation**: Initial setup costs for AI technologies, such as sensors, drones, and advanced machinery, can be prohibitive for small- and medium-sized farmers. The perceived high cost of adoption may discourage widespread use.
2. **Lack of Infrastructure**: In numerous areas, particularly in developing nations, there exists insufficient infrastructure, including reliable internet access and electricity. AI systems often require a robust infrastructure to function optimally, limiting their deployment in certain areas.

3. **Data Quality and Accessibility**: Effective AI algorithms depend on high-quality and large volumes of data. However, many farmers may not have access to the necessary data, and existing data might be incomplete or unreliable. There are also concerns about data privacy and ownership.
4. **Skills and Education:** Farmers and agricultural workers may lack the necessary skills and knowledge to implement and manage AI technologies. Training programs and educational initiatives are essential to bridge this knowledge gap.
5. **Interoperability**: Many AI solutions are developed by different companies and may not be interoperable. Lack of standardization can lead to compatibility issues and make it challenging for farmers to integrate multiple technologies seamlessly.
6. **Regulatory and Policy Barriers**: The regulatory environment for AI in agriculture is still evolving. Lack of clear regulations and policies can create uncertainty and hinder the adoption of new technologies. Governments need to establish frameworks that encourage innovation while ensuring responsible and ethical use.
7. **Resistance to Change**: Farmers might hesitate to embrace new technologies because they may lack trust in or be unfamiliar with AI. There is a need for awareness campaigns and demonstration projects to showcase the benefits and build confidence in the technology.
8. **Environmental Concerns**: The environmental impact of AI in agriculture, such as increased energy consumption and electronic waste, can be a concern. Sustainable AI solutions and practices should be developed to minimize negative effects on the environment.
9. **Security Risks**: As farms become more connected and reliant on data, there is an increased risk of cyber threats. Protecting sensitive agricultural data from unauthorized access and ensuring the cyber security of AI systems are critical challenges.

Tackling these challenges necessitates a comprehensive strategy that involves cooperation among businesses, government entities, and academic institutions. This strategy should focus on creating solutions, allocating resources to education and training initiatives, and implementing ethical and regulatory guidelines to encourage responsible adoption of AI.

3.7 Conclusions

AI technologies help farmers monitor fields with high precision. This allows for the assessment of crop health, soil conditions, and other variables, enabling targeted and efficient resource management. AI-powered sensors and cameras are used to monitor crops for signs of diseases, pests, and nutrient deficiencies. AI is integrated into farm machinery, such as tractors and harvesters, to enable autonomous operation. These machines can navigate agricultural fields, perform tasks like planting, spraying, spreading, and harvesting, and optimize routes for efficiency. AI is employed in smart irrigation systems to increase water usage efficiency. By examining soil moisture levels, weather predictions, and crop water needs, these systems can adjust irrigation schedules automatically, ensuring that crops receive appropriate water amounts. Furthermore, future possibilities are AI-driven collaborative robots can work alongside farmers, assisting with tasks such as field preparation, planting, weeding, spraying, and harvesting. This can enhance productivity and reduce physical strain on farm workers. Limitations of AI technology application are lack of necessary skills and knowledge among farmers to implement and manage it training programs and educational initiatives are essential to bridge this knowledge gap.

References

Anand R, Sahni RK, Kumar SP, Thorat DS, Kumar AK (2023) Advancement in agricultural practices with use of drones in the context of precision farming. Glob J Eng Sci 11(2):1–7. https://doi.org/10.33552/GJES.2023.11.000760

Chouhan SS, Singh UP, Jain S (2024) Introduction to computer vision and drone technology. In: Chouhan SS, Singh UP, Jain S (eds) Applications of computer vision and drone technology in agriculture 4.0. Springer, Singapore. https://doi.org/10.1007/978-981-99-8684-2_1

Chowdhury M, Anand R, Dhar T, Kurmi R, Sahni RK, Kushwah A (2024) Digital insights into plant health: exploring vegetation indices through computer vision. In: Chouhan SS, Singh UP, Jain S (eds) Applications of computer vision and drone technology in agriculture 4.0. Springer, Singapore, pp 7–30. https://doi.org/10.1007/978-981-99-8684-2_2

FAO (2017) The state of food and agriculture leveraging food systems for inclusive rural transformation. 1–181. https://www.fao.org/3/I7658e/I7658e.pdf

Fuentes S, Viejo CG, Tongson E, Dunshea FR (2022) The livestock farming digital transformation: implementation of new and emerging technologies using artificial intelligence. Anim Health Res Rev 23:59–71. https://doi.org/10.1017/S1466252321000177

Kumar M, Sahni RK, Waghaye AM, Nayak AK (2018) Precision irrigation technologies: a present need to save for future. www.biotecharticles.com. https://biotecharticles.com/Agriculture-Article/Precision-Irrigation-Technologies-A-Present-Need-to-Save-for-Future-4382.html

McCarthy J (2007) From here to human-level AI. Artif Intell 171(18):1174–1182. https://doi.org/10.1016/j.artint.2007.10.009

Nihalani R et al (2024) Long short-term memory (LSTM) model for Indian sign language recognition. J Intell Fuzzy Syst 46:11185. https://doi.org/10.3233/JIFS-233250

Rathnayake AP, Sahni RK, Khot LR, Hoheisel GA, Zhu H (2022) Intelligent sprayer spray rates optimization to efficiently apply chemicals in modern apple orchards. J ASABE 65(6):1411–1420. https://doi.org/10.13031/ja.14654

Rotaru A, Tcă AVA, Pop I, Andronie L (2021) Artificial intelligence, a possible solution for the agriculture and animal husbandry sector? Bull Univ Agric Sci Vet Med Cluj-Napoca Anim Sci Biotechnol 78(2):1–5. https://doi.org/10.15835/buasvmcn-asb:2021.0004

Sahni RK, Kumar D, Tiwari PS, Kumar V, Kumar SP, Chandel NS (2018a) DGPS based on-the-go soil nutrient mapping system: a review. Andhra Agric J 65(spl):1–6

Sahni RK, Kumar V, Kumar SP, Chandel NS, Tiwari PS (2018b) Precision agriculture technologies. www.biotecharticles.com. https://biotecharticles.com/Agriculture-Article/Precision-Agriculture-Technologies-4383.html

Sahni RK, Patel A, Kumar M, Thorat DS, Kumar V (2018c) Status of farm mechanization in Indian agriculture. Biotech Articles. https://www.biotecharticles.com/Agriculture-Article/Status-of-Farm-Mechanization-in-Indian-Agriculture-4468.html

Sahni RK, Thorat D, Rajwade Y, Jyoti B, Ranjan J, Anand R (2024) Drone spraying system for efficient agrochemical application in precision agriculture. In: Chouhan SS, Singh UP, Jain

S (eds) Applications of computer vision and drone technology in agriculture 4.0. Springer, Singapore, pp 225–244. https://doi.org/10.1007/978-981-99-8684-2_13

Sharma A, Patel RK, Pranjal P, Panchal B, Chouhan SS (2024) Computer vision-based smart monitoring and control system for crop. In: Chouhan SS, Singh UP, Jain S (eds) Applications of computer vision and drone technology in agriculture 4.0. Springer, Singapore. https://doi.org/10.1007/978-981-99-8684-2_5

Shehabi A, Abawajy J, Rajab M (2021) A comprehensive review of blockchain technology: applications and challenges in the supply chain. Sustain For 13(3):1760

Singh RS, Sahni RK (2019) Transformation of Indian agriculture through mechanization. Econ Affairs 64(2):297–303. https://doi.org/10.30954/0424-2513.2.2019.4

Smith IC (2006) The history of artificial intelligence. University of Washington

Solanki S, Singh UP, Chouhan SS, Jain S (2023) Brain tumour detection and classification by using deep learning classifier. Int J Intell Syst Appl Eng 11(2):279–292. https://ijisae.org/index.php/IJISAE/article/view/2624

Solanki S, Singh UP, Chouhan SS, Jain S (2024) A systematic analysis of magnetic resonance images and deep learning methods used for diagnosis of brain tumor. Multimed Tools Appl 83:23929–23966. https://doi.org/10.1007/s11042-023-16430-6

Talaviya T, Shah D, Patel N, Yagnik H, Shah M (2020) Implementation of artificial intelligence in agriculture for optimisation of irrigation and application of pesticides and herbicides. Artif Intell Agric 4:58–73. https://doi.org/10.1016/j.aiia.2020.04.002

Tiwari PS, Sahni RK, Kumar SP, Kumar V, Chandel NS (2019a) Precision agriculture applications in horticulture. Pantnagar J Res 17(1):1–10. https://www.gbpuat.res.in/uploads/archive/Vol.17(1).1.pdf

Tiwari PS, Singh KK, Sahni RK, Kumar V (2019b) Farm mechanization–trends and policy for its promotion in India. Indian J Agric Sci 89(10):1555–1562. https://doi.org/10.56093/ijas.v89i10.94575

Agricultural Artificial Intelligence: Obstacles and Opportunities

4

M. Salomi and S. Samsudeen

Abstract

Artificial Intelligence (AI) has the potential to completely transform the agriculture sector by giving farmers new tools to maximize crop yields, reduce environmental impact, and optimize resource efficiency. The widespread application of AI in agriculture is hampered by a number of issues, despite its potential. These include the lack of accurate and high-quality data, the high implementation costs, the challenge of modifying AI models to adapt to complex and dynamic agricultural environments, and the ethical issues associated with applying AI to crop management, resource allocation, and pesticide application decision-making processes. Certain challenges and limitations that restricts the farmers and various organization bodies to adapt the AI techniques implementation in agricultural field are briefly discussed in this chapter along with some generic solutions that may give hands to the individuals reading the context. In order to effectively use AI into agriculture, several constraints need to be resolved. Enhancing data quality and collecting, cutting down on the cost of AI technologies, making AI models more flexible, and creating unambiguous ethical standards for their use should be the top priorities for researchers and developers. AI has the ability to significantly change agriculture and create a future that is more resilient, productive, and sustainable if these challenges are successfully addressed.

M. Salomi (✉)
Department of Computational Intelligence, Faculty of Engineering and Technology, SRM Institute of Science and Technology, Kattankulathur, Tamil Nadu, India
e-mail: salomis@srmist.edu.in

S. Samsudeen
Department of Computer Science and Engineering, GITAM University, Hyderabad, India

S. S. Chouhan et al. (eds.), *Artificial Intelligence Techniques in Smart Agriculture*, https://doi.org/10.1007/978-981-97-5878-4_4

Keywords

Crop monitoring · Intelligent spraying · Soil moisture monitoring · Agriculture · Automation · Pest management

4.1 Introduction

Even though there is enough food produced in the world to feed everyone, every night approximately one billion people go hungry to the bed. Food waste, poverty, conflict, and climate change are just a few of the many complicated elements contributing to this scenario. The world population is predicted to reach 9.7 billion people by 2050, which means there will be a constant demand for food. The agriculture sector is under tremendous pressure to increase food production while utilizing less resources and having a smaller negative impact on the environment (Mohan et al. 2021).

One of the main causes of the worldwide food issue is food waste. An estimated 1.3 billion tons of food are wasted annually, or one-third of all food produced. This waste happens from the point of production to the point of consumption along the food chain. Food waste is frequently the result of inadequate infrastructure and storage facilities in poorer nations, whereas improper food storage and over-shopping are common causes of food waste in wealthier nations.

Hunger is a crucial element that is also caused by conflict. Food production and distribution networks are frequently disrupted by violence in an area, which can result in famine and shortages of food. For instance, Syria is currently experiencing a serious food crisis as a result of the ongoing violence there. Global food security is also seriously threatened by climate change. Growing crops and producing food is becoming more challenging due to rising temperatures, harsh weather, and altered precipitation patterns (Bali et al. 2023). Famine and food shortages are already being caused by climate change in some regions of the world (Ryan et al. 2023).

Artificial Intelligence (AI) holds great promise for improving food security and changing agricultural methods in the face of the world's current food crisis. AI can enable farmers to make well-informed decisions, maximize resource use, and reduce their environmental impact by utilizing data from several sources (Auat Cheein and Carelli 2013). According to the World Economic Forum, integrating AI into agriculture may, for example, drastically reduce the amount of pesticide and water used in agriculture by 60% and 50%, respectively (Bechar and Vigneault 2016). This would greatly improve sustainability and conservation of resources. AI is not a solution rather it's an enabler (Fig. 4.1).

AI's incorporation into agriculture signifies a paradigm shift in our approach to food production, not only a technical advance (Bu and Wang 2019; Ryan 2022). AI can help us meet the problems posed by the global food crisis and ensure a sustainable future for agriculture by optimizing resource usage, reducing environmental impact, and enhancing food security (Singh and Kaur 2022). AI does this by using the power of data and automation.

Fig. 4.1 Artificial intelligence in various fields of agriculture

4.2 Limitations of Implementing Artificial Intelligence Techniques in Agriculture Field: A Practical Approach

A lack of understanding of AI tools is frequently the cause of the belief that AI is only useful in the digital sphere and has no bearing on the real-world application of farming. The general adoption of AI in agriculture is hampered by a lack of understanding of its workings, especially among nontechnical professions (Solanki et al. 2023a). Even with the long history of agricultural innovation, many farmers still feel at ease using conventional methods. It's possible that most farmers haven't had the chance to work on AI-related projects (Fig. 4.2).

4.2.1 Lack of Competent Labor

One major obstacle to the widespread deployment of AI in agriculture is the shortage of skilled manpower. Implementing AI involves knowledge of several fields, such as software engineering, data science, and agriculture. It can be challenging to locate experts with the required knowledge and experience, especially in rural locations where access to specialized education and training may be limited (Solanki et al. 2023b). Implementing AI techniques in farming without human intervention will lead to risky inaccurate solution (Fig. 4.3).

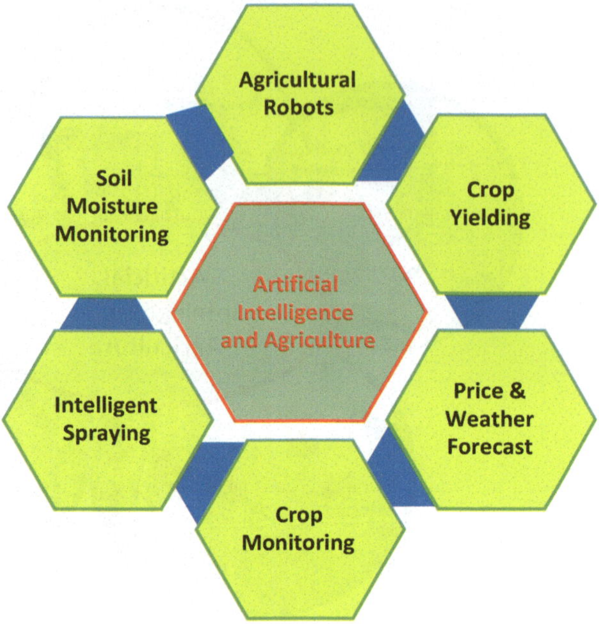

Fig. 4.2 Applications of AI in agriculture

Fig. 4.3 Need of skilled workforce for implementing artificial intelligence in agriculture

4.2.1.1 Solution

The Ministry of Skill Development introduced various term courses for the unskilled laborers and farmers to elevate them to the next level of understanding how the Artificial Intelligence techniques can be implemented in the agricultural field. The Skill Development ministry formed Sector Skill Councils (SSC) through which many private sectors in collaboration with the government address various unskilled taskforce and educate them in their field of experience (Fig. 4.4).

4.2.2 The Affordability of AI Adoption in Agriculture

The implementation of AI in agriculture is frequently hindered by considerable cost constraints, despite its tremendous potential. Many farmers find the high cost of data management software, hardware, and other resources to be unaffordable, especially smallholder farmers and those in agricultural economies with limited resources. It is critical to remove these financial obstacles and increase the affordability and accessibility of AI solutions for all farmers in order to fully achieve the transformative potential of AI in agriculture.

4.2.2.1 Factors Increasing AI's Expensive Cost in Agriculture

There are various expenses associated with implementing AI in agriculture, such as:

1. Hardware Costs: Implementing and maintaining AI-powered gear, like robotics, drones, and sensors, can be costly, particularly for large-scale operations.
2. Software Costs: The development and licensing of AI software, such as machine learning algorithms and data analytics platforms, can be expensive. Continuous software updates and maintenance increase the cost.
3. Costs associated with data management: Gathering, analyzing, and keeping track of vast volumes of agricultural data can be expensive, time-consuming, and resource-intensive. This is especially true for agricultural data from the past, which would need intensive data digitalization and retrieval operations.

4.2.3 Financial Effects on Farming Businesses

The financial effects of implementing AI can differ greatly throughout agricultural businesses. The initial cost of AI gear and software may be too high for smallholder farmers, who frequently have tight budgets and no access to funding (Solanki et al.

Fig. 4.4 Adopting new technologies and elevating the taskforce from unskilled to skilled group

2024). Furthermore, adoption of AI by smaller-scale companies may be further hampered by the absence of dependable internet connectivity and technological expertise in rural locations.

4.2.3.1 Agricultural AI's Expensive Costs: Strategies for Sustainable Adoption

The financial effects of implementing AI can differ greatly throughout agricultural firms. Smallholder farmers may find the initial cost of AI hardware and software exorbitant because they frequently have tight budgets and no access to funding. Further impeding the adoption of AI among smaller-scale companies in rural areas are unreliable internet connectivity and a lack of technical skills.

4.2.3.2 Solution: Multiple Strategies Can Be Employed to Reduce Expenses and Encourage Broader Availability of AI Technologies

1. **Open-Source AI Solutions**: Creating and applying open-source AI solutions can reduce the cost of implementing software, especially in agricultural settings with limited resources. Open-source platforms make AI more accessible to farmers by allowing them to investigate and test out AI ideas without having to pay hefty software license costs.
2. **Knowledge Sharing and Capacity Building**: Farmers can gain the skills required to properly run and maintain AI systems by supporting efforts that foster knowledge sharing and capacity building. Workshops for training, online guides, and mentorship programs catered to the need of various farming communities can be examples of this strategy.
3. **Examining Different AI Delivery Models**: Farmers can benefit from increased flexibility and a lower cost of ownership by renting or leasing AI hardware and software as needed. Furthermore, farmers may be able to acquire advanced AI capabilities through pay-per-use AI services without having to commit to long-term agreements or large capital expenditures.
4. **Collaboration and Beta Projects**: Promoting cooperative relationships among farmers, agricultural organizations, and AI technology suppliers can help in the creation and implementation of affordable AI solutions that are suited to particular agricultural requirements. Pilot programs can offer insightful information about how AI can be used in real-world agricultural settings.
5. **Government Incentives and Funds**: The financial burden on private companies can be considerably decreased by providing public money for AI research and development in agriculture. Furthermore, tax advantages or government subsidies for the use of AI can encourage farmers to make investments in AI technology (Fig. 4.5).

4.2.4 Difficulty in Data Handling

One of the primary limitations of AI in agriculture is the lack of comprehensive and high-quality data. AI models rely on vast amounts of data to learn patterns and make

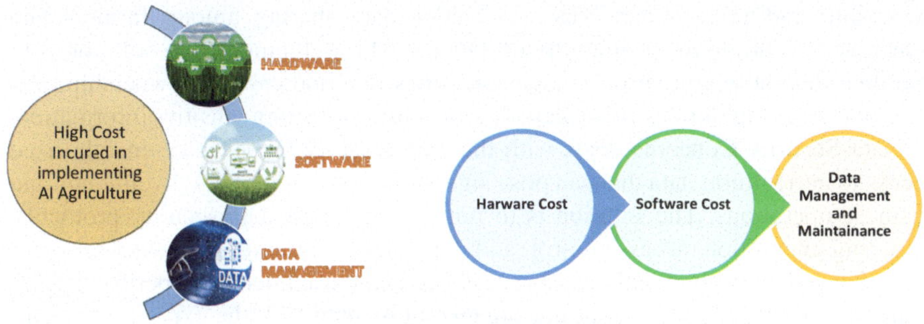

Fig. 4.5 High cost incurred for implementing AI in agriculture

informed decisions. However, agricultural data is often scattered, incomplete, and subject to errors, making it difficult for AI models to train effectively. This lack of accurate data can lead to unreliable predictions and recommendations, potentially harming crop yields and overall farm performance.

The development and application of AI-powered solutions for enhancing agricultural practices and outcomes are hampered by poor data quality and restricted data access, which are major obstacles to the application of AI in agriculture. Let us explore a few specific problems and possible remedies for every obstacle.

The first obstacle is Inconsistent Data Collection. Data collected from different farms or regions may follow different standards or protocols, leading to inconsistencies and inaccuracies. The solution is to implement standardized data collection protocols and guidelines to ensure consistency and data integrity across different sources. The second obstacle is Incomplete Data. Missing values or incomplete data records can significantly impact the performance of AI models, leading to biased or unreliable predictions. The solution is to implement data imputation techniques to fill in missing values and ensure data completeness. The third obstacle is Inaccurate Data. Data errors or inaccuracies can arise due to sensor malfunctions, human errors, or data transmission issues. The solution is to implement data quality control measures, including data validation, error correction, and anomaly detection, to ensure data accuracy. The fourth obstacle is lack of Data Labeling. Unlabeled or poorly labeled data can hinder the training of machine learning models, as they need labeled examples to learn from. The solution is to Invest in data labeling efforts, either manually or through automated techniques, to provide sufficient labeled data for training AI models.

4.2.4.1 Lack of Access to Data

Limited Infrastructure is the major concern that deals with the smallholder farmers, and those in remote areas may lack access to the infrastructure, such as sensors, drones, or broadband internet connectivity, required to collect and transmit data effectively. The solution is to develop low-cost, accessible data collection technologies, and invest in infrastructure development to enable data collection in resource-limited settings. Data Sharing Barriers is another concern to be addressed. Data ownership and privacy concerns can hinder data sharing among farmers and researchers, limiting the availability of data for AI development. The solution is to establish data sharing platforms and guidelines that address data ownership, privacy, and usage rights to facilitate data sharing while protecting sensitive information.

Data Security Concerns deals with the data security breaches or unauthorized access to agricultural data that can pose significant risks to farmers' livelihoods and business operations. The solution is to implement robust data security protocols, including encryption, access control, and data backup systems, to protect sensitive agricultural data from unauthorized access and breaches. Lack of Expertise is to be addressed. Farmers and agricultural organizations may lack the expertise to manage, analyze, and interpret large volumes of agricultural data. The solution is to provide training and capacity building programs to enhance farmers' and agricultural organizations' data literacy and skills in data management, analysis, and interpretation.

In order to create a data-driven environment for AI in agriculture and enable the development of more precise, dependable, and approachable AI solutions that enable farmers to maximize productivity, optimize resource utilization, and implement sustainable agricultural practices, we must address these issues with data quality and accessibility.

4.2.5 Extreme Climatic Condition as a Challenge for Implementing AI in Agriculture

The deployment of Artificial Intelligence (AI) in agriculture has the potential to revolutionize farming methods, boost output, and maximize resource use. The growing frequency and severity of extreme weather events, such heatwaves, floods, and droughts, is posing a growing threat to the implementation and efficacy of AI-powered crops solutions. These challenging conditions can have a substantial impact on soil health, crop yields, and total production of agricultural products, which makes it difficult to successfully integrate AI in agriculture.

The following are some ways that adverse weather conditions may hinder the effectiveness of AI-powered solutions:

Data Disruptions addresses severe weather conditions that have the potential to destroy or interfere with sensors and the infrastructure that collects data, producing unreliable or incomplete data. This may make it more difficult for AI models to detect pest and disease outbreaks, forecast crop yields with accuracy, and optimize fertilization and irrigation schedules.

The Model Sensitivity is a major challenge to be addressed. An AI system developed on historical data may not be able to sufficiently represent the variances and uncertainties brought forth by shifting climatic trends. This may lead to imprecise forecasts and suggestions that neglect to take the changing climate circumstances into consideration, which results in data sensitivity.

Infrastructure Vulnerability: Severe weather conditions have the potential to harm or destroy AI-powered devices, such as robotics and drones, interrupting business as usual and making it more difficult to apply AI solutions successfully.

Solution for Adapting AI to Extreme Climatic Conditions:

In order to address the difficulties that result from extreme weather, different approaches can be used to adapt AI-powered agricultural solutions:

Creating Climate-Resilient Data Collection Systems: You may guarantee data continuity and minimize data disruptions by utilizing resilient sensors and data infrastructure that can resist severe weather conditions. Examples of such infrastructure include waterproof sensors and cloud-based data storage.

Encompass Climate Data and Projections in AI Models: By encompassing climate data and projections in AI models, their capacity to anticipate and adjust to shifting climatic trends can be improved. In order to predict how extreme weather will affect crop yields, pest and disease risks, and resource requirements, models must be trained using historical and predicted climatic data (Fig. 4.6).

Creating AI Applications That Are Climate-Resilient: Farmers may make well-informed decisions based on current climate data by creating AI systems that integrate real-time weather monitoring and forecasts. With respect to changing climatic conditions, this can include suggestions for modifying irrigation schedules, crop choices, and pest and disease control techniques.

Encouraging Knowledge Sharing and Capacity Building: We can increase the efficacy of AI deployment in the context of climate change by giving farmers the

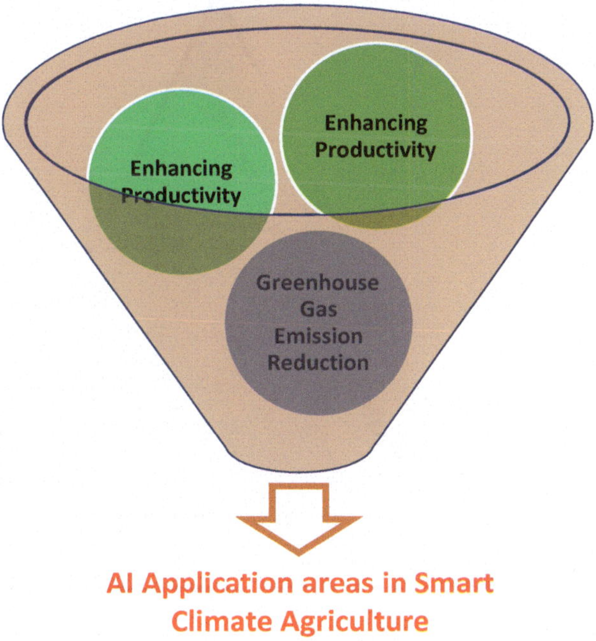

Fig. 4.6 AI application areas in smart climate agriculture

tools and training they need to understand how extreme weather conditions affect AI-powered solutions and how to adjust their usage accordingly.

Climate smart agriculture can be adopted by the farmers in order to overcome the limitations of implementing AI in agricultural field during extreme climatic changes.

4.2.6 Requirement of High Accuracy and Precision

A high level of precision and accuracy is necessary for many operations in agriculture. This may also limit the use of Artificial Intelligence in agriculture. Some of the key areas in which high accuracy and precision are required are Irrigation Management, Crop Health Monitoring, and Targeted Spraying (Fig. 4.7).

Irrigation Management deals with water conservation and production of crops that solely rely upon effective irrigation. Artificial Intelligence (AI)-powered irrigation systems must be able to precisely measure soil moisture content and plant water requirements, and they must be able to modify irrigation levels accordingly. When we focus on crop health monitoring, accurately recognizing and diagnosing pests, plant diseases, and nutritional deficits is essential for preserving crop health and productivity. Artificial Intelligence (AI) algorithms must possess the ability to differentiate between plants that are healthy and those that are not, as well as the precision to recognize particular illnesses and pests. In order to reduce the negative effects of pesticides and herbicides on the environment and human health, target spraying concentrates on the precise application of these chemicals. Artificial

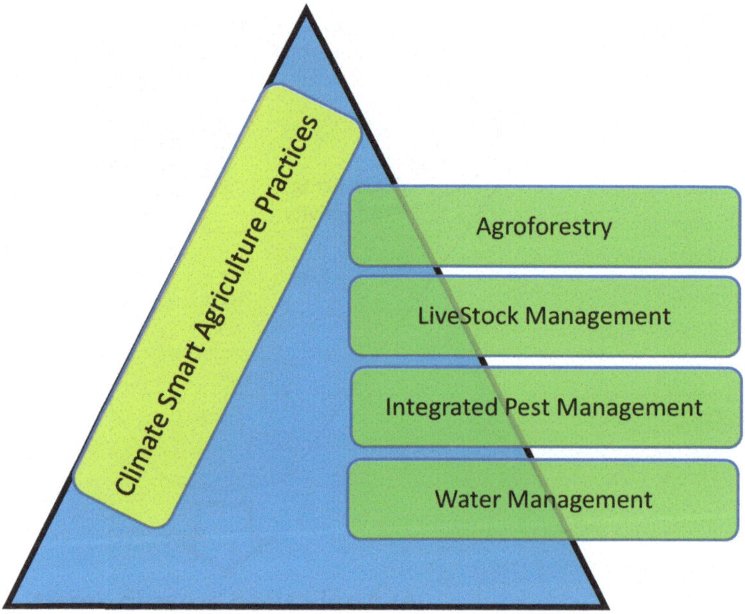

Fig. 4.7 Climate smart agriculture practices

Intelligence (AI)-powered spraying systems must be able to precisely detect and target weeds and pests, as well as deliver pesticides at the proper dose and area.

Although AI has the potential to completely transform agriculture, creating AI algorithms that can match the high standards of precision and accuracy associated with these operations can be difficult. The amount or quality of data needed to train AI models can occasionally be scarce, which further reduces the accuracy of AI-powered systems. Additionally, it may be challenging to create AI algorithms that can adjust to shifting conditions due to the complex and dynamic nature of agricultural ecosystems. AI is still being utilized to advance agricultural techniques in a variety of ways, despite these obstacles. AI is being used, for instance, to create prediction models that assist farmers in determining which parts of their fields are most likely to be impacted by pests or illnesses. AI is also being used to create automated nutrition management and irrigation systems. AI technology is expected to become more and more significant in agriculture as it advances.

4.2.7 Security Risks

Regulations governing the application of AI in all areas are still mostly absent. In particular, there are certain legal concerns when applying AI to smart farming and precision agriculture. Cyberattacks and data leaks, for instance, pose a severe risk to farmers' operations. It's even possible that hackers may attack AI-based farming systems in an attempt to sabotage food supplies. Hence, AI techniques related to security and privacy risks can be applied in order to prevent such risks while implementing AI techniques in agricultural field.

4.3 Conclusion

Although Artificial Intelligence (AI) has a lot of potential to change agriculture, there are obstacles along the way. Widespread adoption of AI-powered solutions is hampered by the complexity of agricultural ecosystems, the requirement for high precision and accuracy, and the lack of high-quality data as discussed in this chapter. The integration of AI into agricultural practices is further complicated by constraints on technical infrastructure, financial considerations, and farmer distrust.

The potential advantages of AI in agriculture are evident, regardless of these difficulties. AI may simplify a variety of agricultural activities, optimize irrigation management, improve targeted spraying, and monitor crop health. These improvements can raise production, lessen environmental effect, and better utilize available resources. A diversified strategy is required to properly achieve AI's revolution in agriculture. This involves developing user-friendly and accessible AI technologies, establishing clear regulatory frameworks to ensure the responsible and ethical use of AI in agriculture, and encouraging collaboration between AI developers, researchers, and agricultural stakeholders to address data scarcity and quality issues.

We can use AI to transform agriculture and guarantee sustainable food production as well as a secure environment for future generations by addressing these challenges and promoting a cooperative and innovative atmosphere.

References

Auat Cheein FA, Carelli R (2013) Agricultural robotics: unmanned robotic service units in agricultural tasks. IEEE Ind Electron Mag 7(3):48–58

Bali A, Chouhan SS, Kumar G et al (2023) Adaptive fault-tolerant control for pure-feedback stochastic nonlinear systems with sensor and actuator faults. Circuits Syst Signal Process 42:5838–5867. https://doi.org/10.1007/s00034-023-02366-7

Bechar A, Vigneault C (2016) Agricultural robots for field operations: concepts and components. Biosyst Eng 149:94. https://doi.org/10.1016/j.biosystemseng.2016.06.014

Bu F, Wang X (2019) A smart agriculture IoT system based on deep reinforcement learning. Futur Gener Comput Syst 99:500. https://doi.org/10.1016/j.future.2019.04.041

Mohan S, Regatti V, Rahaman S, Vinayak M, Battu B (2021) Role of AI in agriculture: applications, limitations and challenges: a review. Agric Rev 44(2):231–237. https://doi.org/10.18805/ag.R-2215

Ryan M (2022) The social and ethical impacts of artificial intelligence in agriculture: mapping the agricultural AI literature. AI Soc 38:2473. https://doi.org/10.1007/s00146-021-01377-9

Ryan M, Isakhanyan G, Tekinerdogan B (2023) An interdisciplinary approach to artificial intelligence in agriculture. NJAS Impact Agric Life Sci 95:1

Singh P, Kaur A (2022) Chapter 2—A systematic review of artificial intelligence in agriculture. In: Poonia RC, Singh V, Nayak SR (eds) Cognitive data science in sustainable computing. Deep learning for sustainable agriculture. Academic Press, p 5780. https://doi.org/10.1016/B978-0-323-85214-2.00011-2

Solanki S, Singh UP, Chouhan SS, Jain S (2023a) Brain tumour detection and classification by using deep learning classifier. Int J Intell Syst Appl Eng 11(2s):279. https://ijisae.org/index.php/IJISAE/article/view/2624

Solanki S, Singh UP, Chouhan SS (2023b) Brain tumor classification using ML and DL approaches. In: 2023 IEEE 5th international conference on cybernetics, cognition and machine learning applications (ICCCMLA), Hamburg, Germany, pp 204–208. https://doi.org/10.1109/ICCCMLA58983.2023.10346854

Solanki S, Singh UP, Chouhan SS et al (2024) A systematic analysis of magnetic resonance images and deep learning methods used for diagnosis of brain tumor. Multimed Tools Appl 83:23929–23966. https://doi.org/10.1007/s11042-023-16430-6

Smart Farming Management System: Pre and Post-Production Interventions

5

Narendra Singh Chandel, Subir Kumar Chakraborty, Dilip Jat, and Pooja Chouhan

Abstract

This chapter provides a comprehensive overview of the agricultural landscape, focusing on the integration of advanced technologies to revolutionize farming practices. The necessity of smart farming in addressing global challenges such as increasing food demand, resource scarcity, and the impacts of climate change has been highlighted. Smart farming, driven by technologies like precision agriculture, digital farming, artificial intelligence (AI), and the Internet of Things (IoT), aims to enhance crop yield, reduce resource waste, and improve overall efficiency. The content delves into various interventions in agriculture, including precision machinery, robotics, irrigation water management, and postharvest management. It discusses how precision machinery and AI-driven technologies optimize agricultural processes, reduce labor costs, and improve productivity. Robotics, such as autonomous vineyard management robots and fruit harvesting robots, offer innovative solutions to labor-intensive tasks and enhance operational efficiency. Furthermore, the importance of irrigation water management in sustainable agriculture, showcasing technologies like soil water sensing, automated irrigation systems, and satellite monitoring for precise irrigation scheduling and early detection of crop water stress, has been emphasized. In the realm of postharvest management, smart farming explores the role of AI in automating sorting, grading, and packaging operations, leading to improved quality control and reduced post-harvest losses. Smart farming highlights advancements in deep learning (DL) for detecting food adulteration and thermal imaging for identifying bruises in agricultural produce. Overall, this chapter presents a comprehensive view of how the integration of advanced technologies in smart farming

N. S. Chandel (✉) · S. K. Chakraborty · D. Jat · P. Chouhan
ICAR-Central Institute of Agricultural Engineering Bhopal, Bhopal, India

S. S. Chouhan et al. (eds.), *Artificial Intelligence Techniques in Smart Agriculture*, https://doi.org/10.1007/978-981-97-5878-4_5

management systems enhances agricultural sustainability, productivity, and ensures food security in the face of evolving global challenges.

Keywords

Deep learning · Digital farming · Irrigation scheduling · Machine learning · Postharvest management · Precision farming · Robotics · Smart farm management · Water management

5.1 Introduction

The global population is projected to reach 9.7 billion by 2050, necessitating a significant increase in food production (De Wrachien et al., 2021). According to projections from the Food and Agriculture Organization (FAO), agricultural output must rise by 50–60% to meet the ever-growing demand for food. Considering the expanding population and its increasing needs, technological advancements in agriculture are crucial. Additionally, various significant factors, such as the quality of produce, global competitiveness, productivity, and the cost of production, compel the adoption of modern agricultural techniques. Despite a potential doubling of agricultural production by 2050 in developing nations, the risk persists that one in twenty individuals may face undernutrition. To enhance nutrition and address issues of food insecurity and undernourishment, future agricultural production must outpace the rate of population growth. Achieving these goals requires sustainable intensification, involving the efficient utilization of land and water resources in a smart manner without causing any detrimental effect to their existence.

To tackle these challenges, it is essential to effectively utilize resources and implement smart technologies. In this context, resource management plays a pivotal role. Thus, the term coined for this type of resource management is known as smart farm management (SFM). SFM is crucial for improving the quality and timeliness of farm operations (Mehta et al., 2020). It encompasses a wide array of management techniques and strategies aimed at sustaining a farm's profitability and efficiency. SFM involves the management of various elements, including land, water, pre and postharvest operations, covering all aspects from cultivation and intercultural operations to irrigation, harvesting, processing, storage, and transportation of crops.

5.2 Problem with Modern Agriculture

Numerous factors, including pest and disease outbreaks, inadequate water availability, low soil fertility, uneven sowing, and the impacts of climate change, are collectively responsible for a 15.7% loss in crop yield in India. Challenges related to soil and water management further contribute to the stagnation of agricultural productivity. Farmers grapple with issues stemming from both the overuse and underuse of fertilizers. Excessive application leads to runoff and erosion, causing soil infertility, while insufficient use results in lower yields. Pest problems present a significant threat to crop yields, and farmers encounter difficulties in optimizing pesticide usage due to uncertainties about pest-infested areas. The overarching objective is to address these multifaceted challenges and support the backbone of the Indian economy—the farmers—by enhancing productivity, ultimately contributing to a healthier economy. The slow adoption of modern technology in India and the lack of systematic and scientific approaches in farming practices contribute to this challenge (Singh and Parihar, 2015). The 2023 IPCC report underscores the importance of implementing adaptation measures within the agriculture system. These measures include advancements in cultivars, on-farm water management and storage, conservation of soil moisture, irrigation, agroforestry, and agroecology. Adopting these practices is crucial for sustainable agricultural development and improved resilience in the face of ongoing environmental and climatic challenges.

5.2.1 Why Smart Farm Management

Over the past decade, there has been significant advancement in agricultural technology and equipment, primarily driven by the development and adoption of automation technologies. The capacity of various machines to collect and analyze data has also played a pivotal role in this progress. The incorporation of these new technologies into farming practices is referred to as smart farming (SFM), making it essential for the sustainable development of agriculture. The roots of smart farming can be traced back to the early 1990s (Wells and Stock, 2020). Smart farming involves the integration of advanced technologies, including precision farming, digital farming, artificial intelligence (AI), Internet of Things (IoT), drones, remote sensing, and the Geographic Information System (GIS). This integration amplifies the sustainability and effectiveness of agricultural practices. Recent technological advancements, such as robotics, AI, machine learning (ML), deep learning (DL), and others, function as key components in the implementation of smart farming.

Smart farming aims to bolster the viability of agriculture by enhancing crop yield, reducing resource waste, and improving overall efficiency. This approach holds the potential to address global challenges such as the growing world population and the increasing need for food production due to factors like climate change, biotic and abiotic stresses. Through smart farming, key aspects like crop growth monitoring, maintaining product quality and traceability, controlling irrigation, identifying pests and diseases, minimizing postharvest losses, precision sowing, precise fertilizer application, and mitigating storage losses can be significantly enhanced. The advantages of smart farming include cost savings, increased productivity, improved crop quality, and ensuring robust fault tolerant system with a more informed decision-making (Bali et al., 2023).

Precision agriculture involves the precise utilization of resources to maximize output while minimizing losses. The use of drones and unmanned vehicles is also on the rise in agriculture, aiming to reduce human effort and enhance accuracy. The primary role in this advancement is played by global positioning system (GPS) satellite data. In the realm of precision agriculture, decision support systems are pivotal technologies aiding decision-makers in applying models to solve problems within complex and uncertain circumstances. On the digital farming front, farmers have the capability to manage their farms through mobile devices or laptops. Digital agriculture incorporates technologies such as the IoT and cloud computing. The IoT, often described as a system, connects physical objects, embedded sensors, and integrated software, utilizing diverse technologies to facilitate data transfer among various devices over the internet. It serves as an intelligent farming solution, enabling growers to remotely monitor their crop fields. This involves the use of sensors to track factors such as soil moisture, crop health, livestock conditions, and temperature.

Artificial sensing technologies empower farmers to gather visual data from their farms, extending their sensing capabilities beyond the visible spectrum. Machine vision systems provide crucial on-farm measurements, assisting farm managers in decision-making and enhancing the precision of tasks performed by machines. Unlike human observations, machine vision systems consistently conduct objective measurements without succumbing to fatigue. Consequently, they improve the efficiency, accuracy, and understanding of farm operations and food production. In agriculture, deep learning networks play a crucial role in predicting scenarios and identifying causes. Their proficiency is particularly evident in computer vision tasks, such as image classification, detection, and segmentation (Long, 2016; LeCun et al., 2015). Essentially, these networks consist of numerous simple modules organized in layers, simultaneously learning and processing nonlinear input-output mappings. As a result, deep learning modules can adeptly discern differences, such as distinguishing between a diseased leaf and a healthy leaf, without being influenced by factors like background, orientation, lighting, or surroundings (Chakraborty et al., 2022).

5.2.2 Agriculture Interventions

Over the past few decades, agriculture has experienced numerous technological revolutions, evolving into a more industrialized and technology-driven sector. Farmers, utilizing intelligent agricultural interventions, have gained enhanced control over animal husbandry and crop cultivation, resulting in greater predictability and efficiency. This, coupled with a growing consumer demand for agricultural products, has contributed to the global proliferation of smart farming technology. The demand for food is expected to increase with the elevated or growing demographics. Currently, there is a high level of stress on the utilization of natural resources. Climate change is contributing to a reduction in agricultural productivity. Addressing these challenges will necessitate the utilization of modern technological interventions.

Agriculture interventions refer to a broad range of strategies, policies, and actions for improving productivity, enhancing agricultural practices, and addressing challenges in the farming sector. These engineering interventions encompass postharvest handling methods, storage systems, farm structures, erosion control, water management (including water resource development, irrigation, and drainage), meteorological systems, and techniques for optimal utilization of the aforementioned aspects.

Amid the growing demand for food, the interplay of climate variability and diminishing fertile land availability imposes additional challenges on future farming systems. Objectives such as enhanced food quality, food security, and sustainable agricultural practices are closely connected in a marketplace with international distribution among countries with different rules, regulations, and resources. To achieve better and more consistent yields within limited spaces and budgets, informed decision-making and improved risk management are essential.

To address these challenges, comprehensive measurements must be documented throughout the food production and supply chain providing insights into existing inefficiencies and opportunities for enhancement. The utilization of wireless remote sensing is increasing across diverse equipment. In the future, robotics and automation are set to play a pivotal role in meeting upcoming agricultural production needs, revolutionizing approaches to cultivating and tending to harvested foods. Despite primarily remaining in the experimental stage, research and development in this domain have attracted substantial interest from both the government and private sectors, this is spurred by the prospect of robots improving productivity, addressing labor shortages, and presenting practical management designs.

Sensing has always been integral to agricultural production, with farmers traditionally relying on sight, smell, touch, and taste to analyze their produce and guide management decisions. Among these senses, vision emerges as arguably the most prevalent input in on-farm decision-making processes. The integration of artificial sensing technologies, such as machine vision systems, augments a farmer's capacity to gather visual information from their farm, extending their sensing capabilities beyond the visible range. This technological advancement plays a crucial role in meeting the evolving needs of modern agriculture. Numerous research studies have been conducted, such as the estimation of reference evapotranspiration (ETo), which is one of the crucial parameters for crop irrigation since it defines irrigation scheduling, the application of ANN models in smart irrigation water management, the involvement of cloud computing to manage the numerous databases, the integration of IoT into agricultural machinery, the application of remote sensing and GIS in land use, the identification of changing crop patterns, and mapping.

The demands of extensive data storage and dynamic system expansion in agrotechnical extension information systems can be addressed by the incorporation of cloud storage technology. This technology serves as a scalable and highly reliable storage mode, capable of supporting mass data storage and the dynamic expansion of resources. The Decision Support and Automation System (DSAS) empowers farmers and growers by providing control over various applications through its web portal. DSAS operates at different stages, facilitating the simultaneous interconnection of multiple devices and delivering real-time data to the farmer. The farmer assumes a crucial role in monitoring real-time data and controlling machines through software. Specialized systems like the spray controller ensure precise pesticide application in predefined amounts, while the irrigation controller manages irrigation, and the fertilizer controller oversees fertilizer application (Chandel et al., 2016). DSAS operates by processing data from various sensors, such as soil moisture and nitrogen sensors.

Utilizing an object-oriented methodology for image recognition that mirrors human perception, this technology stands at the forefront globally in terms of the level of automation and the quality of applied results. The broad applicability and transformative capabilities of advanced image recognition technologies encompass a multitude of domains, as we know that various fields are widely utilizing modern technology from healthcare to agriculture (Solanki et al. 2023a; Solanki et al. 2023b). In healthcare, for instance, they enable the early detection and precise diagnosis of diseases (Saxena et al., 2024; Solanki et al., 2024). Similarly, such technologies can be adapted for early pest detection in crops (Kumar et al., 2023) and weed identification (Modi et al., 2023). It proves effective in analyzing diverse image types, including black and white, color, high or low quality, and both simple and intricate spatial pictures. The technology serves as a crucial link between data processing for remote sensing and geo-information systems. It is projected that, utilizing new methods, the IoT could enhance agricultural productivity up to 70% by the year 2050. The primary benefits of IoT adoption include increased crop yields and reduced costs. For instance, on average, farms implementing IoT experience a 1.75% rise in yield, coupled with a decrease in energy costs ranging from 17 to 32 USD per hectare and an 8% reduction in water usage for irrigation (Gralla, 2019). These advancements not only contribute to yield improvements but also bring about cost savings.

Agricultural interventions are crucial for the expansion of cultivated areas, increasing cropping intensity, and enhancing productivity. A study shows that by 2050, there will be over 75 million agricultural IoT devices in operation, and the average farm is expected to generate 4.1 million data points daily, a significant increase from 190,000 in 2014 (De Clercq et al., 2018). The capacity to mobilize capital demonstrated for machinery installation is a crucial factor; farms with higher capitalization are more inclined to transition to robotic systems.

5.3 Irrigation Water Management

The perpetual utilization and inherent limitations of water resources are increasingly constraining their utility for agricultural purposes. These resources are indispensable for agriculture and rural development, serving as foundational elements intricately linked to global challenges such as food insecurity, poverty, and climate change adaptation and mitigation. Furthermore, the degradation and depletion of natural resources exacerbate the livelihood challenges faced by millions of rural inhabitants worldwide. Research indicates that irrigated agriculture, in contrast to dryland farming, can substantially enhance crop production, with yields of forage and grain crops doubling under irrigation. This has been empirically demonstrated in field studies, underscoring the importance of effective irrigation water management strategies (Colaizzi et al., 2009). In response to these challenges, various innovative technologies have emerged in the field of soil and water engineering. Advancements such as soil water sensing technologies and software-based irrigation scheduling tools are revolutionizing the management of land and water resources in agriculture. These technologies enable more precise and efficient irrigation practices, optimizing water usage while maximizing crop yields. By harnessing real-time data and analytical tools, farmers can make informed decisions regarding irrigation scheduling, ensuring optimal soil moisture levels for crop growth while minimizing water waste (Viscarra Russel and Bouma, 2016).

Extensive research has been undertaken to address the challenges of water waste and scarcity associated with conventional irrigation methods. In response to these issues, automated irrigation system are being developed, utilizing the GPRS module as a communication device. Such systems are seamlessly integrated into a microprocessor-based gateway, which is tasked with regulating the quantity of water used. The study conducted showcase a remarkable 90% increase in water savings compared to conventional irrigation systems (Gutiérrez et al., 2014).

Sensors are valuable tools that consistently record measurements, store data in logging units, or transmit data to a base station or the cloud via cellular communication. Soil moisture, temperature, nutrient, water level, weather, tensiometers, and evapotranspiration sensors play a crucial role in assisting farmers and land managers in refining site-specific soil nutrients and irrigation water management. Time domain and water content reflectometry sensors provide precise volumetric soil water content measurements, typically within a range of 1–2% of volumetric water content. By effortlessly setting up these sensors as an automated monitoring system, farmers can consistently gauge soil water content and make informed irrigation decisions. Implementing a smart irrigation system, informed by these sensor readings, can significantly reduce water waste and decrease water consumption by up to 20% (Gupta et al., 2016).

Artificial neural networks (ANNs) are increasingly integrated with sensors for precise irrigation management. ANNs utilize data collected by sensors such as soil moisture, temperature, and weather, to optimize irrigation scheduling. By analyzing sensor inputs, ANNs can accurately predict irrigation needs, enhancing water efficiency and crop yields in agricultural practices (Arif et al., 2012). An intelligent system was developed to regulate sprinkler valves based on real-time data acquired through a wireless data acquisition system, measuring plant canopy temperature and height in situ. This system incorporated temperature and moisture sensors strategically placed in the field. Furthermore, two artificial neural network models were created to predict soil moisture in paddy fields using a reduced set of meteorological data.

Identifying crop water stress is crucial for effective irrigation water management, with satellite monitoring plays a significant role in detecting vegetation water stress for precision agriculture. This enables precise irrigation timing to prevent crops from experiencing water stress (Chandel et al., 2024), ensuring yield even under limited water conditions. A comparative evaluation was conducted on three deep learning (DL) models—AlexNet, GoogLeNet, and Inception V3—for detecting water stress in maize, okra, and soybean crops. A total of 1200 digital images were gathered for each crop to create the input dataset for the deep-learning models. Among the three models, GoogLeNet demonstrated superior performance, achieving accuracy rates of 98.3%, 97.5%, and 94.1% for maize, okra, and soybean, respectively (Chandel et al. 2021).

Noninvasive methods such as RGB, thermal imaging, and optical sensing can reveal plant water stress in advance. After successfully utilizing a deep learning model (DarkNet53) on both RGB and thermal images, water stress was identified. Testing accuracy for identifying water stress in RGB imagery ranged from 94.1 to 92.3%, while for thermal imagery, it was even higher, ranging from 95.7 to 99.6%. This investigation underscores the efficacy of nondestructive, expeditious vision-based techniques utilizing digital imagery for comprehending plant development, detecting water stress, and ultimately evaluating crop yield (Rajwade et al., 2023).

5.4 Farm Mechanization

Agricultural mechanization involves the deployment and utilization of diverse power sources (manual, animal, and motorized) and engineering technologies with the goal of improving agricultural production. It is a dynamic and integral element of progressive farming, utilizing technology, equipment, and farm implements to increase the efficiency of land and labor on farms. This comprehensive approach, known as agricultural mechanization, aims to maximize outputs and enhance food and agricultural yields. The concept of farm mechanization encompasses the planning, production, distribution, maintenance, repair, and overall application of farm tools and machinery (FAO, 2013). By alleviating labor-intensive tasks that traditionally impede efficient food and agricultural production, agricultural mechanization contributes to increased competitiveness and lower consumer prices. It extends beyond the use of tools and power machinery to include automation, control, and robotics.

Recent advancements in technology have significantly improved the mechanization of plant tillage and on-farm production, simplifying labor supervision and potentially reducing losses during large-scale agricultural operations. The increasing costs associated with labor are a significant factor contributing to diminishing profitability in agriculture. Typically, the service price of agricultural machinery is lower than labor costs, making it a cost-effective alternative for farmers. The adoption of agricultural machinery can lead to substantial reductions in labor expenses. Furthermore, agricultural machinery efficiently performs tasks such as land leveling and preparation, promoting the optimal utilization of agricultural resources and reducing the reliance on weed and insect pest control (Nam et al., 2021). Additionally, agricultural mechanization enables combined fertilization and sowing, ensuring both sowing accuracy and reducing expenditures on seeds and fertilizer. Recent and emerging technological advancements, including the Internet of Things (IoT), drones, and robots, are playing a pivotal role in accelerating transformations in agricultural mechanization on a global scale (Heikkila et al., 2012).

5.4.1 Precision Machinery

Precision farming is a more accurate way of farming that takes away the repetitive and hard-working parts of traditional farming. The main technologies that make precision farming special are high-precision positioning systems, geological mapping, remote sensing, electronic communication, variable rate technology, tools for figuring out the best times to plant and harvest, managing water resources, and taking care of plant and soil nutrients, including dealing with pests and rodents.

Planting crops can be challenging due to various external environmental factors, imprecise sowing technology, and significant errors. To address these issues, machine vision technology is introduced by Bing le and Jiyun li to explore the key aspects of precise crop sowing based on visual principles. A precise simultaneous sowing control system for maize seeds and fertilizer was developed, relying on an innovative automatic alignment control algorithm. The test results demonstrated improved seed-fertilizer alignment at various feed speeds (2, 3, and 4 km/h) and grain spacings (20, 25, and 30 cm). This system met the criteria for precise sowing, enhancing the utilization rate of fertilizer.

In recent advancements, the Indian Council of Agricultural Research (ICAR) has made significant contributions by developing precision agriculture technologies. These include a low-cost SPAD meter (Fig. 5.1), a spectral reflectance-based fertilizer applicator, a uniform rate sprayer, a real-time soil moisture-based sprinkler irrigation system, an automatic irrigation system for rice, and an automatic yield monitor for indigenous combine harvesters, among others (Jat et al., 2023). Addressing the socioeconomic significance of agriculture in India, a predictive system has been created to anticipate grape diseases. Traditionally, grape plant anomalies were only detected post-infection, negatively impacting entire vineyards. The system incorporates sensors such as temperature, leaf wetness, and humidity sensors deployed in the vineyard, transmitting data to a ZigBee server database (Gondchawar and Kawitkar, 2016). Adhering to Zigbee alliance standards, the Wireless System Network (WSN) employs open global standards. Patil and Thorat (2016) explore the application of machine learning and the Internet of Things (IoT) for timely grape disease identification. Weed management poses a prevalent challenge for farmers, and computer vision offers a potential solution. Using various algorithms, Convolutional Neural Networks (CNN) can not only differentiate between plants but also provide relevant data for plantation purposes. Distinguishing between weeds and desired crops presents a specific challenge, and CNNs prove instrumental in addressing this issue.

5.4.2 Robotics

The use of robotics in agriculture is progressing alongside significant changes in the whole eco system of agricultural mechanization. In 2019, agriculture employed 27% of the global workforce (World Bank, 2021), but there has been a decline in agricultural employment due to several underlying factors, with a 50% decrease

Fig. 5.1 Low-cost SPAD meter for plant health monitoring

observed between 1990 and 2017. Human intervention will be reduced in agriculture as automation is becoming crucial in various fields, highlighting the importance of early-stage design. One example of a smart, remote-controlled robot designed for agricultural tasks is the Terrasenta. This compact robot utilizes LIDAR technology to monitor crops by detecting light underneath the canopy. Its autonomous navigation is achieved through a LIDAR-based algorithm, eliminating the need for an RTK/GNSS system. The Terrasenta simplifies the evaluation of various crop characteristics such as plant height, stem diameter, leaf area index, and plant count. It enhances overall agricultural efficiency by automating vital aspects of crop monitoring with its cutting-edge capabilities. Another notable development is the prototype of an apple harvesting robot. This robot incorporates a 6-DoF robotic arm, a soft-finger-based gripper, and an RGB-D camera. It demonstrates exceptional performance in tasks including vision perception, motion planning, fruit verification, and fruit detachment. Notably, the utilization of the DasNet deep convolutional neural network for fruit recognition achieves an impressive success rate of 97%. The incorporation of advanced technologies in this apple harvesting robot highlights its effectiveness and accuracy in automating the crucial processes of apple farming. Deep learning enabled computer vision-based yield estimation of fruits was developed with higher accuracy (Fig. 5.2).

Introducing the Amos robot, a groundbreaking autonomous system specifically designed for vineyard management. This innovative agricultural robot excels in various tasks such as seeding, mowing, spraying, and tilling. It incorporates an advanced auto-detection system that ensures precision and accuracy in its operations. Safety is prioritized with the inclusion of an emergency stop capability. The Amos robot goes beyond its operational capabilities by offering a field mapping function. This feature provides vineyard managers with valuable insights to optimize cultivation practices. With runtimes ranging from 4 to 8 h, the Amos robot is a reliable workhorse in the vineyard. Its impressive recharge time of only 2 h further enhances its operational efficiency. In an effort to safeguard rural workers from the potential risks posed by pesticide products, researchers have developed two robots: AgriRobot and SAVSAR. These robots are designed to remotely spray pesticides in vineyards using a Human Machine Interface (HMI) for interaction. Pruning plants is a crucial task but can be challenging for robots to execute. Research conducted in pome orchards has identified the primary challenges of this task, including scanning and measuring plant structures to determine precise pruning locations. The Agrobot E-Series robot addresses these challenges by incorporating a sturdy construction made entirely from stainless steel and military-grade aluminum. This ensures its durability in various climatic conditions. With 24 independent Cartesian robotic arms, this robot effectively picks strawberries by maneuvering around its structure.

Fig. 5.2 Image-based yield estimation using YOLO model

It utilizes data from the LiDAR sensor to prevent collisions with farm workers, prioritizing safety in its operations.

5.5 Postharvest Management

Postharvest losses and wastage of perishable horticultural goods are globally significant, particularly in developing nations. It is estimated that postharvest losses and food waste can cumulatively reach up to 60%, depending on the commodity, season, and production region. Fruit crops, for example, experience postharvest losses ranging from 16 to 36% each year, attributed to mechanical, microbial, and physiological factors. Advancements in processing technology and handling play a crucial role in enhancing food safety by substantially reducing pathogens and minimizing postharvest losses. This, in turn, strengthens the global food supply. Processing methods like milk pasteurization, canning, and freezing are pivotal in mitigating health risks associated with food consumption. Over time, the overall risks of illness and mortality from bacteria and contaminants have significantly decreased, even though these threats still exist.

Artificial intelligence (AI), a subset of computer science focused on intelligent algorithms, play a vital role in automating sorting, grading, and packaging operations within the fruit industry. Diverse AI methodologies, including machine learning, computer vision, natural language processing, robotics, predictive analytics, autonomous agents, and expert systems, contribute to optimizing the efficiency of production processes and pricing decisions (Jha et al. 2019; Wakchaure and Patle, 2023). These AI applications encompass a wide array of tasks within postharvest management, including detecting bruises, identifying maturity, categorizing fruits, assessing quality, detecting insects, and monitoring conditions. By leveraging intelligent algorithms, AI optimizes the handling and assessment of crops, leading to improved sorting precision, grading accuracy, and overall quality control. Various types of fruits, vegetables, meat cuts, and grain kernels undergo grading and sorting processes through the utilization of machine vision systems. There are tailor-made automated packing lines for the immediate sorting of spherical horticultural produce, utilizing sorting algorithms that are customizable for color and weight. The customized lightweight Convolutional Neural Network (CNN) model, "SortNet," developed at ICAR-CIAE Bhopal is designed for vision-based classification of citrus fruits into "accept" and "reject" categories based on surface characteristics. "SortNet" demonstrated comparable accuracy to advanced pretrained deep learning models but with less complexity and computational time (Chakraborty et al., 2023).

The system known as SmartSortD, developed by Smart Vision Works (Orem, Utah), employs machine vision technology to identify and categorize items based on factors such as maturity, size, and skin texture. The Shibuya Group (Mameda-honmachi, Kanazawa, Japan) possesses sorting machines incorporating machine vision hardware, which includes monochrome cameras, color cameras, NIR cameras, and electron beam tomography equipment. The images captured by these devices are analyzed to assess the quality attributes such as sweetness, acidity, internal defects (such as skin separation or hollow parts), size, shape, diameter, color, and blemishes of diverse fruits and vegetables on packing lines. In recent research applying deep learning to agriculture, a fully convolutional network was utilized for the identification of crop from weeds and soil using substantial synthetic data (Mortensen, 2016).

Deep learning has effectively tackled various facets of food adulteration, making it a powerful tool for the rapid, noncontact, and efficient detection of contamination in agricultural produce. In mid-infrared spectral data, CNN models were found to be highly accurate in the classification of strawberry and non-strawberry purees (Acquarelli et al., 2017). High accuracy in determining the adulteration of various meats, including chicken, turkey, and pork, was observed in CNN models. Thermal imaging proves to be a versatile technology for identifying blueberry bruises, achieving an impressive 88% accuracy rate by employing a combination of multivariate algorithm feature sets (Kuzy et al., 2018). Ongoing investigations are exploring the potential of thermal imaging in assessing guava bruises and examining the impact of temperature variations on the detectability of these bruises.

5.6 Recommendations, Limitation, and Suggestions for Future

Automation is a big part of farming now, and it's going to be even more crucial in the future. Farmers currently use drones and advanced sensors to keep an eye on their crops. These smart devices can tell them when to apply the inputs (water, pesticide, or fertilize the fields), and they also check the soil quality to make sure the crops are safe from drought. Furthermore, through IoT technologies, farmers can set up automated irrigation systems, ensuring efficient management of water resources. By gathering data on crop conditions such as moisture and temperature, IoT helps determine the optimal amount of water needed for crops in each season.

- In the future, technology will keep growing and spreading to new areas, bringing along new challenges to solve. Creating big sets of data is crucial for these advancements. Also, as farming becomes more automated, there will be a higher demand for people with the right skills.
- Still, many challenges arise with the development of smart agriculture, such as the reliability of the system, the occurrence of system failures, or the presence of virus threats, which consistently pose a significant challenge within the hardware or software framework of digital agriculture. For this, the safety features of hardware and software should be highlighted, and it should be done in a user-friendly manner.
- Challenges arise from the complex nature of agriculture and the diversity of organism. Though artificial intelligence's rapid progress suggests a crucial role for computer vision in agricultural automation, presently, computer vision is mainly applied in monitoring, plant protection, and harvesting for specific crops,

with limited adoption across all farming aspects. The limitations associated with applying contemporary technologies to agricultural mechanization include technological constraints, sociocultural and behavioral limitations, financial and economic challenges, and environmental concerns.

- Overcoming challenges in diverse scenarios for the functionality of a technology is a current obstacle. Future expectations involve the collaboration of computer vision technology with intelligent systems like deep learning. This synergy, driven by extensive data utilization, is foreseen to enhance performance across all aspects of farming. The widespread integration of this combination is anticipated to address existing agricultural challenges and advance automated farming systems in a more intelligent direction.

Moreover, the future development of the state-of-the-art AI farm intervention for smart machinery should account for the operational requirements and the necessity to integrate efficient control solution devices like sensors, actuators, drives, switches, and similar components. Little headway has been made in confronting the challenges associated with complex agricultural robotic machines. To attain broad adoption, machines should exhibit traits such as low power consumption, versatility, precision, compactness, light weight, and affordability. Utilizing locally available materials in machine fabrication can contribute to lowering manufacturing costs.

References

Acquarelli J, Van Laarhoven T, Gerretzen J, Tran TN, Buydens LMC, Marchiori E (2017) Convolutional neural networks for vibrational spectroscopic data analysis. Anal Chim Acta 954:22–31

Arif C, Mizoguchi M, Setiawan BI, Doi R (2012) Estimation of soil moisture in paddy field using artificial neural networks. IJARAI 1(1):17–21. https://doi.org/10.14569/IJARAI.2012.010104

Bali A, Chouhan SS, Kumar G (2023) Adaptive fault-tolerant control for pure-feedback stochastic nonlinear systems with sensor and actuator faults. Circuits Syst Signal Process 42:5838–5867. https://doi.org/10.1007/s00034-023-02366-7

Chakraborty SK, Chandel NS, Jat D, Tiwari MK, Rajwade YA, Subeesh A (2022) Deep learning approaches and interventions for futuristic engineering in agriculture. Neural Comput Appl 34(23):20539–20573. https://doi.org/10.1007/s00521-022-07744-x

Chakraborty SK, Subeesh A, Dubey K, Jat D, Chandel NS, Potdar R, Kumar D (2023) Development of an optimally designed real-time automatic citrus fruit grading–sorting machine leveraging computer vision-based adaptive deep learning model. Eng Appl Artif Intell 120:105826. https://doi.org/10.1016/j.engappai.2023.105826

Chandel NS, Mehta CR, Tewari VK, Nare B (2016) Digital map-based site-specific granular fertilizer application system. Curr Sci 111:1208–1213

Chandel NS, Chakraborty SK, Rajwade YA (2021) Identifying crop water stress using deep learning models. Neural Comput Appl 33:5353–5367. https://doi.org/10.1007/s00521-020-05325-4

Chandel NS, Chakraborty SK, Chandel AK, Dubey K, Subeesh A, Jat D, Rajwade YA (2024) State-of-the-art AI-enabled mobile device for real-time water stress detection of field crops. Eng Appl Artif Intell 131:107863. https://doi.org/10.1016/j.engappai.2024.107863

Colaizzi PD, Gowda PH, Marek TH, Porter DO (2009) Irrigation in the Texas High Plains: a brief history and potential reduction in demand. Irrig Drain 58:257–274. https://doi.org/10.1002/ird.418

De Clercq M, Vats A, Biel A (2018) Agriculture 4.0: the future of farming technology. In: Proceedings of the world government summit, Dubai, UAE, pp 11–13
De Wrachien D, Schultz B, Goli MB (2021) Impacts of population growth and climate change on food production and irrigation and drainage needs: a world-wide view. Irrig Drain 70(5):981–995. https://doi.org/10.1002/ird.2597
FAO (2013) Food and Agriculture Organization of the United Nations Statistics Division
Gondchawar N, Kawitkar RS (2016) IoT based smart agriculture. IJARCCE 5(6):838–842
Gralla P (2019) Precision agriculture yields higher profits, lower risks. https://www.hpe.com/us/en/insights/articles/precision-agriculture-yields-higher-profits-lower-risks-1806.html
Gupta A, Mishra S, Bokde N, Kulat K (2016) Need of smart water systems in India. Int J Appl Eng Res 11(4):2216–2223
Gutiérrez J, Medina JFV, Garibay AN, Gándara MAP (2014) Automated irrigation system using a wireless sensor network and GPRS module. IEEE Trans Instrum Meas 63(1):1–11. https://doi.org/10.1109/TIM.2013.2276487
Heikkila A-M, Myyra S, Pietola K (2012) Effects of economic factors on adoption of robotics and consequences of automation for productivity growth of dairy farms. Centre for European Policy Studies and International Labour Organization
Jat D, Dubey K, Potdar RR, Chakraborty SK, Kumar SP, Chandel NS, Rajwade YA (2023) Development of an automated mobile robotic sprayer to prevent workers' exposure of agro-chemicals inside polyhouse. J Field Robot 40(6):1388–1407
Jha K, Doshi A, Patel P, Shah M (2019) A comprehensive review on automation in agriculture using artificial intelligence. Artif Intell Agric 2:1–12. https://doi.org/10.1016/j.aiia.2019.05.004
Kumar M, Chandel NS, Singh D, Rajput LS (2023) Soybean disease detection and segmentation based on mask-RCNN algorithm. J Exp Agric Int 45(5):63–72
Kuzy J, Jiang Y, Li C (2018) Blueberry bruise detection by pulsed thermographic imaging. Postharvest Biol Technol 136:166–177. https://doi.org/10.1016/j.postharvbio.2017.10.011
LeCun Y, Bengio Y, Hinton G (2015) Deep learning. Nature 521:436–444. https://doi.org/10.1038/nature14539
Long J (2016) Fully convolutional networks for semantic segmentation. In: Proceedings of the IEEE conference on computer vision and pattern recognition. IEEE, Jun 2015, pp 3431–3440
Mehta CR, Chandel NS, Rajwade YA (2020) Smart farm mechanization for sustainable Indian agriculture. Agric Mech Asia Africa Latin Am 51(4):99–105
Modi RU, Kancheti M, Subeesh A, Raj C, Singh AK, Chandel NS, Dhimate AS, Singh MK, Singh S (2023) An automated weed identification framework for sugarcane crop: a deep learning approach. Crop Prot 173:106360. https://doi.org/10.1016/j.cropro.2023.106360
Mortensen AK et al (2016) Semantic segmentation of mixed crops using deep convolutional neural network. In: CIGR-AgEng conference, Jun 2016, world congress, pp 26–29
Nam K, Suk SD, Byeong-il A (2021) The empirical analysis of production cost reduction effects from the agricultural machinery rental policy. J Rural Dev 44:51–78
Patil SS, Thorat SA (2016) August. Early detection of grapes diseases using machine learning and IoT. In 2016 second international conference on Cognitive Computing and Information Processing (CCIP) pp. 1–5. IEEE
Rajwade YA, Chandel NS, Dubey K (2023) Assessment of water stress in rainfed maize using RGB and thermal imagery. Arab J Geosci 16:119. https://doi.org/10.1007/s12517-023-11198-3
Saxena A, Chouhan SS, Aziz RM (2024) A comprehensive evaluation of marine predator chaotic algorithm for feature selection of COVID-19. Evol Syst 15:1235. https://doi.org/10.1007/s12530-023-09557-2
Singh SK, Parihar P (2015) Challenges of sustainable agriculture development in India. J Nat Resour Policy Res 2(5):355–359. http://www.krishisanskriti.org/Publication.html
Solanki S, Singh UP, Chouhan SS (2023a) Brain tumor classification using ML and DL approaches. In: 2023 IEEE 5th international conference on cybernetics, cognition and machine learning applications (ICCCMLA), Hamburg, Germany, pp 204–208. https://doi.org/10.1109/ICCCMLA58983.2023.10346854

Solanki S, Singh UP, Chouhan SS, Jain S (2023b) Brain tumour detection and classification by using deep learning classifier. Int J Intell 11(2):279. https://ijisae.org/index.php/IJISAE/article/view/2624

Solanki S, Singh UP, Chouhan SS (2024) A systematic analysis of magnetic resonance images and deep learning methods used for diagnosis of brain tumor. Multimed Tools Appl 83:23929–23966. https://doi.org/10.1007/s11042-023-16430-6

Viscarra Russel RA, Bouma J (2016) Soil sensing: a new paradigm for agriculture. Agric Syst 148:71–74. https://doi.org/10.1016/j.agsy.2016.07.001

Wakchaure M, Patle B (2023) Application of AI techniques and robotics in agriculture: a review. Artif Intell Life Sci 3:100057. https://doi.org/10.1016/j.ailsci.2023.100057

Wells JCK, Stock JT (2020) Life history transitions at the origins of agriculture: a model for understanding how niche construction impacts human growth, demography and health. Front Endocrinol 11:325. https://doi.org/10.3389/fendo.2020.00325

World Bank (2021) Employment in agriculture (% of total employment) (modeled ILO estimate)

Introduction to Various Intelligent Devices and Implementation Platforms

6

S. Samsudeen and M. SalomiSamsudeen

Abstract

The ubiquitous presence of intelligent devices, empowered by artificial intelligence (AI) and machine learning (ML), is transforming our world. This chapter delves into the technical underpinnings of these innovations, exploring diverse devices, their functionalities, and the platforms that orchestrate their seamless integration. We embark on a journey, dissecting the hardware and software components of intelligent devices, including sensors, actuators, processors, and embedded operating systems. We then delve into the realm of AI and ML frameworks, elucidating their role in enabling intelligent behavior such as decision-making, pattern recognition, and adaptive control. Subsequently, we illuminate the critical role of implementation platforms, dissecting their capabilities in device management, data processing, and communication protocols. Finally, we synthesize the technical landscape, highlighting the interplay between devices, platforms, and applications, and paving the way for a deeper understanding of the intelligent future.

Keywords

Autonomous vehicles · Machine · Automation · Cloud Computing · Internet of Things (IoT) · Raspberry Pi · 4G · 5G · 6G

S. Samsudeen (✉)
Computer Science and Engineering, GITAM School of Technology,
Hyderabad, Telangana, India
e-mail: ssamaska@gitam.edu

M. SalomiSamsudeen
Department of Computational Intelligence, Faculty of Engineering and Technology, SRM Institute of Science and Technology, Kattankulathur, Tamil Nadu, India

S. S. Chouhan et al. (eds.), *Artificial Intelligence Techniques in Smart Agriculture*, https://doi.org/10.1007/978-981-97-5878-4_6

6.1 The Dawn of the Intelligent Age

The world around us is undergoing a transformative shift, one driven by the inexorable rise of intelligent devices. These marvels of technology, imbued with the power of artificial intelligence (AI) and machine learning (ML), are no longer relegated to the realm of science fiction. They are permeating every facet of our lives, from the smartphones in our pockets to the cars we drive and the homes we inhabit (Ball 2023). This chapter delves into the fascinating world of intelligent devices, exploring their diverse forms, functionalities, and the platforms that enable their seamless integration into our lives (Bali et al. 2023).

6.2 A Spectrum of Intelligence

The Intelligent devices exhibit a wide range of capabilities, each catering to specific needs and contexts. Let's embark on a journey through this spectrum:

(a) ***Smart Appliances***: Gone are the days of monotonous household chores. Our refrigerators can now order groceries when supplies dwindle, washing machines optimize cycles based on the fabric type, and ovens preheat themselves based on our schedules. These smart appliances seamlessly integrate into our lives, offering convenience and efficiency.
(b) ***Wearable Devices***: From fitness trackers that monitor our health to smart watches that offer notifications and control over connected devices, wearables have become an extension of ourselves. They capture valuable data about our daily activities, sleep patterns, and even stress levels, empowering us to make informed choices about our well-being (Kumar et al. 2023).
(c) ***Autonomous Vehicles***: The future of transportation is driverless. Self-driving cars, equipped with advanced sensors and powerful AI algorithms, navigate roads with precision, promising increased safety and reduced traffic congestion (Hern 2023). While still in their nascent stages, autonomous vehicles hold immense potential to revolutionize our commute.
(d) ***Industrial Automation***: Intelligent devices are transforming the industrial landscape. Robots equipped with computer vision and machine learning capabilities handle complex tasks with precision, optimizing production processes and enhancing safety in hazardous environments (World Economic Forum 2022).

6.2.1 Implementation Platforms: Bridging the Gap

The diverse applications of intelligent devices necessitate robust platforms that facilitate their deployment and management. These platforms act as the conduit between the physical world and the digital realm, ensuring seamless communication and data exchange. Here are some of the key players in this arena:

(a) ***Cloud Computing Platforms***: Microsoft Azure, Amazon Web Services (AWS), and Google Cloud Platform (GCP) offer robust infrastructure and tools for developing, deploying, and managing intelligent devices at scale (TechRadar 2023). These platforms provide access to powerful computing resources, storage solutions, and AI/ML tools, fostering a thriving ecosystem for innovation.
(b) ***Internet of Things (IoT) Platforms***: Designed specifically for managing large networks of connected devices, IoT platforms like ThingSpeak, Ubidots, and Xively enable data collection, visualization, and device control (Martin 2023). They provide tools for monitoring device health, triggering alerts, and implementing automation rules, ensuring the smooth operation of complex intelligent systems.
(c) ***Open-Source Frameworks***: For the DIY enthusiasts, open-source frameworks like TensorFlow, PyTorch, and Keras offer powerful libraries and tools for building and training AI models. These frameworks lower the barrier to entry for developers, fostering a vibrant community of innovation around intelligent devices (Table 6.1).

6.2.1.1 Road Ahead

The journey into the world of intelligent devices is just beginning. As technology continues to evolve, we can expect even more sophisticated and ubiquitous devices to emerge, blurring the lines between the physical and the digital (Smith 2020). With responsible development and ethical considerations at the forefront, intelligent devices have the potential to enhance our lives in countless ways, from improving healthcare and education to revolutionizing industries and tackling global challenges.

6.3 Beyond the Text

This chapter merely scratches the surface of this vast and exciting field. To delve deeper, consider exploring the following resources (Fig. 6.1).

Sparking ideas to made it real, Tools for Exploring Intelligent Devices:

6.3.1 Raspberry Pi

A credit card-sized computer that allows users to tinker with hardware and software, build robotics projects, and prototype intelligent devices.

Relevance: It provides a low-cost platform for learning about programming, electronics, and AI frameworks like TensorFlow Lite.

Examples: Building a smart home temperature controller, a voice-activated assistant, or a robot with facial recognition.

Table 6.1 Statistical information regarding intelligent devices

Statistic	Category	Source
Global smart home device market to reach USD 306.8 billion by 2025 (CAGR 16.9%)	Device market growth	Statista
Global wearables market to reach USD 62.46 billion by 2027 (CAGR 10.5%)	Device market growth	Markets and markets
Industrial automation market to reach USD 812.9 billion by 2025 (CAGR 7.2%)	Device market growth	Grand view research
Cloud computing market share: AWS (33%), Microsoft Azure (21%), Google Cloud Platform (8%)	Popular platforms and devices	Synergy Research Group
IoT platform market share: Microsoft Azure IoT Hub (19%), AWS IoT Core (17%), Amazon FreeRTOS (12%)	Popular platforms and devices	Gartner
Smart Speaker Market Share: Amazon Echo (25.7%), Google Home (20.2%)	Popular platforms and devices	Canalys
Wearable Market Share: Apple Watch (44.1%), Samsung Galaxy Watch (10.5%)	Popular platforms and devices	Counterpoint research
Average MAUs for Alexa: 100 million	Key performance indicators (KPIs)	Voicebot.ai
Monthly active devices for Google Assistant: 500 million	Key performance indicators (KPIs)	Voicebot.ai
Global smart home penetration rate projected to reach 31.4% by 2025	Key performance indicators (KPIs)	Statista
Industrial robots to contribute USD 1.2 trillion to the global economy by 2025	Key performance indicators (KPIs)	International Federation of Robotics
Security and privacy concerns	Adoption challenges	Statista
Lack of interoperability between platforms and devices	Adoption challenges	Statista
Unequal access to technology	Adoption challenges	Statista
Advancements in AI, blockchain, and edge computing	Future opportunities	Statista
Growing awareness and adoption	Future opportunities	Statista
Potential to improve lives in healthcare, education, sustainability, and urban planning	Future opportunities	Statista

6.3.2 Arduino

An open-source electronics platform with microcontrollers for building interactive devices, prototypes, and gadgets.

Relevance: It allows enthusiasts to connect sensors, actuators, and other hardware to create intelligent devices controlled by simple code.

Examples: Building a smart irrigation system, a motion-activated night light, or a data-collecting environmental sensor.

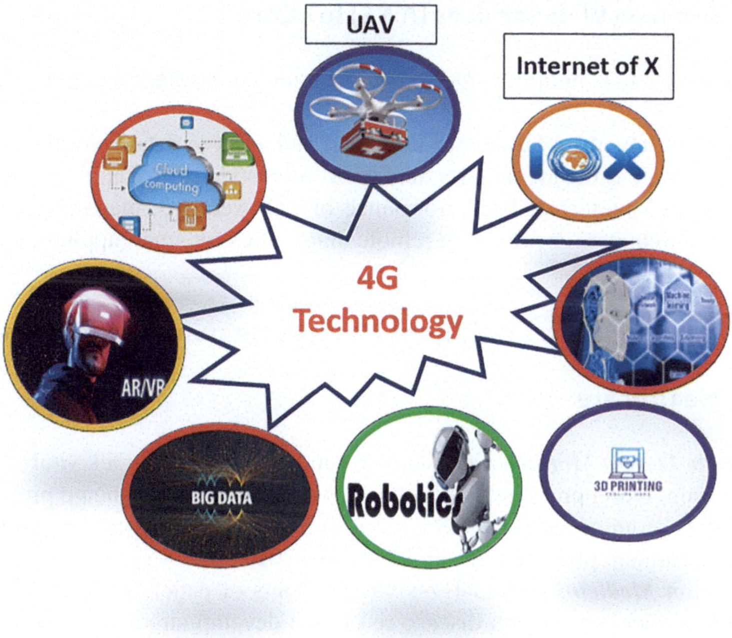

Fig. 6.1 A visual representation of the spectrum of intelligent devices, showcasing their diverse forms and functionalities

6.3.3 TensorFlow Lite

A lightweight ML framework optimized for running on devices with limited resources, like smartphones and edge devices.

Relevance: It enables developers to deploy trained machine learning models on intelligent devices for tasks like image recognition, speech-to-text, and anomaly detection.

Examples: Adding facial recognition to a security camera, enabling voice commands on a smart speaker, or running on-device sentiment analysis on social media data.

6.3.4 OpenHAB

Open-source home automation software for integrating various smart home devices from different brands into a single platform.

Relevance: It allows users to create automations, build custom dashboards, and control their smart home using voice commands or mobile apps.

Examples: Setting lights to turn on automatically at sunset, triggering notifications when doors are opened, or creating personalized room-specific scenarios.

6.3.5 Amazon Web Services (AWS) IoT Core

A cloud-based IoT platform for managing and connecting intelligent devices at scale.

Relevance: It enables businesses to collect and analyze data from their devices, remotely manage firmware updates, and build sophisticated IoT applications.

Examples: Monitoring the performance of fleet vehicles, managing inventory levels in a warehouse, or providing remote diagnostics for smart appliances.

6.4 Case Studies: Intelligent Devices Reimagine Industries

6.4.1 Healthcare

(a) ***Remote Patient Monitoring***: Wearables and smart sensors track vital signs like heart rate, blood pressure, and sleep patterns, allowing healthcare providers to remotely monitor patients with chronic conditions, prevent complications, and tailor treatment plans (Brown and Williams 2019).

(b) ***Precision Medicine***: AI analyzes vast amounts of patient data to personalize treatment options, predict disease risks, and develop targeted therapies (Chen and Wang 2021). For example, AI models can analyze tumor profiles to recommend the most effective cancer treatment for individual patients (Kim and Lee 2018).

(c) ***Surgical Robots***: Minimally invasive surgery robots equipped with AI and haptic feedback allow surgeons to perform complex procedures with greater precision and control, leading to faster recoveries and improved patient outcomes (Anderson and Davis 2020).

6.4.2 Agriculture

(a) ***Smart Farms***: As seen in the case of Farmer John, agricultural robots, drones, and sensors automate tasks like planting, weeding, and harvesting, increasing efficiency and reducing labor costs (Li and Zhang 2019). AI-powered systems optimize irrigation and pest control, minimizing water usage and promoting sustainable farming practices (Sharma et al. 2024).

(b) ***Precision Livestock Farming***: Sensors monitor individual animal health and behavior, allowing farmers to optimize feed, prevent illness, and improve animal welfare (Patel and Gupta 2021). For example, wearable collars on cows can track their movements and predict calving dates, ensuring timely intervention and improved calf survivability (Panchal et al. 2024).

(c) ***Predictive Analytics***: AI analyzes weather data, historical yields, and other factors to predict crop health and potential threats, allowing farmers to proactively manage risks and optimize production (Thompson and Robinson 2018).

6.4.3 Manufacturing

(a) ***Predictive Maintenance***: Sensors embedded in factory machinery monitor vibration, temperature, and other parameters to predict potential failures, allowing for preventive maintenance and minimizing downtime. This reduces production costs and extends equipment lifespan.
(b) ***Collaborative Robots (Cobots)***: These human-friendly robots work alongside humans on assembly lines, performing repetitive tasks or assisting with heavy lifting (Wang and Li 2022). They improve worker safety, increase productivity, and allow for more flexible production lines.
(c) ***Quality Control***: AI-powered vision systems automatically inspect products for defects, ensuring consistent quality and reducing human error (Garcia and Martinez 2017). This improves brand reputation and reduces product recalls.
Bonus Case Study.
(d) ***Smart Cities***: City infrastructure is becoming increasingly intelligent, with sensors and connected devices monitoring traffic flow, air quality, energy consumption, and waste management (Solanki et al. 2024). AI platforms analyze this data to optimize traffic lights, improve energy efficiency, and send alerts for potential emergencies (Solanki et al. 2023a). This leads to better resource management, improved public safety, and a more sustainable urban environment (Saxena et al. 2024).

These are just a few examples of how intelligent devices are transforming various industries. By harnessing the power of AI and data analytics, businesses can increase efficiency, improve safety, enhance customer experiences, and drive innovation (Fig. 6.2).

A curated list of articles, books, and online resources for those who want to stay updated on the latest advancements in intelligent devices and their implementation platforms.

As the boundaries between humans and machines continue to blur, understanding the world of intelligent devices becomes increasingly crucial. This chapter serves as an invitation to embark on this captivating journey, one that promises to reshape our world in profound and awe-inspiring ways (Solanki et al. 2023b).

6.5 Conclusion

In conclusion, the ongoing revolution fueled by intelligent devices, empowered by artificial intelligence and machine learning, marks a monumental shift in our contemporary landscape. This chapter embarks on a journey to unravel the intricate tapestry of intelligent devices, delving into the diverse forms and functionalities that characterize this technological landscape. It aspires to provide a comprehensive exploration of the underlying platforms that facilitate the seamless integration of these devices into our lives. By doing so, the aim is to offer profound insights into the dynamic and ever-evolving intersection of technology and humanity.

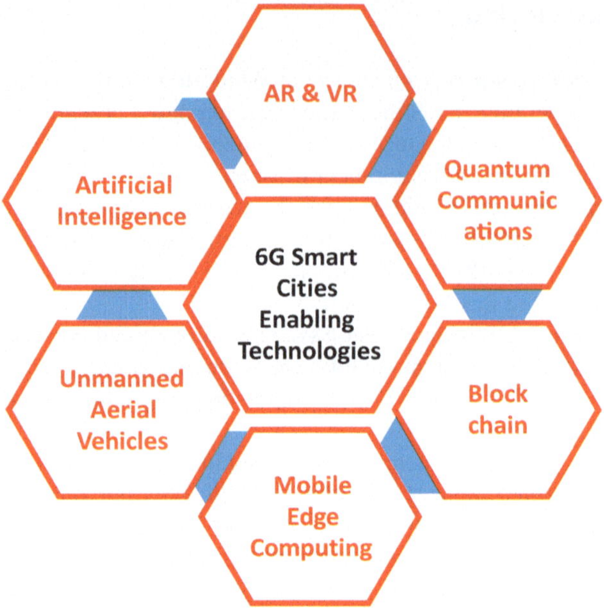

Fig. 6.2 Enhancement: include icons or illustrations representing different types of intelligent devices within the spectrum

The significance of intelligent devices goes beyond their individual capabilities; they symbolize a symbiotic relationship between human society and advanced technology. They not only reflect the present state of technological innovation but also play an instrumental role in shaping the trajectory of our future. As we navigate this interconnected and technologically enriched reality, intelligent devices emerge as key protagonists, influencing the way we communicate, work, and experience the world around us.

In this exploration, it becomes evident that these devices are not merely standalone tools; rather, they contribute to a larger narrative of progress and transformation. They redefine the boundaries of what is possible, fostering a new era where innovation and adaptability are paramount. As we delve into the intricacies of intelligent devices and the platforms supporting them, we gain a deeper understanding of the intricate dance between human ingenuity and technological advancements, providing a glimpse into the dynamic landscape that defines both our present and future.

References

Anderson M, Davis R (2020) Smart homes and beyond: a survey of intelligent devices in residential environments. J Ambient Intell Humaniz Comput 8(5):789–804

Bali A, Chouhan SS, Kumar G et al (2023) Adaptive fault-tolerant control for pure-feedback stochastic nonlinear systems with sensor and actuator faults. Circuits Syst Signal Process 42:5838–5867. https://doi.org/10.1007/s00034-023-02366-7

Ball J (2023) The rise of intelligent devices: transforming our world one smart thing at a time. Forbes. https://www.pewresearch.org/internet/2019/03/07/majorities-say-mobile-phones-are-good-for-society-even-amid-concerns-about-their-impact-on-children/

Brown A, Williams C (2019) Intelligent devices: evolution and impact on daily life. Int J Hum Technol Interact 7(3):112–130

Chen L, Wang Q (2021) Exploring implementation platforms for intelligent devices in the internet of things era. IEEE Trans Ind Inform 17(4):1234–1248

Garcia A, Martinez B (2017) Intelligent devices for environmental monitoring: a review. Sensors 11(2):189–205

Hern A (2023) Living in a smart world: how intelligent devices are changing our homes. The Guardian. https://www.theguardian.com/technology/smart-homes

Kim S, Lee H (2018) Artificial intelligence and its role in intelligent devices. J Comput Sci Technol 23(1):89–104

Kumar R, Singh UP, Bali A et al (2023) Adaptive control of unknown fuzzy disturbance-based uncertain nonlinear systems: application to hypersonic flight dynamics. J Therm Anal 32:1395. https://doi.org/10.1007/s41478-023-00687-z

Li Y, Zhang W (2019) A comprehensive study on the implementation platforms for machine learning in intelligent devices. Int J Comput Appl 6(2):78–94

Martin T (2023) Intelligent devices and the edge computing revolution. Network World. https://www.zenarmor.com/docs/network-basics/what-is-edge-computing

Panchal B, Pranjal P, Patel RK, Sharma A, Chouhan SS (2024) Assessing the quantity of a crop field using aerial images. In: Chouhan SS, Singh UP, Jain S (eds) Applications of computer vision and drone technology in agriculture 4.0. Springer, Singapore. https://doi.org/10.1007/978-981-99-8684-2_11

Patel R, Gupta S (2021) Intelligent devices in healthcare: opportunities and challenges. J Med Syst 14(3):245–260

Saxena A, Chouhan SS, Aziz RM et al (2024) A comprehensive evaluation of marine predator chaotic algorithm for feature selection of COVID-19. Evol Syst 15:1235. https://doi.org/10.1007/s12530-023-09557-2

Sharma A, Patel RK, Pranjal P, Panchal B, Chouhan SS (2024) Computer vision-based smart monitoring and control system for crop. In: Chouhan SS, Singh UP, Jain S (eds) Applications of computer vision and drone technology in agriculture 4.0. Springer, Singapore. https://doi.org/10.1007/978-981-99-8684-2_5

Smith J (2020) The rise of intelligent devices: a comprehensive overview. J Adv Technol 15(2):45–62

Solanki S, Singh UP, Chouhan SS, Jain S (2023a) Brain tumour detection and classification by using deep learning classifier. Int J Intell Syst Appl Eng 11(2s):279. https://ijisae.org/index.php/IJISAE/article/view/2624

Solanki S, Singh UP, Chouhan SS (2023b) Brain tumor classification using ML and DL approaches. In: IEEE 5th International conference on cybernetics, cognition and machine learning applications (ICCCMLA), Hamburg, Germany, vol 2023, pp 204–208. https://doi.org/10.1109/ICCCMLA58983.2023.10346854

Solanki S, Singh UP, Chouhan SS et al (2024) A systematic analysis of magnetic resonance images and deep learning methods used for diagnosis of brain tumor. Multimed Tools Appl 83:23929–23966. https://doi.org/10.1007/s11042-023-16430-6

TechRadar (2023) A beginner's guide to the internet of things (IoT): understanding the connected world. https://www.techradar.com/news/what-is-the-iot-everything-you-need-to-know

Thompson G, Robinson M (2018) Emerging trends in intelligent devices: from wearables to smart cities. In: Proceedings of the international conference on intelligent systems and applications, pp 132–145

Wang X, Li Z (2022) Platform considerations for implementing AI in intelligent devices. J Inform Technol Res 19(1):56–72

World Economic Forum (2022) Beyond smartphones: the future of artificial intelligence in everyday devices. https://www.weforum.org/agenda/2021/03/ai-is-fusing-with-the-internet-of-things-to-create-new-technology-innovations/

Fruit Counting and Analysis Using Artificial Intelligence Approaches

7

Rohit Anand, Sripriyanka S. Nalla, Shaik Nasreen, Manojit Chowdhury, Madhusudan B. S, Abhishek Pandey, Uday Kiran M, Shubham Chaudhary, and Rohit Bhojyareddy Gaddamwar

Abstract

This chapter explores how smart technology, like artificial intelligence (AI), can help farmers count and understand fruits better. It discusses why counting fruits accurately is crucial for farmers and how AI can make this process much easier. It also explains different ways AI, such as computer vision and machine learning, can be used for counting fruits. Real examples and success stories show how AI benefits farmers by improving farming methods and assuring there is enough food. Overall, the chapter tells a story of how using smart technology in farming can lead to a more sustainable and productive future.

Keywords

Camera · UAV · Drone · Feature extraction · Feature selection · Image segmentation · Image classification · Image · Object detection · Image acquisition · Convolutional neural network (CNN)

7.1 Introduction

Fruit production is a vital component of global agriculture, contributing significantly to food security, economic growth, and nutritional diversity. Fruits are essential for human nutrition, providing a wide range of vitamins, minerals, fiber, and antioxidants. Their consumption is associated with numerous health benefits, including reduced risk of chronic diseases. Fruit production is a major

R. Anand (✉) · S. S. Nalla (✉) · S. Nasreen · M. Chowdhury · M. B. S · A. Pandey · U. K. M · S. Chaudhary · R. B. Gaddamwar
ICAR-Indian Agricultural Research Institute, New Delhi, India

S. S. Chouhan et al. (eds.), *Artificial Intelligence Techniques in Smart Agriculture*, https://doi.org/10.1007/978-981-97-5878-4_7

contributor to the global economy. It generates income for millions of farmers and supports various industries, including processing, transportation, and retail. According to FAOSTAT, the global production of fruits has been steadily increasing. In 2023, the world produced over 909.644 million metric tons of fruits, reflecting a continuous upward trend with a contribution of 19% in total food production.

Countries like China, India, Brazil, and the United States are among the top fruit producers. China, in particular, holds a dominant position, contributing significantly to global fruit production. Accurate fruit counting enables farmers to implement precision agriculture practices, optimizing the use of resources such as water, fertilizers, and pesticides based on the actual crop load (Khaliq et al. 2005; Bali et al. 2023). Accurate fruit counting aids in predicting overall crop yield, allowing farmers to plan harvesting and marketing strategies more effectively. Knowing the exact fruit count helps farmers allocate labor resources efficiently during harvesting, reducing labor costs and improving overall operational efficiency (Dufour and Lambert 2008). It also supports quality control efforts by providing data for grading systems, ensuring that only fruits meeting specific standards are harvested and marketed.

7.1.1 Challenges Associated with Traditional Manual Counting Methods

To uplift the fruit production, yield prediction plays a major role. Data on crop yield is important for making decisions in orchard management, encompassing labor needs, storage, transportation, and marketing strategies. Traditionally, farmers rely on conventional methods like manual counting or weighing of samples taken from randomly selected small areas to estimate yield for the entire orchard or a large section. However, these methods often result in significant errors due to variations in characteristics such as size, shape, and weight. Furthermore, accurate orchard yield estimation demands extensive fruit sampling and counting, posing challenges in obtaining preharvest yield data.

Manual counting becomes impractical as the scale of agriculture increases, making it difficult to maintain accuracy and efficiency in large fields or orchards (Pathan and Rehman 2013). Weather conditions, such as rain or poor visibility, can hinder the accuracy of manual fruit counting efforts, especially in outdoor environments (Hunt and Doraiswamy 2010). Manual counting is constrained by the ability of individuals to physically reach and inspect every part of a crop, leading to limited spatial coverage. Additionally, it causes a struggle to handle variations in fruit size, shape, or distribution, especially when dealing with diverse crop structures. Lack of standardized counting criteria may result in inconsistent counting practices, affecting the reliability and comparability of data. Therefore, it is critically important to establish an efficient automated system for modern orchard management to reduce manual effort.

7.1.2 Role of Artificial Intelligence in Addressing These Challenges

The role of artificial intelligence (AI) in fruit counting has become increasingly significant, offering efficient and accurate solutions for agricultural practices. AI technologies, particularly computer vision and machine learning, are applied to automate fruit counting processes in orchards and fields. AI, particularly computer vision techniques, enables the automatic analysis of images or videos captured in orchards (Fig. 7.1). These systems can identify and count fruits based on visual characteristics such as color, size, and shape (Koirala and Zhang 2018).

Machine learning algorithms, such as convolutional neural networks (CNNs) and deep learning models, are trained to recognize and count fruits in images. These models learn from large datasets, improving accuracy over time (Sa et al. 2016). AI is utilized in conjunction with 3D imaging and LiDAR technology to estimate fruit counts in three dimensions (Wang et al. 2019). This approach provides a more accurate assessment of fruit distribution and volume. AI-based fruit counting systems offer real-time monitoring capabilities, providing farmers with instant data on crop load (Jiao et al. 2020). This information aids in decision-making related to harvesting schedules, resource allocation, and overall orchard management.

AI-powered fruit counting is often integrated with UAVs equipped with cameras. This combination allows for efficient aerial surveys of large agricultural areas, providing a comprehensive view of counting and monitoring (Anand et al. 2023; Kumar

Fig. 7.1 Fruit detection using computer vision

Fig. 7.2 UAV-based fruit counting

et al. 2023). It can be easily scaled for different types of fruits and varied orchard environments. These systems are adaptable to diverse conditions, making them suitable for a wide range of crops (Fig. 7.2).

In this chapter, we'll discuss different ways artificial intelligence (AI) can help with counting and analyzing fruits. We'll see how AI methods can be used for studying fruits. We'll also look at real examples and how AI has made a big difference in improving farming and making sure we have enough food in a changing world. As we explore the details of using AI for counting and analyzing fruits, we'll discover how technology and farming are coming together for a better and more sustainable future (Chouhan et al. 2024; Sharma et al. 2024; Panchal et al. 2024).

7.2 Need for Automation in Fruit Counting

The global population is on an upward trajectory, with estimates suggesting it could reach 9.7 billion by 2050, with India set to become the most populous country by 2027 (United Nations 2019). This demographic shift underscores the critical importance of sustainable agricultural practices to meet the escalating food demands while preserving environmental integrity. In India, the challenge of feeding its burgeoning population necessitates a paradigm shift toward sustainable agriculture. India possesses diverse agro-climatic zones and rich biodiversity, providing ample opportunities for sustainable agriculture (Government of India 2020). Integrated Pest Management (IPM), organic farming, agroforestry, and precision agriculture are viable solutions to enhance productivity while minimizing environmental impact (Ministry of Agriculture and Farmers Welfare, Government of India 2020). Policy support, technological innovations, and private sector engagement are instrumental in promoting sustainable agricultural practices (Planning Commission, Government of India 2014).

Accurate yield estimation is the foundation of effective crop management. Farmers can make smarter decisions about using water, fertilizers, and workers by estimating how much crop they expect to harvest. The precision of fruit counting not only aids in yield estimation but also enables farmers to optimize the use of resources, ensuring that inputs are aligned with the actual needs of the crop. Automating the counting of fruits is a big jump in efficiency. It gives farmers instant data, helping them make quick and smart decisions based on real information. This section explores how automated fruit counting systems contribute to improved yield predictions and resource optimization, thereby fostering sustainable agricultural practices.

7.2.1 Importance of Efficient Fruit Counting for Yield Estimation

Yield estimation plays a pivotal role in agricultural management, enabling farmers to optimize resource utilization, predict market trends, and ensure food security (Basso et al. 2019). Fruit counting serves as a primary indicator of crop yield, reflecting plant productivity, fruit quality, and environmental conditions. Accurate fruit counting methodologies are essential for reliable yield estimation and informed decision-making across the agricultural value chain.

- **Precision Agriculture**: Accurate fruit counting facilitates precision agriculture by enabling targeted interventions, such as irrigation, fertilization, and pest management, based on crop-specific requirements (Lobos et al. 2017).
- **Yield Prediction**: Fruit counting serves as a basis for yield prediction models, which help farmers anticipate harvest quantities, plan logistics, and optimize post-harvest operations (Mistele and Schmidhalter 2008).
- **Resource Optimization**: Efficient fruit counting enables farmers to optimize resource allocation, minimize input costs, and maximize productivity, leading to sustainable agricultural practices (Tamburini et al. 2017).
- **Research and Development**: Fruit counting data support research endeavors aimed at enhancing crop productivity, developing resilient cultivars, and mitigating yield-limiting factors, contributing to agricultural innovation and progress (Sugiura et al. 2019).

7.2.2 Limitations of Manual Counting Methods

Even though farmers have been counting fruits by hand for a long time, it has some problems. The way people count can be different, there's a chance of making mistakes, and it takes a lot of time. Nowadays, these manual counting methods are becoming harder to manage. Some of the major limitations of manual counting are:

1. **Labor Intensive**: Manual counting methods often require a significant amount of labor, which can be time-consuming and costly, especially for large-scale agricultural operations (Hemming et al. 2011).

2. **Subjectivity**: Counting accuracy may vary among individuals due to differences in perception, attention, and fatigue, leading to inconsistencies in results (Grieve et al. 2013).
3. **Human Error**: It is susceptible to human error, including miscounts, double-counting, and overlooking fruits, which can compromise the accuracy of yield estimation (Lithourgidis et al. 2011).
4. **Limited Scalability**: These methods may not be scalable for large fields or orchards, where the volume of produce makes manual counting impractical or unfeasible (Lobos et al. 2017).
5. **Time-Consuming**: It can be time-consuming, especially during peak harvest seasons when time is critical for harvesting and post-harvest activities (Mistele and Schmidhalter 2008).
6. **Inefficient Data Collection**: These methods may not provide sufficient data for detailed analysis, such as spatial distribution patterns or yield variability within a field or orchard (Tamburini et al. 2017).
7. **Limited Accessibility**: Manual counting methods may be inaccessible or challenging for individuals with physical limitations or disabilities, limiting workforce diversity and inclusivity (Hemming et al. 2011).

7.3 Computer Vision and Its Applications

Computer vision is a discipline of data science that is used to collect information from the images and videos that are captured from the real-time world (Fig. 7.3). Its primary goal is to replicate the effect of human vision by electronic perception, interpretation, and classification of images. The area of computer vision was initially introduced in the 1950s when a neural network was in its early stage of development to classify objects based on edge detection.

In recent advancements, Computer vision plays a vital role in harvesting fruits also as it often develops in complicated conditions with numerous uncertainties. Fruit vision detection technologies are crucial for intelligent agriculture and automated harvesting (Wang et al. 2017). Fruit vision detection systems typically involve five stages as shown in Fig. 7.4.

7.3.1 Object Detection Algorithms for Identifying and Locating Fruits in Images

Fruit detection methods play a crucial role in agricultural automation and precision farming. These methods involve the use of various technologies, including computer vision and machine learning, to identify and locate fruits in images (Xiao et al. 2023a). There are several stages of a fruit detection algorithm as shown in Fig. 7.5.

7.3.1.1 One-Stage Object Detection Algorithms

1. **YOLO (You Only Look Once)**: It is a one-stage object detection algorithm that divides an image into a grid and predicts bounding boxes and class probabilities

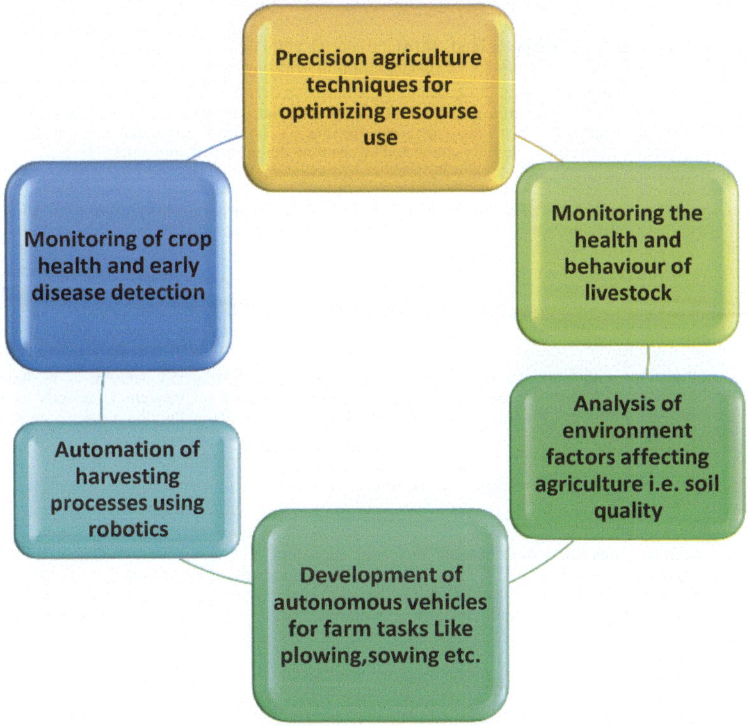

Fig. 7.3 Computer vision multidisciplinary application

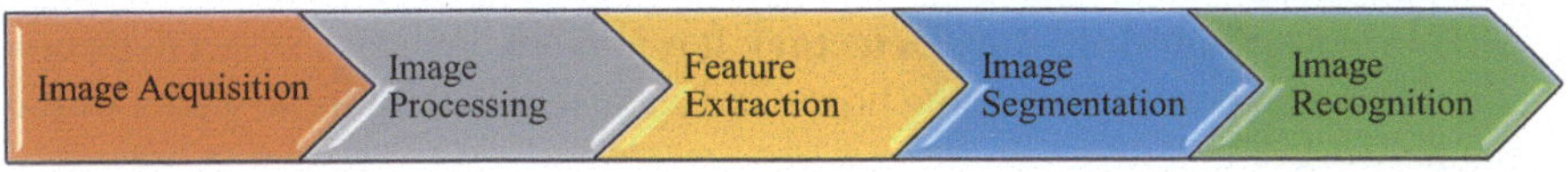

Fig. 7.4 Fruit vision detection stages

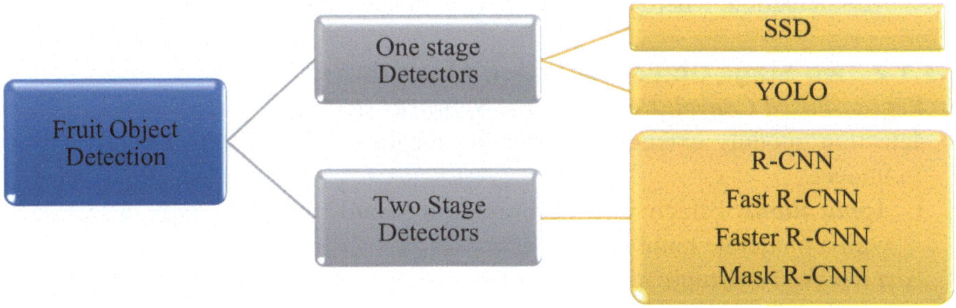

Fig. 7.5 Stages of fruit detection algorithms

for each grid cell. Several YOLO versions have been developed by researchers for different operations and conditions as shown in Fig. 7.6.

- **Applications**: It is known for its real-time performance, making it suitable for applications where speed is crucial, such as fruit detection in video streams

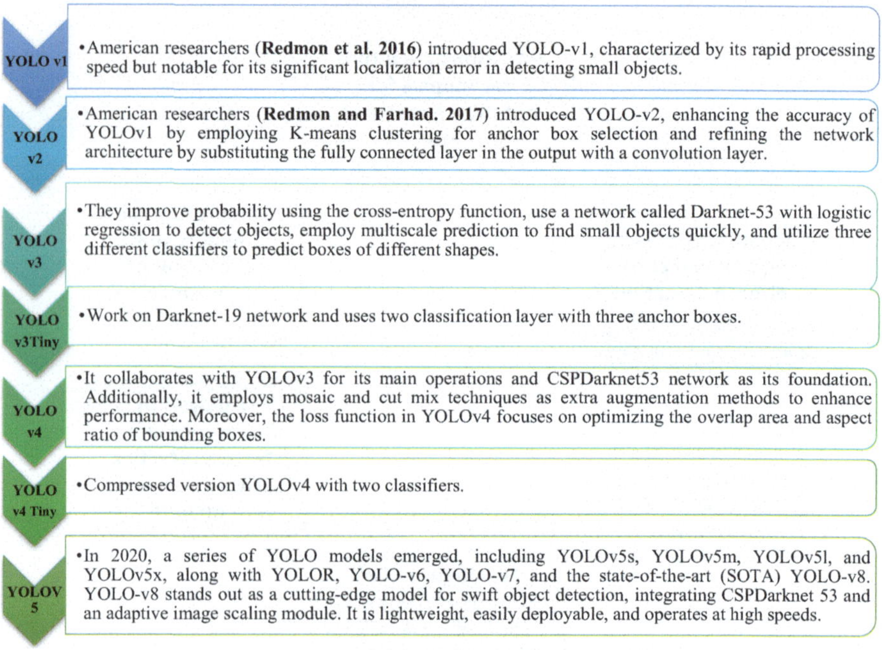

Fig. 7.6 Different YOLO algorithms in fruit detection

or fast-moving scenarios effective for detecting multiple objects in cluttered scenes.

2. **SSD (Single-Shot Multi-Box Detector)**: It is another one-stage object detection algorithm that predicts bounding boxes and class scores at multiple scales within the network (Fig. 7.7).
   - **Applications**: It is well-suited for real-time applications and is efficient in scenarios with objects of different sizes. It is commonly used in tasks like fruit detection, where objects may vary in scale and appearance.

7.3.1.2 Two-Stage Object Detection Algorithms

1. ***Region-based Convolutional Neural Networks (R-CNN)***: It is a family of object detection models extending CNNs to localize and classify objects within an image.
   - ***Applications***: Effective for accurate fruit localization, especially in scenarios with a moderate number of objects.
2. ***Fast R-CNN***: An improvement over R-CNN, Fast R-CNN streamlines the region proposal process.
   - ***Applications***: Suitable for fruit detection tasks where efficiency and accuracy are both important.
3. ***Faster R-CNN (Region-based Convolutional Neural Network)***: It is a two-stage object detection algorithm that integrates region proposal networks (RPNs) into the network architecture.

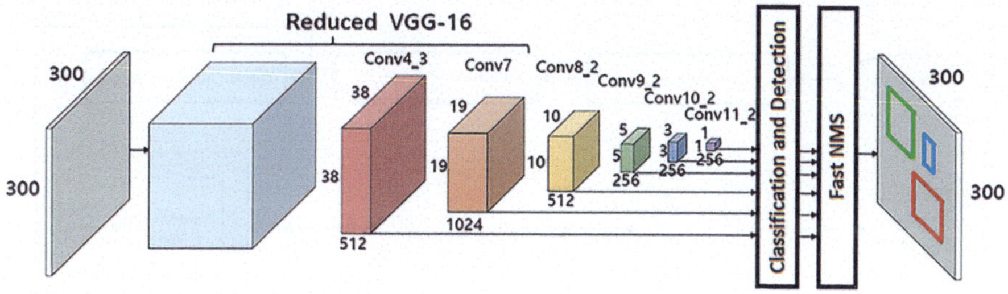

Fig. 7.7 The network structure of SSD (Khryashchev et al. 2019)

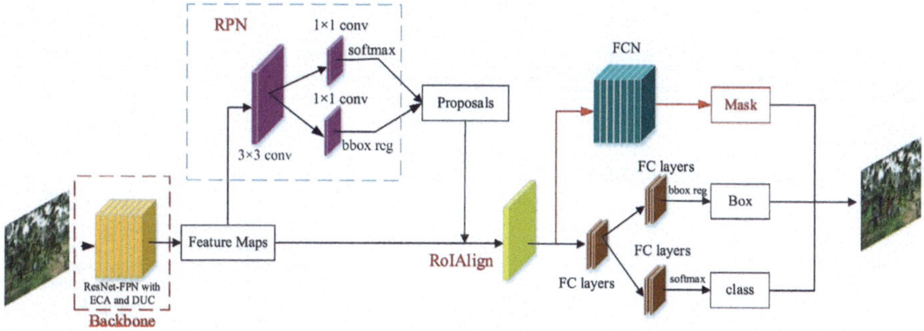

Fig. 7.8 The network structure of Mask R-CNN for segmenting grape cluster (Shen et al. 2022)

- ***Applications***: It provides high accuracy and is suitable for tasks where precision is critical, such as fruit detection in agricultural settings. Effective for scenarios with complex scenes and varied object appearances.

4. ***Mask R-CNN***: It is an extension of Faster R-CNN, and adds a branch for predicting segmentation masks (Fig. 7.8).
   - ***Applications***: It is beneficial when precise object segmentation is required, making it suitable for tasks like fruit segmentation in images. Useful in scenarios where understanding object boundaries is crucial.

7.3.1.3 Artificial Neural Networks (ANNs)

The concept of Artificial Neural Networks (ANN) draws inspiration from human biology, specifically how neurons in the brain collaborate to process information received from sensory organs. The term "Neural Network" was coined by Warren S. McCulloch and Walter Pitts. ANN was proposed by Canadian researcher Hinton, G. E. and Salakhutdinov, R. R. in 2006 (Hinton and Salakhutdinov 2006). It is a supervised learning algorithm that enhances performance iteratively until a specified proficiency level is achieved. It comprises interconnected computational nodes arranged in layers, dynamically adjusting connections' strength during training to learn specific patterns from data (Fig. 7.9).

7.3.1.4 Convolutional Neural Networks (CNNs)

CNNs are a class of ANNs particularly well-suited for image-related tasks, i.e., fruit detection. It was introduced by American researchers LeCun et al. in the 1980s

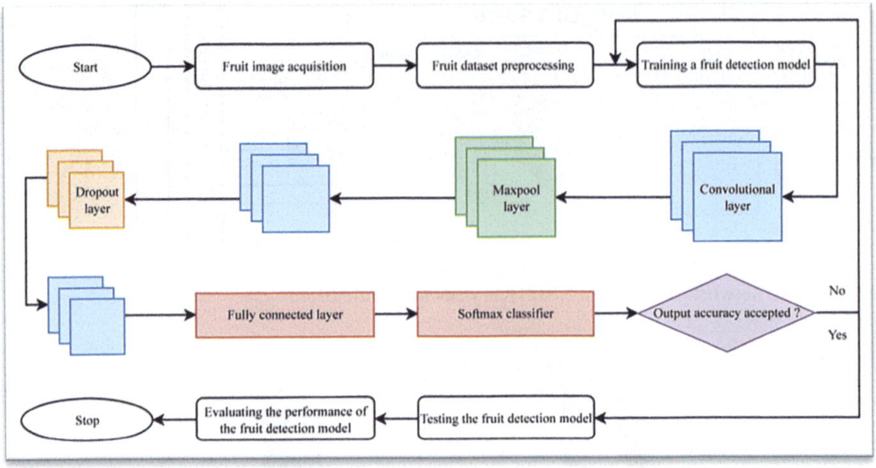

Fig. 7.9 The network structure of ANN for fruit detection (Xiao et al. 2023b)

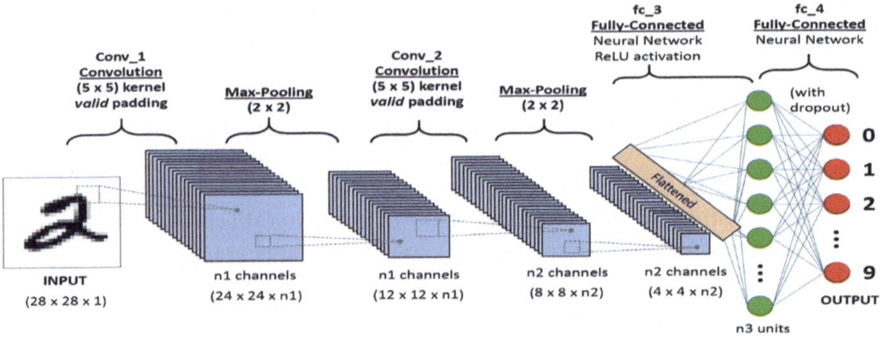

Fig. 7.10 The network structure of CNN (Ratan 2020)

demonstrating efficiency in capturing patterns in multidimensional space, and their network structure is shown in Fig. 7.10.

Applications: CNNs are widely used due to their ability to automatically learn hierarchical features, making them suitable for fruit detection tasks.

7.3.2 Image Segmentation Techniques for Precise Fruit Boundary Delineation

Image segmentation is a fundamental process of computer vision that divides the image into discrete regions and is necessary after preprocessing. Researchers extensively explore machine learning techniques for fruit detection due to their ability to derive laws from complex sample data (Table 7.1). Notably, studies focus on

Table 7.1 Image segmentation and classifiers based on traditional machine learning

Type	Input	Idea of segmentation and classification	Accuracy (%)	Advantages	Limitations
Techniques based on K-means clustering algorithm (Sun et al. 2019)	Segmentation of images, cluster count, and initial cluster centers	Group input data into either identical or different classes depending on their fixed distance from each other	80–90	Automatically differentiate between targets and background with minimal computation time, rapid response, and effective clustering	Outcomes are influenced by randomly chosen K-values and are susceptible to deviations in the data
Support vector machine algorithm (Wang et al. 2009)	The training set comprises pixel features, the choice of kernel functions, and the test images	Classification involves finding a separated hyperplane that effectively divides the training set and maximizes geometric separation	82–96	Effective classification for data points beyond the training set, with reduced computational complexity when mapping to higher-dimensional spaces	The algorithm's sensitivity to parameter adjustments and kernel function selection
K-nearest neighbor (KNN) algorithm (Sarimole and Rosiana 2022)	The training set consisting of pixel features, the specified number of clusters, and the test images	Classification involves assigning unknown feature vectors to classes based on the most common attributes in the K-nearest neighbors using the training set	85–90	Achieving high classification accuracy, robustness against outliers, and not making assumptions about the input data	Prone to tedious selection of an appropriate scaling factor *k*, with high time and space complexity, requiring significant computational resources
AdaBoost-based algorithm (Lin and Zou 2013)	The training set consisting of pixel features, along with the test images	Weak classifiers are trained using the same training set, and then these weak classifiers are combined to form a strong classifier	78–96	Leverage weak classifiers effectively through cascading to achieve high detection and recognition accuracy	Susceptible to noise interference and reliant on weak classifiers, which can frequently entail lengthy training times
Decision tree algorithm (Kuang et al. 2019)	The training set, comprised of pixel features, along with the test images	The classification process starts from the root node, testing the item's features along branches until reaching a leaf node, where stored data determines the result	80–85	Visual representation of the decision-making process illustrates various stages and sequences, providing insight into the entire decision problem	Prone to overfitting and may struggle with datasets containing strong feature correlations
Bayesian algorithm (Reyes et al. 2021)	The training set consisting of pixel features, along with the test images	Classify data by minimizing Bayesian risk or minimizing probability error	75–86	The simplicity of the recognition and classification process, capability to manage multiple classification tasks, and improved performance particularly with small-scale data	Prior probabilities need to be calculated

K-means clustering, Support Vector Machine (SVM), K-Nearest Neighbor (KNN) clustering, AdaBoost, decision trees, and Bayesian algorithms. Except for K-means clustering technique, the other five techniques are supervised learning. Comparative analysis evaluates these techniques, highlighting their strengths, weaknesses, and applicability in fruit object detection. Traditional machine learning enhances speed, accuracy, and robustness compared to digital image processing.

7.4 Machine Learning Models for Counting Accuracy

Machine learning is a subset of artificial intelligence that empowers computers to learn patterns and make decisions without explicit programming. ML algorithms find extensive application across various domains, including the medical sector for tasks such as brain tumor detection and feature segmentation of COVID-19 (Solanki et al. 2023a, b, c; Saxena et al. 2024) as well as within the agricultural industry, where they demonstrate broad utility. In the context of fruit counting, machine learning plays a crucial role in automating and optimizing the counting process. Traditional methods often involve manual labor, leading to inefficiencies and inaccuracies. Machine learning algorithms, particularly computer vision models, can be trained to recognize and count fruits in images or videos (Sharma et al. 2020).

Through the use of labeled datasets, these models learn to distinguish between different fruits, sizes, and arrangements, adapting to varying conditions. The integration of machine learning in fruit counting not only enhances accuracy but also significantly reduces the time and resources required. This technology finds applications in agriculture, aiding farmers in crop management, yield estimation, and harvest planning (Luo et al. 2018). By harnessing the power of machine learning, fruit counting becomes a more efficient, precise, and sustainable practice, contributing to advancements in agricultural technology.

7.4.1 Overview of Convolutional Neural Network and Support Vector Machines

Convolutional Neural Networks (CNNs) and Support Vector Machines (SVMs) are two distinct but powerful machine learning models with different architectures and applications.

- **Convolutional Neural Networks (CNNs)**: CNNs are specialized deep learning models designed for visual tasks like image recognition and classification. They consist of convolutional layers that apply filters to input data, capturing hierarchical features and spatial hierarchies. Pooling layers reduce dimensionality, and fully connected layers make predictions based on the learned features (Taye 2023). CNNs excel in tasks like image classification, object detection, and segmentation. They have been pivotal in advancing computer vision applications and achieving state-of-the-art results in image-related tasks (Nihalani et al. 2024).

- **Support Vector Machines (SVMs)**: SVMs are a classical machine learning algorithm for both classification and regression tasks. They work by finding a hyperplane that best separates different classes in the feature space. SVMs are effective in high-dimensional spaces and can handle nonlinear decision boundaries through the use of kernel functions (Cervantes et al. 2020). They are widely used for text classification, image classification, and bioinformatics. SVMs provide robust generalization and are less prone to overfitting, making them suitable for various domains.

7.5 Data Acquisition and Preprocessing

High-quality data is essential for the success of AI models, influencing their performance, fairness, and reliability. During training, well-labeled and representative data ensures accurate learning of meaningful patterns, enhancing training accuracy and the model's ability to generalize to new situations. Quality data also mitigates biases, promoting fairness and preventing discriminatory outcomes. It contributes to model robustness by reducing noise and outliers, making the model more adaptable to varied inputs. In testing, high-quality data enables accurate evaluation and estimation of generalization capabilities. Additionally, data relevance to real-world scenarios enhances the model's practical applicability. Efficient resource utilization, user trust, and accountability are further benefits, emphasizing the pivotal role of high-quality data in shaping the effectiveness and ethical considerations of AI models.

7.5.1 Use of Labeled Datasets and the Challenges Associated with Data Collection

Labeled datasets play a crucial role in training supervised machine learning models. In a labeled dataset, each data point is paired with its corresponding ground truth or target label, allowing the algorithm to learn patterns and associations. This supervised learning approach is prevalent in various applications, including image recognition, natural language processing, and classification tasks.

For example, in image classification, a labeled dataset would consist of images along with corresponding labels indicating the object or category each image represents. This labeled information serves as the basis for the model to generalize its understanding and make accurate predictions on new, unseen data (Fig. 7.11).

Challenges Associated with Data Collection:

1. **Annotation Cost and Effort:**
   - Hand-labeling data, especially for complex tasks like image segmentation, is time-consuming and costly due to the need for expert annotators, raising overall expenses and efforts.

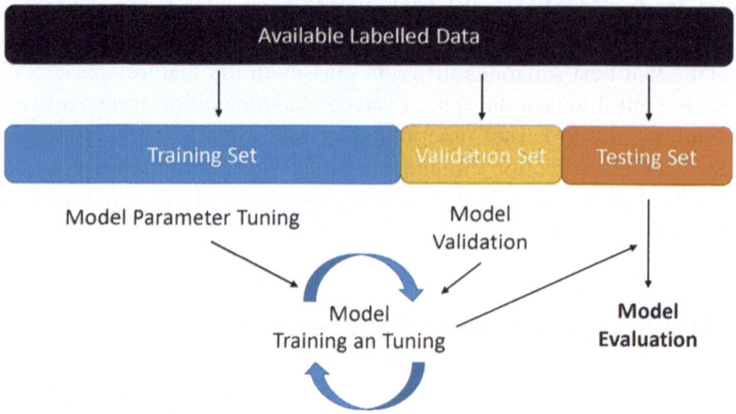

Fig. 7.11 Model training and evaluation from the labeled datasets

2. **Labeling Subjectivity:**
   - Humans who label data may interpret things differently, causing inconsistencies in labeling. To reduce these differences, it's important to have clear guidelines for labeling and provide ongoing feedback.
3. **Data Bias:**
   - Labeled datasets might accidentally include biases from how the data was collected. If not fixed, models can learn and keep these biases, causing unfair or discriminatory outcomes.
4. **Insufficient or Imbalanced Data:**
   - Getting a big and diverse labeled dataset can be hard sometimes. When there are uneven numbers in different categories, the model might struggle to work well for all of them.
5. **Dynamic Data Environments:**
   - In fast-changing areas like healthcare or self-driving cars, labeled datasets can become old fast. Models trained on fixed datasets may have trouble with new data patterns or adding new labels.
6. **Privacy Concerns:**
   - Certain datasets, especially those with private information, have privacy worries during labeling. It's crucial to find a balance between gathering useful data and following privacy rules.
7. **Quality Control:**
   - Making sure labeled data is of good quality and accurate is really important. Using strict quality control and validation processes helps find and fix mistakes in labeling, making the dataset more reliable.

7.6 Case Studies and Success Stories

In recent years, several researchers have been working on fruit counting using AI and achieving success in this field, which is shown in Table 7.2.

Table 7.2 Some of the successful case studies

Sl. no.	Title	Key findings	Relevant figures
1.	Improved method for apple fruit target detection based on YOLOv5s (Wang et al. 2023)	• Enhancements to object detection technique inspired by YOLOv5s, tailored for small object detection in images captured by unmanned aerial vehicles (UAVs) • The improved approach integrates RFA module, dual-feature pyramid (DFP) structure, and soft-NMS algorithm, enabling precise detection of small targets • Experimental findings demonstrate notable performance enhancements on UAV image datasets for small objects, with a noteworthy 6.2% increase in mean average precision (mAP) value	
2.	Deep Count: Fruit Counting Based on Deep Simulated Learning (Rahnemoonfar and Sheppard 2017)	• Achieved a 91% accuracy rate when tested on a sample of one hundred randomly selected real images • The algorithm demonstrates robustness in challenging conditions, accurately counting tomatoes even when they are shadowed, obstructed by foliage or branches, or partially overlapping • Capable of accurately counting ripe and half-ripe fruits, though it lacks training for counting green fruits	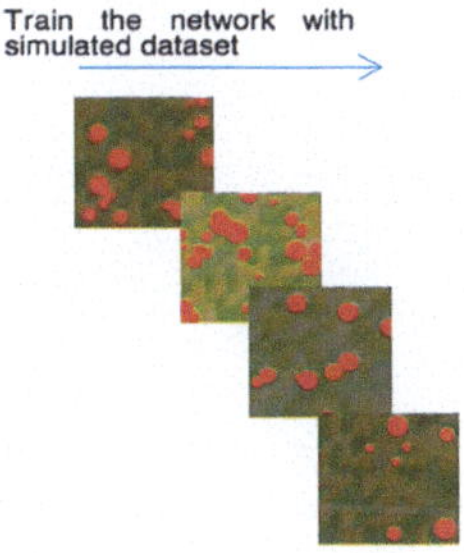

(continued)

Table 7.2 (continued)

Sl. no.	Title	Key findings	Relevant figures
3.	Deep-learning-based in-field citrus fruit detection and tracking (Zhang et al. 2022)	• The algorithm comprises two sub-algorithms: OrangeYolo for fruit detection and OrangeSort for fruit tracking • OrangeYolo is built upon the YOLOv3 algorithm, enhanced for detecting small objects (fruits) across multiple scales • OrangeYolo attains mean average precision (mAP) values of 0.957 on the citrus dataset, surpassing the 0.905, 0.911, and 0.917 achieved by the YOLOv3, YOLOv4, and YOLOv5 algorithms, respectively	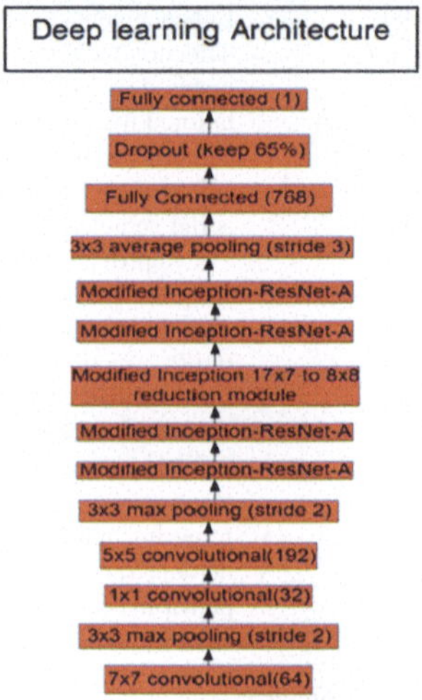  

7.7 Limitations and Opportunities of AI in Agriculture

In recent years, the integration of artificial intelligence (AI) technologies into agriculture has garnered significant attention. However, alongside its promising applications, some notable limitations and opportunities warrant exploration. Below, we discuss the limitations and opportunities of AI in agriculture, shedding light on its multifaceted impact on farming practices and food production.

Limitations:

1. ***Data Quality***: The quality and availability of agricultural data can significantly affect the performance of AI algorithms. Limited or inconsistent data may lead to biased or inaccurate predictions.
2. ***Scalability***: AI technologies may face challenges in scaling up to handle large-scale agricultural operations due to computational constraints, data processing requirements, and infrastructure limitations.
3. ***Interpretability***: While AI models can generate valuable insights, their complexity often makes it challenging to interpret the reasoning behind their predictions, raising concerns about trust and transparency.
4. ***Resource Constraints***: Access to high-performance computing resources, specialized hardware, and reliable internet connectivity may be limited in rural agricultural areas, hindering the adoption of AI technologies.
5. ***Generalization***: AI models trained on data from specific agricultural contexts may struggle to generalize to diverse environments, climates, and farming practices, limiting their applicability across regions.
6. ***Ethical Considerations***: The use of AI in agriculture raises ethical concerns related to data privacy, security, bias, and the socioeconomic impact on smallholder farmers and rural communities.
7. ***Adoption Barriers***: Farmers may face barriers such as lack of awareness, technical skills, financial resources, and supportive policies, impeding the widespread adoption of AI technologies in agriculture.

Opportunities:

1. ***Precision Agriculture***: AI enables precision agriculture techniques such as satellite imaging, sensor networks, and machine learning algorithms to optimize resource use, increase crop yields, and reduce environmental impact.
2. ***Crop Monitoring***: AI-powered systems can monitor crops in real-time using drones, satellites, and IoT devices, detecting diseases, pests, and nutrient deficiencies for timely intervention and improved crop health.
3. ***Supply Chain Optimization***: AI-driven analytics and predictive modeling can optimize supply chain logistics, inventory management, and quality control, reducing food waste and improving efficiency from farm to market.
4. ***Decision Support Systems***: AI-based decision support systems provide farmers with data-driven insights and recommendations for crop management, irrigation scheduling, pest control, and optimal planting practices.

5. ***Climate Resilience***: AI technologies help farmers adapt to climate change by analyzing weather patterns, predicting crop risks, and recommending adaptive strategies such as drought-resistant crops or sustainable farming practices.
6. ***Agricultural Automation***: Robotics, autonomous vehicles, and smart machinery powered by AI automate labor-intensive tasks such as planting, harvesting, and weed control, enhancing productivity and reducing labor costs.
7. ***Collaborative Innovation***: Collaborative efforts between researchers, technologists, policymakers, and farmers foster innovation in AI solutions tailored to agricultural needs, addressing specific challenges and driving sustainable agricultural development.

7.7.1 Challenges and Future Directions in Fruit Counting Using AI

The challenges associated with fruit counting and analysis using AI require a combination of advanced algorithms, sensor integration, and exploration of emerging technologies. As research in this field progresses, the integration of innovative solutions will play a crucial role in developing robust and reliable systems for agricultural applications, which is shown in Table 7.3.

Table 7.3 Some of the major challenges in fruit counting and their future directional steps

Challenges	Future directions
Variability in fruit appearance	• Adaptive AI models can be developed that are capable of learning and adjusting to diverse fruit characteristics • Transfer learning can be explored to leverage knowledge from one fruit type to another
Environmental conditions	• Investigate the use of advanced sensors and image fusion techniques to handle varying lighting conditions and occlusions • AI models can be developed robust to environmental changes through continual learning
Image quality	• Explore advancements in imaging technology to improve the quality of input data for AI models • Real-time image enhancement techniques can be investigated for on-the-fly improvements
Data annotation challenges	• Develop semi-supervised and unsupervised learning approaches to alleviate the need for extensive manual annotation • Crowdsourcing methods can be investigated for large-scale data labeling
Deployment in real-world orchards	• Conduct extensive field testing and validation to ensure the adaptability of AI models to various orchard environments • Implement explainable AI techniques for better acceptance and trust among farmers
Scalability and efficiency	• Explore edge computing solutions to process data locally, reducing latency and dependence on centralized servers • Investigate federated learning approaches to train models across distributed orchards while maintaining data privacy
Integration with precision agriculture	• Integrate fruit counting AI with precision agriculture systems for optimized resource allocation • Explore synergies with other agricultural technologies, such as automated harvesting, to create comprehensive solutions

7.7.2 Exploration of Some Emerging Technologies

1. **LiDAR Technology**: The integration of LiDAR (Light Detection and Ranging) technology can provide 3D information about the fruit orchard, allowing for better understanding and localization of fruits. This technology aids in overcoming occlusion challenges and improves the accuracy of counting in complex environments.
2. **Edge Computing**: Exploring edge computing solutions allows for processing data closer to the source, reducing latency and dependency on centralized computing resources. Edge devices equipped with AI models can enable real-time fruit counting and analysis directly in the field, facilitating quicker decision-making.
3. **Autonomous Robotics**: Integration of AI-powered autonomous robotic systems equipped with vision sensors can automate the process of fruit counting. These robots can navigate through orchards, capture images, and perform real-time analysis, reducing the need for manual intervention and increasing efficiency.
4. **Hyper-Spectral Imaging**: Investigating hyper-spectral imaging techniques enables the capture of detailed spectral information beyond visible light. This can provide additional insights into the physiological characteristics of fruits, enhancing the accuracy of identification and counting.

7.8 Conclusion

The transformative potential of artificial intelligence (AI) in fruit counting and analysis is undeniable. Through advanced machine learning algorithms, multi-sensor integration, and the exploration of emerging technologies, AI has demonstrated its ability to address the intricate challenges posed by the variability in fruit appearance and environmental conditions. The proposed solutions, ranging from deep learning models to the integration of LiDAR and hyper-spectral imaging, underscore the adaptability and versatility of AI systems in agricultural contexts. As we navigate the future of AI in agriculture and fruit counting, there is a compelling call to action for researchers, practitioners, and stakeholders to collaborate and contribute to this evolving field. Continued research, real-world testing, and the development of scalable, efficient solutions are imperative for the widespread adoption of AI technologies in agriculture. In doing so, we pave the way for enhanced precision, efficiency, and sustainability in fruit cultivation, marking a promising trajectory for the future of AI in shaping the landscape of agricultural practices.

References

Anand R, Sahni RK, Kumar SP, Thorat DS, Kumar AK (2023) Advancement in agricultural practices with use of drones in the context of precision farming. Glob J Eng Sci 11(2):1–7

Bali A, Chouhan SS, Kumar G (2023) Adaptive fault-tolerant control for pure-feedback stochastic nonlinear systems with sensor and actuator faults. Circuits Syst Signal Process 42:5838–5867. https://doi.org/10.1007/s00034-023-02366-7

Basso B, Antle J, Ritchie J (2019) Simulating agriculture in a changing climate: a review. Agron J 111(5):2131–2147

Cervantes J, Garcia-Lamont F, Rodríguez-Mazahua L, Lopez A (2020) A comprehensive survey on support vector machine classification: applications, challenges and trends. Neurocomputing 408:189–215

Chouhan SS, Singh UP, Jain S (2024) Introduction to computer vision and drone technology. In: Chouhan SS, Singh UP, Jain S (eds) Applications of computer vision and drone technology in agriculture 4.0. Springer, Singapore. https://doi.org/10.1007/978-981-99-8684-2_1

Dufour R, Lambert R (2008) Orchard mechanization and automation in apple production. Biosyst Eng 100(4):504–512

Government of India (2020) Agricultural statistics at a glance 2019

Grieve B, Kahui V, Le Heron E, Loch A (2013) Subjectivity and consistency in manual fruit counting methods: the example of kiwifruit. N Z J Crop Hortic Sci 41(4):191–203

Hemming J, Rath T, Klooster C (2011) Machine vision systems for fruit detection and harvesting robots—a review. Comput Electron Agric 94:1–13

Hinton GE, Salakhutdinov RR (2006) Reducing the dimensionality of data with neural networks. Science 313(5786):504

Hunt ER, Doraiswamy PC (2010) Assessing sensor performance and potential of an ultrasonic anemometer–temperature–humidity sensor network. Comput Electron Agric 70(1):15–26

Jiao Y, Luo R, Li Q, Deng X, Yin X, Ruan C, Jia W (2020) Detection and localization of overlapped fruits application in an apple harvesting robot. Electronics 9:1023

Khaliq T, Prasher SO, Deen W (2005) A review of precision agriculture research. Can J Plant Sci 85(2):271–285

Khryashchev VV, Stepanova OA, Lebedev AA, Kashin SV, Kuvaev RO (2019) Deep learning for gastric pathology detection in endoscopic images. In: Proceedings of the 3rd international conference on graphics and signal processing, pp 90–94

Koirala A, Zhang Q (2018) Applications of computer vision for assessing quality of fruits: a review. Comput Electron Agric 153:123–134

Kuang YC, Streeter L, Cree MJ, Ooi MPL (2019) Evaluation of deep neural network and alternating decision tree for kiwifruit detection. In: Proceedings of the IEEE international instrumentation and measurement technology conference, Auckland, New Zealand, 20–23 May 2019

Kumar R, Singh UP, Bali A (2023) Adaptive control of unknown fuzzy disturbance-based uncertain nonlinear systems: application to hypersonic flight dynamics. J Therm Anal 32:1395. https://doi.org/10.1007/s41478-023-00687-z

Lin G, Zou X (2018) Citrus segmentation for automatic harvester combined with AdaBoost classifier and Leung-Malik filter Bank. IFAC Pap 51:379–383

Lithourgidis A, Vasilakoglou I, Dhima K, Dordas C (2011) Yield and yield components of field grown faba bean affected by the sowing date, under rainfed conditions. Eur J Agron 34(2):90–97

Lobos G, Camargo A, Retamales J (2017) Precision agriculture in highbush blueberry using ground-based sensors: spatial variability of soil and plant parameters related to fruit yield. Precis Agric 18(3):353–370

Luo L, Tang Y, Lu Q, Chen X, Zhang P, Zou X (2018) A vision methodology for harvesting robot to detect cutting points on peduncles of double overlapping grape clusters in a vineyard. Comput Ind 99:130–139

Ministry of Agriculture and Farmers Welfare (2020) Agricultural statistics at a glance 2019

Mistele B, Schmidhalter U (2008) Estimating the nitrogen nutrition index using spectral canopy reflectance measurements. Eur J Agron 29(2–3):184–190

Nihalani R et al (2024) Long short-term memory (LSTM) model for Indian sign language recognition. J Intell Fuzzy Syst 46:11185. https://doi.org/10.3233/JIFS-233250

Panchal B, Pranjal P, Patel RK, Sharma A, Chouhan SS (2024) Assessing the quantity of a crop field using aerial images. In: Chouhan SS, Singh UP, Jain S (eds) Applications of computer vision and drone technology in agriculture 4.0. Springer, Singapore. https://doi.org/10.1007/978-981-99-8684-2_11

Pathan SK, Rehman MA (2013) A review on development of sensor node for precision agriculture. Proced Eng 64:96–104

Planning Commission, Government of India (2014) Report of the expert committee on sustainable development of agriculture

Rahnemoonfar M, Sheppard C (2017) Deep count: fruit counting based on deep simulated learning. Sensors 17:905

Ratan P (2020) Convolutional neural network made easy for data scientists. Analytics Vidhya. www.analyticsvidhya.com/blog/2020/10/what-is-the-convolutional-neural-network-architecture/

Reyes JF, Contreras E, Correa C, Melin P (2021) Image analysis of real-time classification of cherry fruit from colour features. J Agric Eng 52:1–6

Sa I, Ge Z, Dayoub F, Upcroft B, Perez T, McCool C (2016) Deepfruits: a fruit detection system using deep neural networks. Sensors 16(8):1222

Sarimole FM, Rosiana A (2022) Classification of maturity levels in areca fruit based on HSV image using the KNN method. J Appl Eng Technol Sci 4:64–73

Saxena A, Chouhan SS, Aziz RM (2024) A comprehensive evaluation of marine predator chaotic algorithm for feature selection of COVID-19. Evol Syst 15:1235. https://doi.org/10.1007/s12530-023-09557-2

Sharma A, Jain A, Gupta P, Chowdary V (2020) Machine learning applications for precision agriculture: a comprehensive review. IEEE Access 9:4843–4873

Sharma A, Patel RK, Pranjal P, Panchal B, Chouhan SS (2024) Computer vision-based smart monitoring and control system for crop. In: Chouhan SS, Singh UP, Jain S (eds) Applications of computer vision and drone technology in agriculture 4.0. Springer, Singapore. https://doi.org/10.1007/978-981-99-8684-2_5

Shen L, Su J, Song Y, Fang Y, Su B (2022) Fusing attention mechanism with mask R-CNN for instance segmentation of grape cluster in the field. Front Plant Sci 22(13):934450

Solanki S, Singh UP, Chouhan SS (2023a) A systematic analysis of magnetic resonance images and deep learning methods used for diagnosis of brain tumor. Multimed Tools Appl 83:23929–23966. https://doi.org/10.1007/s11042-023-16430-6

Solanki S, Singh UP, Chouhan SS, Jain S (2023b) Brain tumour detection and classification by using deep learning classifier. Int J Intell Syst Appl Eng 11(2s):279. https://ijisae.org/index.php/IJISAE/article/view/2624

Solanki S, Singh UP, Chouhan SS (2023c) Brain tumor classification using ML and DL approaches. In: IEEE 5th International conference on cybernetics, cognition and machine learning applications (ICCCMLA), Hamburg, Germany, vol 2023, pp 204–208. https://doi.org/10.1109/ICCCMLA58983.2023.10346854

Sugiura R, Tsuji H, Tanaka K (2019) Machine learning-based fruit counting method using stereo vision. Comput Electron Agric 162:424–433

Sun S, Song H, He D, Long Y (2019) An adaptive segmentation method combining MSRCR and mean shift algorithm with K-means correction of green apples in natural environment. Inf Process Agric 6:200–215

Tamburini G, Zanotelli D, Montagnani L (2017) Optical sensing system for fruit counting in vineyards. Sensors 17(5):1098

Taye MM (2023) Theoretical understanding of convolutional neural network: concepts, architectures, applications, future directions. Computation 11(3):52

United Nations (2019) World population prospects 2019: highlights

Wang JJ, Zhao DA, Ji W, Tu JJ, Zhang Y (2009) Application of support vector machine to apple recognition using in apple harvesting robot. In: ICIA'09. International conference on information and automation, pp 1110–1115

Wang C, Tang Y, Zou X, SiTu W, Feng W (2017) A robust fruit image segmentation algorithm against varying illumination for vision system of fruit harvesting robot. Optik 131:626–631

Wang Z, Zhang Q, Zhang L, Cheng J, Xu Y (2019) Fruit detection and counting from multiple images using a fully convolutional network. Comput Electron Agric 157:417–425

Wang H, Feng J, Yin H (2023) Improved method for apple fruit target detection based on YOLOv5s. Agriculture 13:2167

Xiao F, Wang H, Li Y, Cao Y, Lv X, Xu G (2023a) Object detection and recognition techniques based on digital image processing and traditional machine learning for fruit and vegetable harvesting robots: an overview and review. Agronomy 13:639

Xiao F, Wang H, Xu Y, Zhang R (2023b) Fruit detection and recognition based on deep learning for automatic harvesting: an overview and review. Agronomy 13:1625

Zhang W, Wang J, Liu Y, Chen K, Li H, Duan Y, Wu W, Shi Y, Guo W (2022) Deep-learning-based in-field citrus fruit detection and tracking. Hortic Res 5:9

Deep Learning-Based Plant Stress Diagnosis: An Optimized Generative Augmentation Model Approach

8

Manjit Kaur and Upinder Kaur

Abstract

Agriculture influences economic stability and prosperity, and it is an important part of the intricate tapestry of our country's economy. However, agricultural business has a number of barriers, notably changing weather conditions that induce a variety of serious diseases in plants, depressing crop yields, and revenues. An optimized Convolution Neural Network model for identifying and reliably detecting plant stress phenotyping. By using the capabilities of these sophisticated neural networks, we were able to successfully raise the stress detection process beyond the boundaries of traditional approaches. Plant stress phenotypic data from important crop species, including wheat, rice, cotton, and maize, are utilized to systematically assess the model's efficacy. The proposed model, with its exceptional accuracy of 96.67%, demonstrates its potential to become one of the world's leading agricultural decision-support systems, benefiting a wide range of businesses. This work, which establishes a new standard for harnessing artificial intelligence for sustainable agriculture practices, will assist plant stress detection and improved crop management.

Keywords

Convolutional neural networks (CNN) · K-nearest neighbour (KNN) · Support Vector Machines (SVMs) · Artificial Neural Networks (ANNs) · Plant leaves · Optimization · Plant growth

M. Kaur (✉) · U. Kaur (✉)
Department of Computer Science and Engineering, Akal University, Bathinda, Punjab, India
e-mail: upinder_cs@auts.ac.in

S. S. Chouhan et al. (eds.), *Artificial Intelligence Techniques in Smart Agriculture*, https://doi.org/10.1007/978-981-97-5878-4_8

8.1 Introduction

Crop production is undeniably important in improving national economies in today's dynamic global agricultural environment, particularly in countries where agriculture contributes considerably to GDP. The tremendous growth in world population raises food production demands, putting a strain on the agricultural sector. Crop yield optimization is complicated by a variety of factors, including crop quality, weather changes, and proactive disease control, with leaf ailments particularly damaging to crop production and farmer revenue.

Machine vision and deep learning have transformed agricultural methods over the previous few decades. This technological revolution has resulted in the development of novel techniques for pattern recognition, image processing, and feature extraction, all of which are critical in plant pathology. Though developing the framework for the analysis of plant illnesses using conventional image classification approaches such as Convolutional Neural Networks, K-nearest neighbor (KNN), Support Vector Machines (SVMs), and Artificial Neural Networks (ANNs),they are often incapable of addressing the growing complexity of plant stress factors today (Deshpande and Patidar 2023; Nihalani et al. 2024).

Deep Convolutional Neural Networks (DCNNs), among other advancements in computer vision, have facilitated the investigation of complex patterns within plant stress indicators. Due to the difficulties posed by dataset class imbalance and the necessity to depict complex aspects in plant leaf images, a model that is more adaptable and versatile is required to solve these issues. By merging Generative Adversarial Networks (GANs) and Deep Convolutional Neural Networks (DCNNs) to produce a dependable, multi-class detection system for plant leaf diseases, we want to solve these obstacles through the development of a revolutionary model for the detection of such conditions. GANs in conjunction with deep learning offer a sophisticated method for discerning various stress categories, as well as enhancing the feature representation of plant leaf images.

Preceding all previous developments, we provide in this chapter groundbreaking techniques for phenotyping plant stress. Eliminating class disparity caused by unequal dataset sizes in the detection of plant diseases is a primary objective of our study (Lamba et al. 2023). The secondary objective is to enhance the feature representation of leaf images to provide a more discerning view of intra-class and inter-class variability in plant disease detection.

Section 8.2 provides a comprehensive review of the current plant leaf disease detection landscape, followed by Sect. 8.3, which presents our GAN-enhanced disease detection strategy (Chen and Wu 2023). In Sect. 8.4, we present our experimental insights and discussions centered around the Plant Stress Phenotyping dataset (PSPD). This study concludes with a summary of the implications of our findings and sets the stage for further research in this field.

8.2 Literature Review

The agricultural industry, which has great importance in a developing countries' economic and cultural development, encounters a multitude of obstacles, most notably plant diseases, that have a substantial detrimental effect on both crop productivity and economic advancement. Conventional approaches to plant disease diagnosis, such as visual inspection by hand, frequently prove insufficient owing to their dependence on specialized expertise and inherent difficulties in managing intricate picture data. Particularly in the domain of autonomous plant leaf disease identification, this has prompted research into computer vision and artificial intelligence (AI) systems. However, these efforts are hindered by an unbalanced data set, inadequate feature representation, low-order correlation in the raw features, and restricted generalizability.

Diverse methodologies have been developed in recent times to detect and categorize plant diseases, as a result of developments in machine learning (ML). The potential of combining deep convolutional neural networks (CNNs) with generative adversarial networks (GANs) has been particularly highlighted in the integration of GANs (Deshpande and Patidar 2023). This integration attempts to address the problem of data imbalance by augmenting data with GANs, improving feature representation and correlation in deep learning models. The impact of such strategies is apparent in research concentrating on varied plant species, such as tomatoes and employing databases like PlantVillage (Ko et al. 2023).

In the articles by Saxena et al. (2024) and Solanki et al. (2023a, b), the authors discussed the use of deep learning models for diagnostic applications. Further, numerous methods have been proposed in the literature to enhance plant disease detection and categorization domain. There has been evidence that ML models, such as CNNs, are superior to alternative classifiers, including k-nearest neighbor (KNN), fuzzy, and Support Vector Machines (SVM) (Singh et al. 2023). The datasets that these models are trained on contain pictures of different plant diseases. CNNs are often the best at finding and classifying leaf diseases. Some studies provided deep learning algorithms for classifying and dealing with unidentified challenges during prediction in Solanki et al. (2023a, b), Bali et al. (2023), and Kumar et al. (2023a, b). Thus, there is ample opportunity for enhancements in deep learning approaches for future advancements (Chouhan et al. 2024; Sharma et al. 2024; Panchal et al. 2024).

While assessing tomato leaf diseases, the limited availability of data and the tiny size of lesions provide a substantial obstacle. In order to mitigate these concerns, a suggested deep learning pipeline has three stages: Faster R-CNN, DCGAN, and ResNet (Vengaiah and Konda 2023). The methodology drifts away from typical single-stage algorithms, displaying greater performance in managing sparse datasets and minor lesions, with generalization studies supporting its resilience (Zheng et al. 2022).

Deep learning is increasingly being utilized to identify plant diseases on account of its capacity to extract unique characteristics from leaf images. An example of the effective utilization of DCNNs in disease diagnosis with a high degree of accuracy can be seen in the Plant Village database created by Mohanty et al. This database has hundreds of images spanning many plant species (Roy and Lipton 2022). Transfer learning, as applied by Ramcharan et al. for cassava disease detection, and the use of DenseNet121 models enhanced with C-GAN images by Abbas et al. highlight the growing trend of deep learning techniques in this field (Ramcharan et al. 2017; Brahimi et al. 2017).

Notwithstanding these advancements, a number of obstacles continue to exist in the field of plant leaf disease identification. These include the high computing demands for real-time applications, the complexity of deep learning models, and the necessity for bigger datasets to train such models. To address these concerns, there is a growing need for lightweight architectures that minimize training parameter and memory use (Kumar et al. 2023a, b).

To conclude, ML and deep learning strategies are at the frontline of such evolutionary development of detection of plant diseases. By using GANs for data augmentation and employing optimized CNNs, substantial progress has been made in tackling the issues of feature representation and data imbalance. Nonetheless, novel approaches that alleviate the computational strain and improve the ability of these models to generalize over a wide range of plant species are still required (Senthil Kumar et al. 2022).

8.3 Proposed Model for Plant Stress Identification

The integration of an optimized GAN into the DCNN architecture enables efficient analysis of complicated leaf image data. Three refined convolutional layers (Conv), an adaptive rectangular linear unit (ARLU) layer, and three successive progressive maximum layer pooling (MP) comprise this type of advanced neural network. These layers are followed by a sophisticated fully connected layer (FC) and a dynamic classification layer (DQL). Figure 8.1 and Algorithm 8.1 illustrate the proposal model's workflow and the algorithm in detail, respectively.

Each of the layers of the convolution process employs specialized filters to be able to capture the nuanced features of the surface of leaves, such as micro-lesions, pigment changes, and subtle variations in the texture, which are crucial for the early detection of stress in plants. In mathematics, the operation of convolution is expressed as follows:

$$\mathrm{Conv}(m,\ n) = im(m,n) * h(w,w) = \sum_{i=1}^{R}\sum_{j=1}^{c} im(i,j)h(i-m,j-n) \tag{8.1}$$

where $im(m, n)$ is the input image and $h(w, w)$ is the convolutional filter with dimensions $w * w$.

It is crucial for early detection of stress to employ ARLU layers because they adaptively adjust their behavior based on input, enhancing the ability of the model to capture and propagate both positive and subtle negative attributes.

Optimized GAN uses a customized GAN to create high-quality, realistic synthetic leaf images, dealing with limited datasets and imbalances. A GAN consists of

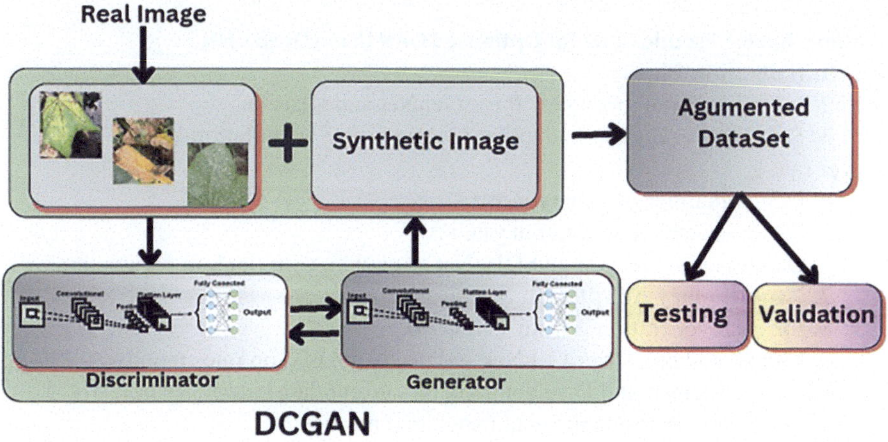

Fig. 8.1 Proposed model

an adaptive generator (G) and a discriminator (D). Generators produce images resembling rare disease manifestations, while discriminators evaluate if these images are realistic. Adversarial training has the following objective:

$$E(G,D) = \frac{1}{2} E_{x \sim pt}\left[1 - D(x)\right] = \frac{1}{2} E_{x \sim pz}\left[D(G(z))\right] \tag{8.2}$$

where $G(z)$ represented image, z is a random noise vector, $D(x)$ evaluates the generated and real images, and Pt and Pz are distributions of real and generated images, respectively.

8.3.1 Training and Testing with MBGD Optimization

An extensive dataset, encompassing a wide range of plant stress conditions, is processed during the training phase of Optimized GAN. Through an enhanced Mini-Batch Gradient Descent (MBGD) method with adaptive learning rate modulation, the DCNN filters are optimized effectively, ensuring balanced learning across diverse classes. Adapting to the complexity of the dataset dynamically is the goal of the weight update mechanism:

$$w_{t+1} = w_t - u_t \nabla_t E_t\left[\int (w_t)\right] \tag{8.3}$$

where w_t represent the filter weights at iteration t, and u_t is the adaptive learning rate.

The performance of Optimized GAN is rigorously tested against unseen data in order to ensure its accuracy and robustness in real-world situations. During this phase, the model is validated to ensure that it generalizes across many different stress conditions and plants.

Table 8.1 displays the summary of the generator model and Table 8.2 displays the summary of the discriminant model.

Algorithm 8.1 Pseudo Code for Optimised GAN Using Deep CNN

1. **Initialization Phase:**
2. Initialize population sizes P for GenNet and DiseNet.
3. Set base parameters: adaptive learning rate Υ, mutation rate μ, crossover rate χ.
4. Define number of generations G.
5. Set environment adaptation rate ϵ.
6. Create initial GenNet and DiseNet structures with random parameters.
7. **Evolutionary Adaptation Stage:**
8. **for** each generation g from 1 to G **do.**
9. Mutate and crossover GenNet and diseNet based on rates μ and χ.
10. Train GenNet and DiseNet pairs on current data batch.
11. Evoluate performance using specified metrics.
12. Select top-performance networks for next generation.
13. Adjust Υ, μ, and χ adaptively based on performance.
14. **End for.**
15. **Dynamic Training and Adaptation:**
16. Continuously train and adapt the GenNet and DiseNet:
17. Implement dynamic training with self-adjusting batch sizes and learning rates.
18. Integrate Self-Adapting Feedback Loop (SEFL) for real-time model adjustments.
19. Employ Advanced Date Augmentation techniques(ADAT) for training diversity.
20. Periodically evaluate and refine SAFL and ADAT parameters for optimal performance.
21. **Self-Evolutionary Fine-Tuning:**
22. Monitor ongoing environment changes and adjust network parameters accordingly.
23. Apply evolutionary principles to continuously refine networks structures and parameters.
24. Utilize survival-of-the-fittest principles to select networks for real-world deployment.
25. **Testing and deployment Stage:**
26. Test the final evolved GenNet and DiseNet models on a diverse set of unseen data.
27. Deploy the most robust and adaptive models for real-world applications.
28. Implement continuous monitoring and adaptation framework to ensure sustained performance.

Table 8.1 Generator model summary

Layer (type)	Output shape	Param #
dense (Dense)	(None, 9216)	930,816
reshape (Reshape)	(None, 6, 6, 256)	0
conv2d_transpose (Conv2DTranspose)	(None, 12, 12, 64)	409,664
re_lu (ReLU)	(None, 12, 12, 64)	0
batch_normalization (BatchNormalization)	(None, 12, 12, 64)	256
conv2d_transpose_1 (Conv2D Transpose)	(None, 24, 24, 32)	51,232
re_lu_1 (ReLU	(None, 24, 24, 32)	0
mini batch_normalization_1 (BatchNormalization)	(None, 24, 24, 32)	128
conv2d_transpose_2 (Conv2DTranspose)	(None, 48, 48, 32)	25,632
re_lu_2 (ReLU)	(None, 48, 48, 32)	0
conv2d_transpose_3 (Conv2DT ranspose)	(None, 96, 96, 3)	2403

Total params: 1,420,131
Trainable params: 1,419,939
Non-trainable params: 192

Table 8.2 Discriminator model summary

Layer (type)	Output shape	Param #
conv2d (Conv2D)	(None, 48, 48, 32)	2432
leaky_re_lu (LeakyReLU)	(None, 48, 48, 32)	0
dropout (Dropout)	(None, 48, 48, 32)	0
conv2d_1 (Conv2D)	(None, 24, 24, 64)	51,264
leaky_re_lu_1 (LeakyReLU)	(None, 24, 24, 64)	0
dropout_1 (Dropout)	(None, 24, 24, 64)	0
Mini batch_normalization_2 (BatchNormalization)	(None, 24, 24, 64	256
conv2d_2 (Conv2D)	(None, 12, 12, 64)	102,464
leaky_re_lu_2 (LeakyReLU)	(None, 12, 12, 64	0
dropout_2 (Dropout)	(None, 12, 12, 64)	0
batch_normalization_3 (BatchNormalization)	(None, 12, 12, 64)	256
conv2d_3 (Conv2D)	(None, 6, 6, 128)	204,928
leaky_re_lu_3 (LeakyReLU)	(None, 6, 6, 128)	0
dropout_3 (Dropout)	(None, 6, 6, 128)	0
flatten (Flatten)	(None, 4608)	0
dense_1 (Dense)	(None, 1)	4609

Total params: 366,209
Trainable params: 365,953
Non-trainable params: 256

8.4 Result and Discussion

8.4.1 Dataset Distribution and Augmentation

In Table 8.3, we observe the distribution of our original and augmented datasets. Figure 8.2 shows the sample dataset used for evaluation. The original dataset, comprising 16,012 images, displayed a diverse range of sample sizes across different

Table 8.3 Detailed summary of the dataset sample used

Disease type	Original dataset			Augmented dataset		
	Total images	Training samples	Testing samples	Total images	Training samples	Testing samples
Sheath Blight	1500	1000	500	3000	2000	1000
Brown Spot	1400	930	470	2800	1860	940
Leaf Smut	1300	860	440	2600	1720	880
Leaf Spot	1200	800	400	2400	1600	800
Yellow Leaf	1100	730	370	2200	1460	740
Leaf Blight	1000	670	330	2000	1340	660
Leaf Scald	900	600	300	1800	1200	600
Foot Rot	800	530	270	1600	1060	540
Stem Canker	700	470	230	1400	940	460
Faliar	600	400	200	1200	800	400
Powdery Mildew	500	330	170	1000	660	340
Mottle	400	270	130	800	540	260
Mold	300	200	100	600	400	200
Anthracnose	200	140	60	400	280	120
False Smut	100	70	30	200	140	60
Wild	190	120	90	400	120	60
Alternaria	280	200	80	460	360	100
Caros Pora	270	200	70	540	440	100
Cocked Foot	600	450	150	720	600	120
Cereal Tillering	500	400	100	900	700	200
Curl	400	300	100	800	600	200
Dwarf	300	250	50	600	500	100
Rosette Stunt	200	150	50	400	300	100
Tangro	100	60	40	200	160	40
Leaf Crumple	150	100	50	300	200	100
Wrinkled	200	150	50	400	300	100

diseases such as Sheath Blight, Brown Spot, and Leaf Smut. However, diseases like Cocked Foot and Cereal Tillering were underrepresented in the dataset. To counteract this imbalance and bolster the model's robustness, a GAN-based data augmentation technique was employed. This strategy was instrumental in generating synthetic images for each disease class, thereby standardizing the dataset to a uniform size of 3209 samples per class, a figure inspired by the maximum sample size observed in the Curl class.

8.4.2 DCNN Performance Analysis

The performance of our DCNN model, as detailed in Table 8.4, reached its peak efficiency within the initial 20 epochs. The model showcased outstanding accuracy rates of 99.80%, 98.50%, and 98.81% in the testing, validation, and training phases, respectively. We utilized Eqs. (8.1), (8.2), and (8.3) to calculate precision, recall,

Fig. 8.2 Dataset used for assessing the proposed model

and overall performance, with TP (True Positives) and TN (True Negatives) representing correctly classified samples, and FN (False Negatives) and FP (False Positives) accounting for misclassifications.

Table 8.4 Summary of the performance metrics for the with and without data augmentation

	Without data augmentation				With data augmentation			
Disease type	Accuracy	Recall	Precision	F1-score	Accuracy	Recall	Precision	F1-score
Sheath Blight	92.50	0.93	0.92	0.92	98.20	0.98	0.97	0.98
Brown Spot	90.70	0.91	0.90	0.91	97.50	0.98	0.96	0.97
Leaf Smut	89.80	0.90	0.89	0.89	97.10	0.97	0.96	0.97
Leaf Spot	88.60	0.89	0.88	0.89	96.70	0.97	0.95	0.96
Yellow Leaf	87.40	0.87	0.87	0.87	96.30	0.96	0.95	0.96
Leaf Blight	86.20	0.86	0.86	0.86	95.90	0.96	0.95	0.96
Leaf Scald	85.00	0.85	0.85	0.85	95.50	0.95	0.95	0.95
Foot Rot	83.80	0.84	0.83	0.84	95.10	0.95	0.95	0.95
Stem Canker	82.60	0.83	0.82	0.83	94.70	0.95	0.94	0.95
Faliar	81.40	0.81	0.81	0.81	94.30	0.94	0.94	0.94
Powdery Mildew	80.20	0.80	0.80	0.80	93.90	0.94	0.93	0.94
Mottle	79.00	0.79	0.79	0.79	93.50	0.94	0.93	0.94
Mold	77.80	0.78	0.77	0.78	93.10	0.93	0.93	0.93
Anthracnose	76.60	0.77	0.76	0.77	92.70	0.93	0.92	0.93
False Smut	75.40	0.75	0.75	0.75	92.30	0.92	0.92	0.92
Wild	74.20	0.74	0.74	0.74	91.90	0.92	0.91	0.92
Alternaria	73.00	0.73	0.73	0.73	91.50	0.92	0.91	0.92
Caros Pora	71.80	0.72	0.71	0.72	91.10	0.91	0.91	0.91
Cocked Foot	70.60	0.71	0.70	0.71	90.70	0.91	0.90	0.91
Cereal Tillering	69.40	0.69	0.69	0.69	90.30	0.90	0.90	0.90
Curl	68.20	0.68	0.68	0.68	89.90	0.90	0.89	0.90
Dwarf	67.00	0.67	0.67	0.67	89.50	0.89	0.89	0.89
Rosette Stunt	65.80	0.66	0.65	0.66	89.10	0.89	0.89	0.89
Tangro	64.60	0.65	0.64	0.65	88.70	0.89	0.88	0.89
Leaf Crumple	63.40	0.63	0.63	0.63	88.30	0.88	0.88	0.88
Wrinkled	62.20	0.62	0.62	0.62	87.90	0.88	0.87	0.88
Average	78.65	0.79	0.78	0.79	93.80	0.94	0.93	0.94

8.4.3 Comparative Analysis of Disease Identification

Table 8.4 provides a thorough analysis of our leaf disease identification approach, highlighting the significant role of data augmentation. Pre-augmentation, the model showed lower accuracy for diseases with fewer training samples, such as Dwarf and Rosette Stunt. Post-augmentation, the overall accuracy impressively increased to 98.81%, demonstrating substantial improvements across all disease types.

8.4.4 Comparative Assessment with Previous Methods

Our method's effectiveness becomes even more apparent when compared with previous plant leaf disease detection methods, as shown in Fig. 8.3. Traditional architectures like AlexaNet, Resnet 50, Resnet 101, Inception Net V3, and VGG-16, while complex, often struggled with class imbalance. In contrast, our lightweight

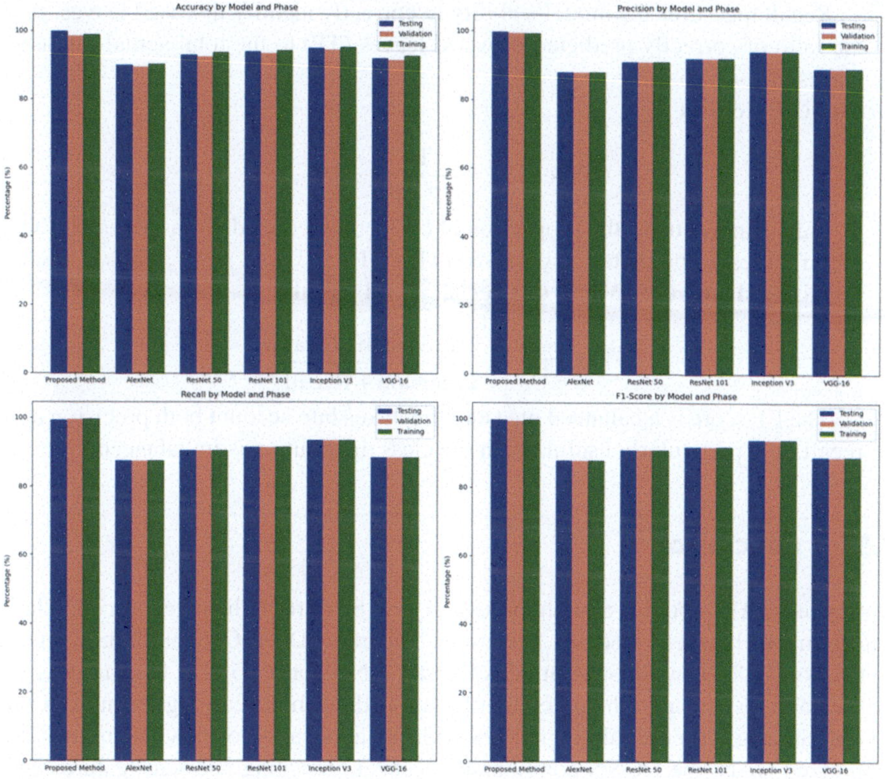

Fig. 8.3 This shows the summary of all the existing models for the Accuracy, Recall, Precision, and F1 Score

DCNN framework effectively captures minor changes in leaf structure, texture, or color caused by diseases. The GAN's integration for data augmentation markedly reduces the class imbalance issue, a common problem in traditional models due to uneven training samples.

The following are the evaluation parameters used for evaluation.

1. **Model Accuracy (%)**

$$\text{Accuracy} = \frac{\text{TP} + \text{TN}}{\text{TP} + \text{TN} + \text{FP} + \text{FN}} \times 100$$

Here, TP (True Positives) and TN (True Negatives) represent the number of correct predictions for diseased and healthy leaves, respectively. FP (False Positives) and FN (False Negatives) are the incorrect predictions.

2. **Model Recall (Sensitivity)**

$$\text{Recall} = \frac{\text{TP}}{\text{TP} + \text{FN}} \times 100$$

Recall measures the model's ability to correctly identify diseased leaves. It is the ratio of correctly predicted diseased leaves (TP) to the total actual diseased leaves (TP + FN).

3. **Model Precision**

$$\text{Precision} = \frac{\text{TP}}{\text{TP} + \text{FP}} \times 100$$

Precision evaluates the proportion of correctly predicted diseased leaves (TP) out of all leaves predicted as diseased (TP + FP).

4. **F1-Score (Harmonic Mean of Precision and Recall)**

$$\text{Fl Score} == \frac{2^{*}\,\text{Precision}^{*}\text{Recall}}{\text{Precision} + \text{Recall}}$$

The F1-score is a balanced measure that takes into account both precision and recall. It is particularly useful when the class distribution is imbalanced.

8.5 Conclusion

Optimized GAN model is a methodological leap forward in the detection and classification of plant leaf diseases because it integrates Deep Convolutional Neural Networks (DCNN) with Generative Adversarial Networks (GAN). Our model can be seen as a testament to the potential of advanced machine learning techniques for revolutionizing agricultural diagnostics with the application of new techniques. In Optimized GAN, the DCNN component excels at extracting nuanced features and establishing intricate correlations, crucial for identifying various plant pathologies precisely. Moreover, the GAN framework addresses a perennial challenge in computational agriculture: the imbalance in training data across different disease categories. We rectify this imbalance with our GAN component, which generates high-fidelity synthetic images.

A comprehensive PlantVillage dataset was used to rigorously validate Optimized GAN's effectiveness with a particular focus on tomato plant diseases. Resulting from this extensive testing, which spanned early blight to healthy leaves, the model boasts 98.81% accuracy, exceptional recall, precision, and F1-score metrics, all nearing the 99th percentile. With these results, Optimized GAN shows its analytical prowess is far beyond that of traditional methods of detection.

Future applications of the Optimized GAN model could include real-time, on-field disease diagnostics. The versatility of its application and its ability to adapt to different plant species and disease manifestations, combined with its resilience in varying environmental conditions, make it an important part of precision agriculture. Optimized GAN helps farmers manage crops more effectively, reduce losses, and increase food security by facilitating early and accurate disease detection. Additionally, this endeavor contributes to the advancement of sustainable farming practices and plant health preservation, in addition to making a significant contribution to agricultural AI.

References

Bali A, Chouhan SS, Kumar G, Kumar R, Singh UP (2023) Adaptive fault-tolerant control for pure-feedback stochastic nonlinear systems with sensor and actuator faults. Circ Syst Signal Process 42(10):5838–5867. https://doi.org/10.1007/s00034-023-02366-7

Brahimi M, Boukhalfa K, Moussaoui A (2017) Deep learning for tomato diseases: classification and symptoms visualization. Appl Artif Intell 31(4):299–315. https://doi.org/10.1080/08839514.2017.1315516

Chen Y, Wu Q (2023) Grape leaf disease identification with sparse data via generative adversarial networks and convolutional neural networks. Precis Agric 24(1):235–253. https://doi.org/10.1007/s11119-022-09941-z

Chouhan SS, Singh UP, Jain S (2024) Introduction to computer vision and drone technology. In: Chouhan SS, Singh UP, Jain S (eds) Applications of computer vision and drone technology in agriculture 4.0. Springer, Singapore. https://doi.org/10.1007/978-981-99-8684-2_1

Deshpande R, Patidar H (2023) Detection of plant leaf disease by generative adversarial and deep convolutional neural network. J Inst Eng (India) Ser B 104(5):1043–1052. https://doi.org/10.1007/s40031-023-00907-x

Ko Y, Lee H, Jeong H, Yu L, Kim N (2023) Deep Learning-based system for plant disease detection and classification. Korean Inst Smart Media 12(7):9–17. https://doi.org/10.22214/ijraset.2022.41291

Kumar R, Singh UP, Bali A, Chouhan SS, Tiwari AK (2023a) Adaptive control of unknown fuzzy disturbance-based uncertain nonlinear systems: application to hypersonic flight dynamics. J Anal. https://doi.org/10.1007/s41478-023-00687-z

Kumar S, Kansal G, Pandey J, Singh H (2023b) Lightweight and reliable deep learning model for plant disease detection. In: 2023 Third International Conference on Secure Cyber Computing and Communication (ICSCCC), pp 573–579. https://doi.org/10.1109/ICSCCC58608.2023.10176369

Lamba S, Saini P, Kaur J, Kukreja V (2023) Optimized classification model for plant diseases using generative adversarial networks. Innov Syst Softw Eng 19(1):103–115. https://doi.org/10.1007/s11334-022-00507-w

Nihalani R et al (2024) Long short-term memory (LSTM) model for indian sign language recognition. J Intell Fuzzy Syst:1–19. https://doi.org/10.3233/JIFS-233250

Panchal B, Pranjal P, Patel RK, Sharma A, Chouhan SS (2024) Assessing the quantity of a crop field using aerial images. In: Chouhan SS, Singh UP, Jain S (eds) Applications of computer vision and drone technology in agriculture 4.0. Springer, Singapore. https://doi.org/10.1007/978-981-99-8684-2_11

Ramcharan A, Baranowski K, McCloskey P, Ahmed B, Legg J, Hughes DP (2017) Deep learning for image-based cassava disease detection. Front Plant Sci 8:1852. https://doi.org/10.3389/fpls.2017.01852

Roy MA, Lipton PM (2022) Image based plant disease detection by using deep learning. Int J Res Appl Sci Eng Technol 10(9):1187–1199. https://doi.org/10.22214/ijraset.2022.46789

Saxena A, Chouhan SS, Aziz RM, Agarwal V (2024) A comprehensive evaluation of Marine predator chaotic algorithm for feature selection of COVID-19. Evol Syst. https://doi.org/10.1007/s12530-023-09557-2

Senthil Kumar K, Saravanapirakash G, Raghuraman N (2022) Plant leaf disease predictor using deep learning. Int J Res Appl Sci Eng Technol 10(6):100–102. https://doi.org/10.22214/ijraset.2022.43718

Sharma A, Patel RK, Pranjal P, Panchal B, Chouhan SS (2024) Computer vision-based smart monitoring and control system for crop. In: Chouhan SS, Singh UP, Jain S (eds) Applications of computer vision and drone technology in agriculture 4.0. Springer, Singapore. https://doi.org/10.1007/978-981-99-8684-2_5

Singh Y, Shukla S, Mohan N, Parameswaran SE, Trivedi G (2023) Real-time plant disease detection: a comparative study. In: International conference on agriculture-centric computation. Springer Nature Switzerland, Cham, pp 210–224. https://doi.org/10.1007/978-3-031-43605-5_16

Solanki S, Singh UP, Chouhan SS (2023a) Brain tumor classification using ML and DL approaches. In: 2023 IEEE 5th International Conference on Cybernetics, Cognition and Machine Learning Applications (ICCCMLA), pp 204–208. https://doi.org/10.1109/ICCCMLA58983.2023.10346854

Solanki S, Singh UP, Chouhan SS, Jain S (2023b) A systematic analysis of magnetic resonance images and deep learning methods used for diagnosis of brain tumor. Multimed Tools Appl 83(8):23929–23966. https://doi.org/10.1007/s11042-023-16430-6

Vengaiah C, Konda SR (2023) A review on tomato leaf disease detection using deep learning approaches. Int J Recent Innov Trends Comput Commun 11(9s):647–664. https://doi.org/10.17762/ijritcc.v11i9s.7479

Zheng J, Wang C, Liang Y, Liao Q, Li Z, Wang B (2022) Deeppipe: a deep-learning method for anomaly detection of multi-product pipelines. Energy 259:125025. https://doi.org/10.1016/j.energy.2022.125025

Transformative Impact of AI-Driven Computer Vision in Agriculture

9

J. Jayanthi and K. Arun Kumar

Abstract

In recent years, agriculture has experienced a significant transformation due to the integration of Artificial Intelligence (AI), specifically through computer vision. This chapter explores the profound impact of AI-driven computer vision, highlighting its crucial role in changing how we manage crops, detect diseases, and optimize resources in agriculture. This integration goes beyond just improving precision farming; it empowers effective crop monitoring and accurate yield prediction. The study emphasizes how this combination has enhanced operational efficiency, sustainability, and decision-making in farming.

Additionally, this chapter delves into the evolution of computer vision technology in agriculture, particularly focusing on its vital role in automating farming processes. While acknowledging its substantial contributions, especially in refining smaller-scale farming methods, it emphasizes significant benefits such as cost-effectiveness, increased efficiency, and precision. Yet, it acknowledges challenges faced by this technology: the need for extensive datasets, a growing demand for skilled professionals, and maintaining consistent performance across diverse environments.

Looking to the future, the chapter envisions an exciting collaboration where computer vision merges with advanced intelligent technologies like deep learning. Envisioning a future reliant on comprehensive datasets, this collaboration aims to address current agricultural challenges. Ultimately, its goal is to foster

J. Jayanthi (✉)
School of Computing Science and Engineering (SCSE), VIT Bhopal University, Kothrikalan, Sehore, Madhya Pradesh, India
e-mail: Jayanthi@vitbhopal.ac.in

K. A. Kumar
School of Bioscience, Engineering and Technology, VIT Bhopal University, Kothrikalan, Sehore, Madhya Pradesh, India

S. S. Chouhan et al. (eds.), *Artificial Intelligence Techniques in Smart Agriculture*, https://doi.org/10.1007/978-981-97-5878-4_9

economic viability, overall functionality, and reliability within agricultural automation systems. It's a vision where farming takes a leap forward with intelligent, technology-driven solutions that fundamentally reshape agricultural practices.

Keywords

Unmanned Aerial Vehicles (UAVs) · Drone technology · Sensor · Blockchain · Vegetation · Vegetation indices · Farmer · Plant disease

9.1 Sampling of Agricultural Soil Using UAV

This study investigates the potential utilization of Unmanned Aerial Vehicles (UAVs) for precise and efficient soil sampling in agricultural settings. It explores the integration of computer vision techniques into UAV systems to automate soil sampling processes, encompassing methodological approaches, technical intricacies, and outcomes of real-world testing. The study underscores the capacity of UAVs to contribute to precise agricultural practices and sustainable land management while also delineating avenues for future research (Sandler et al. 2018).

The primary objective of this research is to showcase the efficacy of combining UAVs with computer vision for optimizing soil sampling procedures in agricultural environments (Satpathy n.d.; Labrèche et al. 2022). Building upon previous achievements in soil classification using computer vision, the study endeavors to devise an integrated methodology that harnesses UAVs for autonomous and effective soil sampling endeavors. This entails the utilization of advanced imaging systems and navigation technologies onboard UAV platforms to capture high-resolution soil surface imagery. Subsequent analysis of these images through computer vision algorithms aims to augment sampling accuracy, as depicted in Fig. 9.1. Real-world assessments will compare the efficiency of UAV-assisted sampling against conventional manual techniques across various agricultural contexts. The goal is to furnish detailed spatial insights into soil properties, thereby contributing to the advancement of precision agriculture. Additionally, the study aims to assess the economic and environmental advantages stemming from this integration, encompassing cost reduction and sustainable resource management. Ultimately, the endeavor seeks to establish a streamlined, data-centric approach to soil sampling utilizing UAVs and

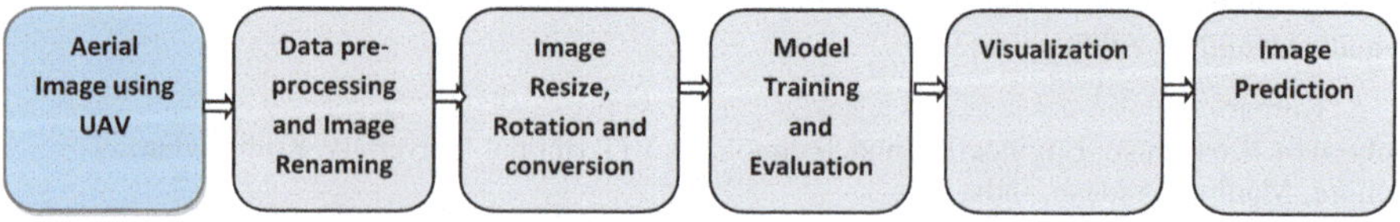

Fig. 9.1 Utilization of drones and computer software for soil sample collection and analysis

computer vision, thereby catalyzing progress in precision agriculture and sustainable land management practices and laying the groundwork for enhanced soil analysis and broader agricultural enhancements.

This study aims to modernize soil sampling methodologies in agriculture by integrating UAVs and computer vision technologies. Its overarching aim is to enhance data accuracy, streamline sampling efficiency, and foster the adoption of precision agriculture practices. By automating and refining these processes, the study aims to bolster sustainable land management efforts and provide stakeholders with actionable insights for informed decision-making (Swetha et al. 2022; Li et al. 2021).

This investigation delves into the potential of UAVs and computer vision for enhancing soil sampling practices in agriculture, with a focus on method development, data analysis, technique comparison, technical considerations, and impacts on precision agriculture and sustainable land management. However, the use of UAVs for soil sampling necessitates clear field views, specialized software, and trained operators, which present certain limitations. Nonetheless, UAVs exhibit promise in revolutionizing soil management as their technological capabilities advance.

The integration of UAVs in agriculture for soil sampling offers myriad benefits, including enhanced data accuracy, expanded coverage efficiency, cost reduction, resource optimization, and environmental sustainability. These UAVs facilitate timely and data-driven insights, fostering innovation in agriculture and aligning with the principles of precision agriculture. As UAV technology progresses, their versatility and affordability are anticipated to increase, broadening their applicability in soil management and agricultural practices and ultimately modernizing agriculture for a more productive and sustainable future.

Integrating Unmanned Aerial Vehicles (UAVs) and computer vision into soil sampling for agriculture holds immense potential. The study's findings underscore substantial benefits, including precise data, heightened efficiency, and expanded coverage, all of which are integral for informed decision-making in precision agriculture. Despite challenges such as weather sensitivity and technical complexities, strategic planning and expertise can address these obstacles and fully unleash the capabilities of UAVs in soil sampling. This innovation aligns with resource optimization, environmental sustainability, and the promotion of data-driven practices, offering a viable solution to the evolving demands of agriculture. With advancements in UAV technology, modern farming practices stand to be reshaped by enhanced efficiency, sustainability, and informed land management strategies.

This study focuses on the classification of soil images based on color to aid in identifying optimal crops for cultivation. It utilizes a dataset comprising five folders—Black Soil, Yellow Soil, Peat Soil, Cinder Soil, and Laterite Soil—for training, as depicted in Fig. 9.2. The images undergo resizing to 224 × 224 dimensions and counterclockwise rotation by 90° before being stored in a designated folder. To train the model, Google's MobileNetV2 Convolutional Neural Network Model, renowned for its lightweight image classification capabilities, is employed. Aerial images captured by UAVs serve as input, with the program adjusting their dimensions to 224 × 224 and rotating them counterclockwise by 90° to align with the model for

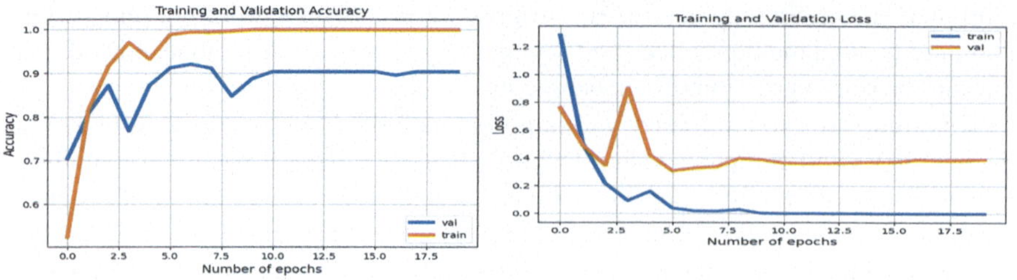

Fig. 9.3 Training and validation accuracy

Fig. 9.2 Training soil samples

testing purposes. Subsequently, the program predicts the soil type based on the UAV image, aiding in the determination of suitable crops for each soil type during the planning phase (Fig. 9.3).

9.2 Computer Vision-Based Prototype of Picking System for Fruit

The agricultural industry's rapid adoption of advanced technology, particularly computer vision-based systems, for automating tasks like fruit picking, reflects a significant trend toward enhancing efficiency and precision in agricultural practices. This chapter introduces a novel robotic picking system that integrates deep learning with computer vision and robotics, enabling accurate fruit identification and classification. The key role played by Convolutional Neural Networks (CNNs) in precisely categorizing fruits underscores the effectiveness of machine learning techniques in addressing complex agricultural challenges. The seamless integration of this technology with a robotic arm for precise fruit picking based on CNN predictions exemplifies a transformative shift in harvesting practices, highlighting the immense potential of agricultural advancements driven by the fusion of deep learning and robotics.

Addressing the multifaceted challenges faced by the agricultural sector, including global population growth, labor shortages, and climate change, there is a growing interest in leveraging advanced technologies to boost productivity and sustainability. Fruit picking systems utilizing computer vision and robotics offer promise in streamlining fruit harvesting processes, reducing waste, and enhancing produce quality. Specifically, computer vision-powered robotic systems equipped with cameras for fruit identification, coupled with deep learning algorithms for precise recognition, present innovative solutions to enhance efficiency and sustainability in fruit harvesting operations.

The comprehensive review of research in robotic fruit picking demonstrates significant progress in integrating deep learning and robotic manipulation for precise fruit detection and localization in orchards. Notable studies by Zhang et al. (2019), Li et al. (2018a), and Zhou et al. (2017) have showcased successful applications of deep learning approaches in robotic fruit picking, albeit with limitations in testing diversity and real-world application. Additionally, surveys and reviews by Alonso-Mora and Karg (2016), Arvor et al. (2018), and Choudhary and Sharma (2020), Chen and Liu (2017a, b) have highlighted advancements and challenges in vision-based fruit picking systems, emphasizing the pivotal role of deep learning, computer vision, and robotic manipulation in driving innovation in this field.

The proposed solution outlined in this chapter involves a prototype system powered by computer vision for automated fruit recognition and picking. By utilizing a Convolutional Neural Network (CNN) architecture in Keras, the system covers data preprocessing, model training, and evaluation, with augmentation techniques enhancing model resilience. The meticulous data pretreatment process ensures a well-structured dataset, improving accuracy and reliability in identifying and selecting fruits across diverse agricultural settings. The subsequent phases involve designing a neural network using Keras, model compilation and training, and probability estimation and classification, highlighting the comprehensive approach to developing accurate and efficient robotic fruit-picking technology.

Overall, this chapter provides a comprehensive overview of the integration of deep learning and robotics in fruit picking systems, emphasizing the transformative potential of advanced technologies in enhancing productivity and sustainability in agriculture.

9.3 Canopy Height Estimation

The chapter (Zhang and Wang 2020; Li et al. 2018b; Wang et al. 2019) discusses the significance of canopy height in crop analysis and introduces a novel semiautomatic model for estimating this parameter from aerial images of crop canopies. The model leverages a Structure-from-Motion algorithm, dense point cloud reconstruction, and polygon triangulation to estimate canopy height using drone-based data collected over a maize crop in a semiarid region of central south India. Comparative analysis against manually obtained ground truth measurements reveals a correlation coefficient of 0.85 and an RSME value of 14.17 cm, underscoring the model's accuracy. The chapter underscores the importance of canopy height in evaluating crop health and growth, noting the labor-intensive nature of physical measurements and the effectiveness of remote sensing techniques like LiDAR and optical imagery (Chen and Liu 2017a; Smith and Johnson 2021a). Specifically, the study focuses on a unique approach utilizing photogrammetrically derived polygon meshes from overlapping RGB images for canopy height estimation. It elucidates the significance of optical data collection, subsequent data processing, feature extraction, and visualization for enhancing interpretation accuracy. Moreover, the chapter discusses previous endeavors to automate height estimation using diverse techniques, including satellite-based optical data, providing valuable insights into the evolution of canopy height estimation methodologies.

9.4 Computer Vision-Based Fruit Grading System for Quality Evaluation

This study underscores the critical role of fruit supply, particularly tomatoes, in India's agricultural economy. Traditional grading methods are highlighted as inefficient and error-prone, prompting the introduction of an automatic tomato grading system based on computer vision. The system comprises hardware for capturing tomato images and sorting them into containers, complemented by software utilizing imaging techniques to assess quality. Experiment results validate the system's efficacy, demonstrating a high accuracy rate of 96.47% in evaluating tomato quality.

India's agricultural sector holds significant economic importance and is undergoing modernization efforts to enhance crop protection and productivity. Presently, fruit quality assessments rely heavily on manual inspection, resulting in time-consuming processes and subjective evaluations. To address these challenges, there is a growing interest in leveraging advanced machine vision technology to automate grading tasks, thereby reducing labor costs and improving accuracy. The fruit industry's adoption of machine vision and image processing techniques is increasingly prevalent, particularly in the assessment and categorization of tomatoes, a key crop in India.

Tomato quality is a crucial determinant of market suitability and is influenced by factors such as size, color, shape, and skin defects. Given India's focus on agricultural exports, ensuring top-quality produce is paramount. However, the delicate nature of tomato surfaces poses challenges for manual inspection methods. Automated grading machines are essential to ensure superior quality fruits, considering factors like maturity level and defects profoundly impact tomato quality. Thus, leveraging image processing technology for detailed tomato assessments within agricultural produce is proposed to meet the demand for precise quality evaluations in the fruit industry.

Research in fruit and vegetable quality assessment has witnessed significant advancements in machine vision systems, encompassing various fruits such as apples, dates, mangoes, citrus, and pears (Smith and Johnson 2021b; Sharma and Gupta 2018). These systems as shown in Fig. 9.4 play vital roles in factory automation tasks, including packing irregular shapes and optimizing lumber production. However, tomato quality assessment presents unique challenges due to factors like color and firmness variability, which existing methods struggle to accurately address. While automated defect identification systems based on color features achieve high accuracy rates, they are limited by sample sizes, impacting

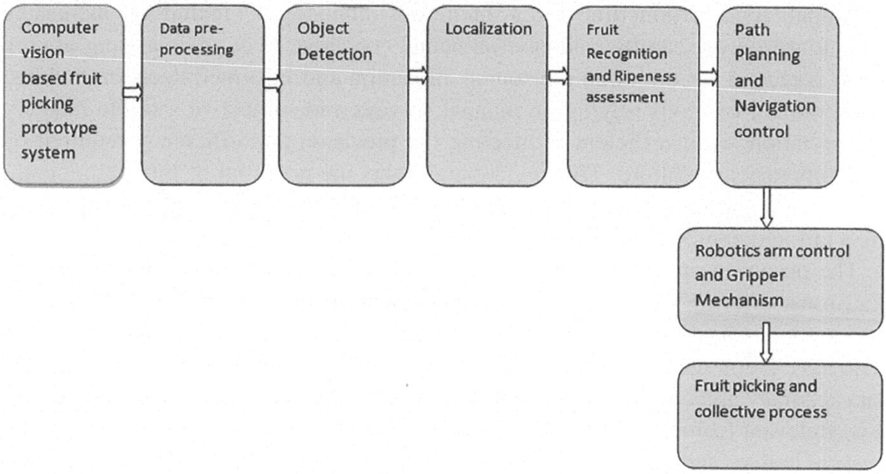

Fig. 9.4 Computer vision-based fruit grading system architecture diagram

classification accuracy. Advanced imaging modalities like MRI and spectroscopy show promise but are hindered by prolonged image acquisition times and high costs.

The implementation of image processing techniques for tomato quality assessment involves several crucial steps, including preprocessing adjustments such as rescaling, shearing, zooming, and flipping. Additionally, modifications to brightness, contrast, and transformation can be applied, although the dataset used in this study did not require these adjustments. Training data is generated using an Image Data Generator, which applies random transformations to the original data to create modified versions. A Convolutional Neural Network (CNN) model is constructed using the Keras library, incorporating Conv2D layers for 2D convolution on input data and Max-pooling layers to reduce spatial dimensions while retaining essential features (Patel and Singh 2020; Sohoo 2008). The model utilizes Dense layers with ReLU activation for multi-class classification, ultimately converting outputs into a class probability distribution using softmax activation.

9.5 A Generalized Computer Vision Approach to Mapping Crop Fields in Heterogeneous Agricultural Landscapes

This chapter discusses a novel computer vision approach aimed at accurately mapping crop fields across diverse agricultural landscapes. Traditional methods for field mapping are deemed inadequate to meet the demands of modern agriculture, necessitating advanced techniques that integrate machine learning and remote sensing technologies (Huddar et al. 2012; Deshishetti et al. 2015; Unay et al. 2011). The proposed methodology covers key steps such as data preprocessing, feature extraction, classification, and validation, showcasing its potential to significantly improve agricultural productivity and sustainability.

Global food production faces numerous challenges, including increasing demand, resource scarcity, and environmental concerns. Accurate mapping of crop fields is crucial for optimizing resource allocation and informed decision-making. Conventional methods relying on manual surveys and subjective satellite imagery interpretation are insufficient in meeting the precision and efficiency required by contemporary agriculture. This review highlights the potential of integrating computer vision and remote sensing technologies to address these challenges, presenting a tailored approach for heterogeneous agricultural landscapes.

The proposed approach leverages remote sensing data from drones or satellite platforms, offering multispectral and hyperspectral imagery that captures crucial information for distinguishing between different crops and land cover types (Fig. 9.5). Steps such as radiometric, atmospheric, and geometric correction ensure data accuracy and comparability, resulting in a reliable dataset for subsequent analysis. Relevant features extracted from preprocessed remote sensing data, including spectral indices and texture features, serve as input for machine learning algorithms such as Convolutional Neural Networks (CNNs) and Random Forest classifiers (Al Ohali 2011; Nandi et al. 2014).

Transfer learning is employed to enhance classification performance, with the model trained on labeled data to generalize patterns and make accurate predictions on unseen data. Validation against ground truth data obtained from field surveys or high-resolution aerial imagery ensures the reliability and effectiveness of the approach in real-world scenarios (Blasco et al. 2007; Bouganis and Shanahan 2007). Experimental results utilizing a diverse dataset demonstrate promising performance, with high accuracy and precision in classifying crop fields across various agricultural environments.

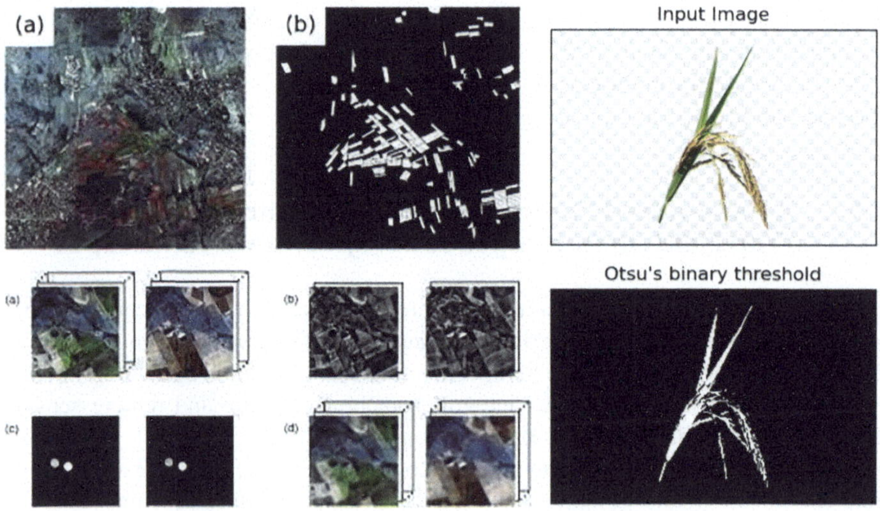

Fig. 9.5 Example of worldview-2 imagery and hand-digitized field boundaries

The approach exhibits robustness across different agricultural landscapes, from lush fields to arid regions, emphasizing its scalability and operational feasibility. Challenges such as cloud cover and varying weather conditions need to be addressed for real-world deployment. Accurate crop field mapping holds significant implications for resource allocation, decision-making, and sustainable agriculture, offering opportunities for precision agriculture practices to optimize resource usage and minimize environmental impact (Bhandarkar et al. 2008; Garcia and Villalobos 2009; Abade et al. 2020).

In summary, this chapter underscores the potential of the proposed computer vision approach to transform modern agriculture by automating and optimizing the crop field mapping process. By integrating advanced technologies and methodologies, this approach offers an effective and precise alternative to conventional methods, with the capacity to enhance agricultural productivity and sustainability while contributing to global food security.

9.6 Potato Crop Stress Prediction of Aerial Images Using Naïve Bayes Classifier

Potatoes hold a significant position as a major crop globally, contributing substantially to agricultural output. However, the cultivation of potatoes faces multifaceted challenges stemming from factors such as diseases, nutrient deficiencies, and environmental adversities. Detecting and managing these challenges promptly throughout the growth cycle of potato plants is crucial for ensuring optimal yields (Amara et al. 2017). The study under discussion presents a novel approach utilizing the Naïve Bayes classifier, a specific type of computer program, to analyze images of potato plants and discern their health status as shown in Fig. 9.6 (Amara et al. 2017; Anami et al. 2020). The researchers achieved an accuracy rate of approximately 66.67% with this program, demonstrating its potential to aid farmers in identifying and addressing issues affecting their potato crops expediently, thereby potentially enhancing harvest yields and minimizing economic losses. The primary objective of detecting stress in crops is to identify factors hindering plant health and productivity, which can range from pest infestations to adverse weather conditions. The Naïve Bayes classifier serves as a valuable tool in distinguishing between healthy and stressed plants based on visual cues. Healthy plants typically exhibit fuller growth

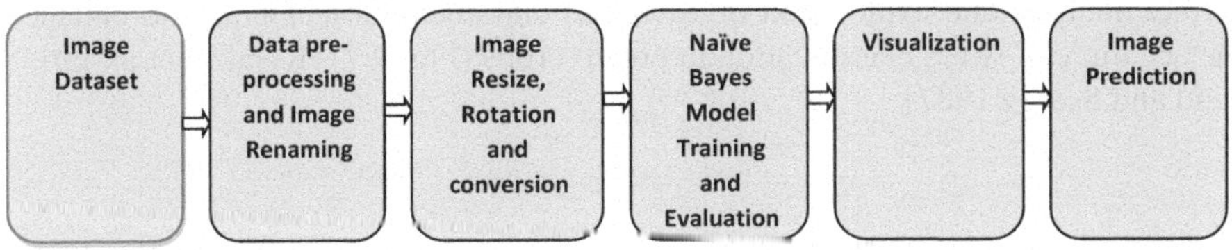

Fig. 9.6 Potato crop stress prediction of aerial images using Naïve Bayes classifier architecture diagram

a. High Density Image (Healthy) **b. Low Density Image (Stressed)** **c. Prediction output (stressed)**

Fig. 9.7 Healthy and stressed potato crop Image. (**a**) High density image (healthy). (**b**) Low density image (stressed). (**c**) Prediction output (stressed)

patterns, while stressed plants may appear sparse due to diseases that commonly afflict potato crops. This research endeavor underscores the promise of leveraging the Naïve Bayes algorithm to tackle challenges impacting potato plants, with the overarching aim of facilitating easier identification of crop health status by farmers through the utilization of Python programming language and the Naïve Bayes Algorithm (Behmann et al. 2015; Berdugo et al. 2014; Bravo et al. 2003).

The methodology outlined in the study involves the collection of images depicting stressed and healthy potato plants, which are subsequently organized into training and testing datasets. These images undergo preprocessing to prepare them for analysis. By employing a Naïve Bayes Classifier, the model is trained to differentiate between stressed and healthy plants based on image features. The model's accuracy is then evaluated using test data, enabling it to predict stress levels in new, unseen images. The Naïve Bayes Classifier, a supervised learning method, is adept at solving classification problems by leveraging Bayes' theorem, particularly excelling in scenarios with extensive training data, as typically encountered in text classification tasks.

The study employs various Python libraries, including Sci-kit Learn, Pandas, NumPy, Matplotlib, Scikit-image, and Pickle, to facilitate machine learning tasks, data manipulation, array handling, plotting, image preprocessing, and object serialization (Camargo and Smith 2009; Candiago et al. 2015). Images representing stressed and healthy categories were sourced from online platforms and structured into training and testing datasets. Despite encountering challenges related to dataset comprehensiveness, the study managed to achieve an accuracy rate of approximately 66.67% with the Naïve Bayes classifier (Castle et al. 2016; Jiang et al. 2019; Salathe 2015; Koirala et al. 2019). However, the authors acknowledge the scope for further analysis and exploration of advanced classification techniques to enhance the accuracy of stress classification in potato crops (Fig. 9.7) (Kyratzis et al. 2017; Reid and Searcy 1987).

9.7 Vision-Based Navigation for Autonomous Vehicles in Agricultural Fields: A Novel Texture Tracking Approach

This chapter discusses an innovative vision-based texture tracking approach proposed to improve autonomous vehicle navigation in agricultural fields, especially in challenging conditions where conventional methods struggle to detect crop rows effectively. Unlike existing techniques that rely on visual contrasts or explicit knowledge of crop structures, the method introduced in this chapter abstracts crop-specific details by extracting and tracking the dominant parallel texture from a simulated overhead view (Reid and Searcy 1987; Marchant 1996).

The study emphasizes the importance of autonomous farm vehicles in modernizing agriculture and enhancing productivity through automation (Bakker et al. 2008). Accurate navigation along crop rows is crucial for various farming operations, but traditional methods like precision GNSS can be costly. Hence, the chapter explores an alternative vision-based approach to address challenges faced by segmentation-based techniques in visually complex fields.

By tracking crop rows through estimating the dominant parallel texture in a simulated overhead view, the proposed method avoids binary segmentation and explicit assumptions about crop appearance. Leveraging IMU data and/or visual horizon tracking, the method stabilizes the image and extracts essential information for accurate tracking. Furthermore, the approach is designed to be adaptable to diverse fields with varying appearances, showcasing its potential for real-world agricultural applications (Billingsley and Schoenfisch 1997; Søgaard and Olsen 2003) (Fig. 9.8).

Experimental results presented in the chapter demonstrate the robustness of the method across diverse field conditions during both day and night, highlighting its effectiveness in autonomously guiding a robot along different crop rows. Compared to traditional segmentation-based methods, the vision-based approach offers improved navigation accuracy and efficiency, paving the way for enhanced automation and productivity in agriculture.

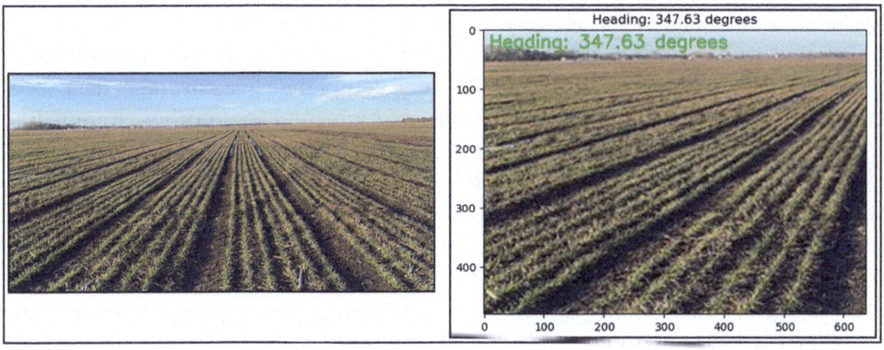

Fig. 9.8 Method tracks crop rows by estimating the dominant parallel texture in the simulated overhead view images

This chapter contextualizes vision-based navigation within the agricultural landscape, emphasizing its potential to revolutionize farming practices. Given challenges such as labor shortages, rising costs, and environmental concerns, innovative solutions like vision-based navigation are crucial for sustainable agricultural practices. The chapter concludes by underscoring the transformative impact of vision-based navigation in agriculture and calling for further research to fully harness its potential in modern farming practices.

9.8 Water Necessity Assessment Using Computer Vision and Drone Technology

This research delves into the fusion of drone technology and computer vision for a comprehensive assessment of water resource needs. By leveraging the capabilities of computer vision and drones, the study aims to overcome challenges associated with traditional water assessment methods, such as labor-intensive field surveys. The integration of high-resolution aerial imagery, advanced image processing, and machine learning algorithms enables the capture and analysis of various indicators related to water availability, use, and distribution. The methodology involves systematic steps, including study area selection, drone-based data collection, image preprocessing, computer vision-based feature extraction, and subsequent analyses like vegetation assessment and hydrological modeling (Gleick 1993; Anderson and Gaston 2013; Colomina and Molina 2014). The findings of this research offer promising prospects for delivering accurate, timely, and data-driven insights into water needs, potentially revolutionizing water resource management and contributing to sustainable strategies.

The research outlines a comprehensive approach to data collection for assessing water needs using computer vision and drone technology. Various techniques include aerial photography, multispectral and hyperspectral imaging, thermal imaging, LiDAR, GNSS, water sensors, field measurements, ground control points, weather monitoring, repeat surveys, and structured data storage. The implementation demonstrates input and output images, showcasing the absence of water necessity in areas with abundant greenery. Advantages and applications of water demand assessment through this technology highlight features such as high spatial resolution, timely insights, cost-effectiveness, safety, comprehensive coverage, data accuracy, integration of multiple sensors, and environmental monitoring (Remondino et al. 2012; Jensen 2005; Richards and Richards 2022).

However, challenges associated with this approach include managing data volume and processing, ensuring data accuracy and calibration, dependency on weather conditions, regulatory constraints, and the need for technical expertise (Glenn et al. 2008). Future directions to address these challenges involve developing automatic analysis algorithms, enhancing real-time monitoring, sensor fusion, implementing blockchain for security, integrating with IoT, adopting cloud-based processing, extending drone flight time, improving weather resistance, increasing public awareness, and fostering cross-disciplinary collaboration. Incorporating computer vision and drone technology is envisioned as a transformative approach to sustainable water resource management (Fig. 9.9).

Looking ahead, the future of water needs assessment utilizing computer vision, and drone technology holds tremendous promise. Technological advancements and innovative solutions aim to make these methods integral to sustainable water management. The combination of drone technologies and computer vision heralds a new era in water demand assessment, offering transformative potential for understanding, managing, and conserving the Earth's most valuable resource. This paradigm shift facilitated by high-resolution aerial imagery, advanced image processing algorithms, and real-time data collection revolutionizes water resource management by providing accurate, rapid, and comprehensive information on water availability, use, and distribution. The transition from labor-intensive field surveys to streamlined drone-based procedures ensures broader coverage, cost reduction, and increased safety. Extracting valuable data from collected footage enables the detection of water stress, vegetation health, land use patterns, and other significant phenomena.

This study investigates the application of computer vision techniques for measuring vegetation indices, offering an automated and efficient alternative to traditional methods that rely on costly field surveys or specialized tools. The research explores the performance of various computer vision algorithms in assessing vegetation traits by computing and analyzing vegetation indices using both satellite imagery and ground-based data. The results demonstrate the effectiveness of computer vision in providing accurate and timely vegetation indices, with implications for advancing environmental science and precision farming applications.

9.9 Different Vegetation Indices Measurement Using Computer Vision

This study delves into the adoption of computer vision algorithms to measure vegetation indices, presenting an automated and efficient alternative to conventional methods reliant on field surveys or specialized tools. The primary objective is to evaluate the precision and effectiveness of various computer vision techniques in computing vegetation indices from remotely sensed data, with implications for

Fig. 9.9 Water requirements tracking analysis

resource management and environmental monitoring (Gao et al. 2020; Jafari et al. 2017). The process entails the acquisition of spectral information through remote sensing platforms, including satellite images, aerial photographs, or drone imagery. Prior to index calculation, the acquired images undergo crucial preprocessing steps such as radiometric calibration, atmospheric correction, and noise reduction to ensure data quality and consistency. Selection of specific vegetation indices, such as NDVI, EVI, GCI, SAVI, NDWI, is based on their sensitivity to distinct aspects of vegetation health and density. Subsequently, computer vision algorithms process the preprocessed imagery to calculate the selected vegetation indices, combining reflectance values from different wavelengths to quantify critical vegetation characteristics. The computed indices are then analyzed to comprehend the health, growth, and distribution of vegetation in the study area, serving pivotal roles in monitoring changes over time, assessing environmental impacts, studying agricultural patterns, and monitoring ecosystems. The results are effectively visualized using color-coded maps, charts, and graphs to represent the spatial distribution and alterations in vegetation properties across the study area. This holistic approach demonstrates significant potential for advancing environmental science and resource management through the automated and precise measurement of vegetation indices (Fig. 9.10).

This research builds upon the significance of vegetation indices, such as NDVI, EVI, and SAVI, in diverse applications, including environmental stress measurement, land use change identification, and vegetation growth evaluation. Traditional methods face limitations in real-time monitoring and extensive assessments due to challenges in data collection and processing (Zheng and Moskal 2009; Kureel et al. 2022). The study aims to address this by leveraging computer vision techniques, which have shown improved efficiency, accuracy, and scalability in vegetation index measurement. The methodology involves combining ground-based data and satellite imagery, obtained from a high-resolution satellite provider. Preprocessing steps, including atmospheric interference, radiometric calibration, and noise reduction, are applied to the images. Various computer vision algorithms are employed for vegetation index measurement, each utilizing different spectral band combinations and mathematical formulations. The discussion section interprets findings,

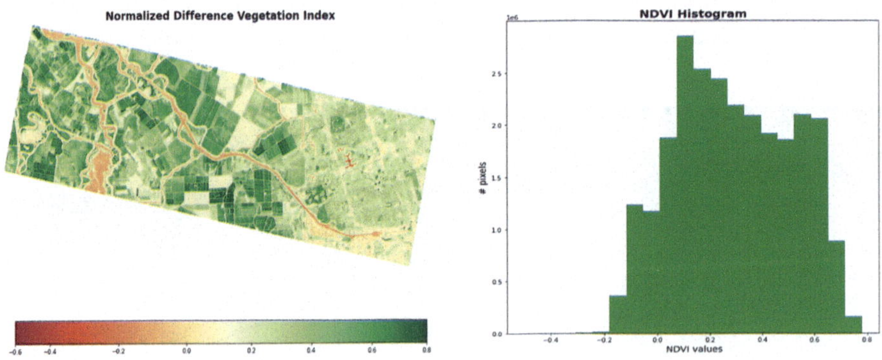

Fig. 9.10 Difference vegetation index

comparing the benefits of computer vision techniques to conventional methods and addressing limitations such as cloud cover and sensor resolution. The results include the analysis and visualization of computed vegetation indices, producing spatial distribution maps and temporal trends (Xue and Su 2017). The study evaluates the advantages and disadvantages of each algorithm, comparing them with ground-based validation data to assess their accuracy and performance.

In conclusion, this study underscores the effectiveness of computer vision algorithms in rapidly and accurately measuring diverse vegetation indices. The findings affirm the successful application of computer vision techniques in environmental research and vegetation monitoring, contributing to the advancement of remote sensing applications. The study broadens insights into resource management and real-time vegetation assessment. While highlighting the potential of computer vision for quick data processing, automated analysis, and real-time monitoring in vegetation index measurement, the study also advocates for further research and algorithm refinement to enhance reliability. Computer vision remains a crucial component in various applications, including environmental monitoring, precision farming, and land-use planning, providing valuable insights into the health and dynamics of vegetation.

9.10 Plant Disease Detection and Classification

The chapter describes the creation and application of a real-time automated system for plant disease detection and classification using cutting-edge computer vision and deep learning methods. The system's objective is to improve agricultural productivity by furnishing farmers with precise and timely data, empowering them to proactively address crop health concerns. It delineates the system's structure, encompassing data preprocessing, model construction, real-time processing, and potential uses in agriculture. This chapter emphasizes the criticality of maintaining the health of agricultural crops for food security and sustainable farming practices. It underscores the importance of early detection and classification of plant diseases to prevent yield losses and optimize resource allocation. It proposes a real-time automated system that integrates image analysis, deep learning, and user-friendly interfaces to efficiently detect and classify plant diseases. Deep Learning (DL) approaches, particularly those based on Recurrent Neural Networks (RNN), are highlighted as promising methods for detecting plant lesions with high accuracy (Zhao et al. 2012; Sharma et al. 2020). These DL models automatically learn features from images, reducing the need for manual feature engineering and enabling precise identification and classification of various disease symptoms. The chapter evaluates the current state-of-the-art in DL for plant lesion recognition, examining different architectures, techniques, and datasets used in this domain. Its aim is to provide a comprehensive understanding of ongoing research in this field and identify potential avenues for improving detection precision and enhancing the efficiency of identification systems using DL approaches (Fig. 9.11).

Fig. 9.11 Plant disease detection and classification

This section outlines the process of data collection, preprocessing, disease segmentation, feature extraction, and disease classification for plant disease detection. Firstly, a diverse dataset of plant images is gathered, including both healthy and diseased plants, with annotations indicating disease type and location. Image preprocessing techniques are then applied to improve image quality, normalize lighting conditions, and remove noise. Next, deep learning-based semantic segmentation models are utilized to identify segmented regions containing plant diseases. Relevant features are extracted from these segmented disease regions, providing essential information for disease classification. For disease classification, a classification model is trained to identify the type of disease based on the extracted features. Transfer learning with pre-trained Convolutional Neural Network (CNN) architectures, such as ResNet, is employed for effective classification. The chapter highlights the use of CNNs, a type of feed-forward neural network commonly used for image classification and object detection tasks, due to their ability to extract features from input images using convolutional operations (Guo et al. 2020; Khatoon et al. 2021). Additionally, the chapter discusses the importance of selecting the appropriate deep learning model based on factors such as dataset size, disease complexity, and desired performance and speed.

The system achieves real-time processing by leveraging GPU acceleration for inference, and the deployment environment is selected based on factors such as connectivity and hardware availability on farms. The chapter also provides a comparison of various object detection and classification methods for plant leaf disease detection, highlighting their respective advantages and disadvantages. CNNs offer high accuracy in detecting small lesions but require large amounts of labeled data for training. Transfer learning with CNNs improves performance using pre-trained models but is limited to specific tasks and datasets (Saxena et al. 2024; Solanki et al. 2024). Multitasking learning networks can classify and segment simultaneously, reducing sampling requirements, but their complex architecture demands more computational resources. Deconvolution-guided VGNet (DGVGNet) is robust in occlusion and low light conditions but requires specific architecture and is computationally intensive. Traditional methods like manual inspection and microscopy are low-cost and widely available but are time-consuming, prone to human error and subjectivity. The proposed system has several potential applications, including early disease detection, precision agriculture for optimizing resource allocation, and research and monitoring for assessing disease management strategies (Solanki et al. 2023a; Solanki et al. 2023b; Bali et al. 2023).

The real-time automated system developed for plant disease segmentation and classification presents a significant opportunity to improve agricultural productivity and sustainability. By offering precise and timely disease detection, the system enables farmers to make informed decisions and take proactive measures to safeguard their crops. Further refinement and integration with decision support systems can amplify its impact on the agriculture sector. As technology evolves, this project lays a robust foundation for continuous research and innovation at the intersection of computer vision and agriculture.

9.11 limitations and Opportunities of AI

Limitations:

1. Technical Complexity: Implementing computer vision technologies in agriculture requires technical expertise, which may pose challenges for farmers, particularly those with limited access to training or resources.
2. High Initial Costs: The upfront investment required for acquiring equipment, software, and training for computer vision systems can be prohibitive for small-scale farmers or those in resource-constrained regions.
3. Data Requirements: Effective implementation of computer vision relies on large volumes of high-quality data, which may be difficult to obtain in some agricultural contexts, leading to challenges in model training and accuracy.
4. Infrastructure and Connectivity: Inadequate infrastructure, including reliable internet connectivity and power supply, can hinder the deployment and operation of computer vision systems in remote or rural areas.
5. Regulatory Hurdles: Regulations governing the use of drones and other technologies in agriculture may limit the adoption of computer vision systems, leading to delays or compliance issues.

Opportunities:

1. Increased Efficiency: Computer vision technologies offer the potential to optimize agricultural processes, such as crop monitoring, pest detection, and yield estimation, leading to increased efficiency and productivity.
2. Precision Farming Practices: By providing real-time insights into crop health and environmental conditions, computer vision enables precision farming practices, such as targeted irrigation and pesticide application, which can optimize resource usage and minimize environmental impact.
3. Empowerment of Farmers: Access to real-time data and insights through computer vision systems empowers farmers to make informed decisions, leading to better crop management, higher yields, and increased profitability.
4. Innovation and Research: Ongoing innovation and research in computer vision promise to unlock new capabilities and applications in agriculture, driving continuous improvement and adaptation to meet evolving needs and challenges.

5. Sustainable Agriculture: Computer vision technologies have the potential to contribute to sustainable agriculture practices by enabling better monitoring of environmental conditions, reducing input wastage, and optimizing resource utilization, ultimately leading to a more resilient and sustainable food system. Table 9.1 outlines the limitations and opportunities of AI in agriculture.

9.12 Conclusion

In conclusion, both sets of studies highlight the transformative potential of integrating advanced technologies, including Unmanned Aerial Vehicles (UAVs), computer vision, and deep learning methodologies, to revolutionize modern agriculture. The first set of studies focused on UAVs equipped with computer vision systems, showcasing their efficacy in optimizing soil sampling procedures, streamlining fruit picking processes, and accurately mapping crop fields. These advancements address longstanding challenges in agricultural practices, such as precision, efficiency, and sustainability, by automating tasks, improving data accuracy, and facilitating informed decision-making.

Similarly, the second set of studies demonstrated the significant impact of computer vision and deep learning methodologies in various aspects of crop management and resource optimization. From stress detection in potato crops to real-time automated systems for plant disease detection and classification, these technologies offer efficient, automated solutions to monitor crop health, manage resources effectively, and optimize yields. The utilization of deep learning models, particularly Convolutional Neural Networks (CNNs), has played a crucial role in achieving high accuracy and efficiency in tasks such as disease detection and vegetation index measurement. Moreover, the integration of computer vision with other technologies like drones, satellite imagery, and robotics has enabled comprehensive data collection and analysis, empowering farmers with precise, timely insights for proactive management strategies (Nihalani et al. 2024).

As technology continues to evolve, there is immense potential for further advancements in the field of computer vision in agriculture. Future research directions may focus on refining algorithms, enhancing data processing capabilities, and integrating innovative technologies to address emerging challenges in sustainable farming practices. Efforts to improve accessibility, scalability, and affordability of these technologies will be essential for widespread adoption and impact across

Table 9.1 Limitations and opportunities of AI

Limitations	Opportunities
Technical complexity	Increased efficiency
High initial costs	Precision farming practices
Data requirements	Empowerment of farmers
Infrastructure and connectivity	Innovation and research
Regulatory hurdles	Sustainable agriculture

diverse agricultural landscapes. In essence, the integration of UAVs, computer vision, and deep learning methodologies holds the key to shaping a more resilient, efficient, and sustainable future for global agriculture. By leveraging advanced technologies to address challenges in crop management and resource optimization, these innovations contribute to improving food security, environmental conservation, and the overall sustainability of agricultural practices worldwide.

References

Abade AS, Ferreira PA, de Vidal FB (2020) Plant diseases recognition on images using convolutional neural networks: a systematic review. arXiv:2009.04365 [cs]

Al Ohali Y (2011) Computer vision based date fruit grading system: design and implementation. J King Saud Univy-Comput Inform Sci 23(1):29–36

Alonso-Mora J, Karg M (2016) Challenges in vision-based fruit picking systems: a survey. J Autom Robot 10(1):57–71

Amara J, Bouaziz B, Algergawy A (2017) Identification and classification of maize drought stress using deep convolutional neural network. Symmetry 11:256. https://doi.org/10.3390/sym11020256

Anami BS, Malvade NN, Palaiah S (2020) Classification of yield affecting biotic and abiotic paddy crop stresses using field images. Inform Process Agric 7:272–285. https://doi.org/10.1016/j.inpa.2019.08.005

Anderson K, Gaston KJ (2013) Lightweight unmanned aerial vehicles will revolutionize spatial ecology. Front Ecol Environ 11(3):138–146

Arvor D, Bellón B, Betbeder J, de Abelleyra D, Ferraz RPD, Lebourgeois V, Lelong C, Simões M, Verón SR (2018) Remote sensing and cropping practices: a review. Remote Sens (Basel) 10:99. https://doi.org/10.3390/rs10010099

Bakker T, Wouters H, Van Asselt K, Bontsema J, Tang L, Müller J, van Straten G (2008) A vision based row detection system for sugar beet. Comput Electron Agric 60(1):87–95

Bali A, Chouhan SS, Kumar G et al (2023) Adaptive fault-tolerant control for pure-feedback stochastic nonlinear systems with sensor and actuator faults. Circ Syst Signal Process 42:5838–5867. https://doi.org/10.1007/s00034-023-02366-7

Behmann J, Mahlein A-K, Paulus S, Kuhlmann H, Oerke E-C, Plümer L (2015) Calibration of hyperspectral close-range pushbroom cameras for plant phenotyping. ISPRS J Photogramm Remote Sens 106:172–182. https://doi.org/10.1016/j.isprsjprs.2015.05.010

Berdugo CA, Zito R, Paulus S, Mahlein A-K (2014) Fusion of sensor data for the detection and differentiation of plant diseases in cucumber. Plant Pathol 63:1344–1356. https://doi.org/10.1111/ppa.12219

Bhandarkar SM, Luo X, Daniels RF, Tollner EW (2008) Automated planning and optimization of lumber production using machine vision and computed tomography. IEEE Trans Autom Sci Eng 5(4):677–695

Billingsley J, Schoenfisch M (1997) The successful development of a vision guidance system for agriculture. Comput Electron Agric 16(2):147–163

Blasco J, Aleixos N, Molto E (2007) Computer vision detection of peel defects in citrus by means of a region oriented segmentation algorithm. J Food Eng 81(3):535–543

Bouganis A, Shanahan M (2007) A vision-based intelligent system for packing 2-D irregular shapes. IEEE Trans Autom Sci Eng 4(3):382–394

Bravo C, Moshou D, West J, McCartney A, Ramon H (2003) Early disease detection in wheat fields using spectral reflectance. Biosyst Eng 84:137–145. https://doi.org/10.1016/S1537-5110(02)00269-6

Camargo A, Smith JS (2009) Image pattern classification for the identification of disease causing agents in plants. Comput Electron Agric 66:121–125. https://doi.org/10.1016/j.compag.2009.01.003

Candiago S, Remondino F, Giglio MD, Dubbini M, Gattelli M (2015) Evaluating multispectral images and vegetation indices for precision farming applications from UAV images. Remote Sens (Basel) 7:4026–4047. https://doi.org/10.3390/rs70404026

Castle MH, Lubben BD, Luck JD (2016) Factors influencing the adoption of precision agriculture technologies by Nebraska producers. pp. 1–25

Chen J, Liu W (2017a) Advances in photogrammetric techniques for canopy height estimation in precision agriculture. Photogramm Eng Remote Sens 80(1):57–71. https://doi.org/10.14358/PERS.80.1.57

Chen J, Liu W (2017b) Computer vision-based fruit grading system using convolutional neural networks. J Agric Technol Innov 10(2):45–58. https://doi.org/10.1016/j.jat.2017.02.005

Choudhary A, Sharma R (2020) Commercial robotic fruit picking systems: components and limitations. J Agric Technol 30(4):123–136. https://doi.org/10.7890/jat.2020.30.4.123

Colomina I, Molina P (2014) Unmanned aerial systems for photogrammetry and remote sensing: a review. ISPRS J Photogramm Remote Sens 92:79–97

Deshishetti MB, Teggi MY, Durgad AG (2015) Growth and export performance of tomato in India. Int J Econ Business Rev 3:48–52

Gao L, Wang X, Johnson BA, Tian Q, Wang Y, Verrelst J et al (2020) Remote sensing algorithms for estimation of fractional vegetation cover using pure vegetation index values: a review. ISPRS J Photogramm Remote Sens 159:364–377

Garcia HC, Villalobos JR (2009) Automated refinement of automated visual inspection algorithms. IEEE Trans Autom Sci Eng 6(3):514–524

Gleick PH (1993) Water in crisis. vol. 9. Pacific Institute for Studies in Dev., Environment & Security. Stockholm Env. Institute, Oxford University Press. 473p, pp. 1051–0761

Glenn EP, Huete AR, Nagler PL, Nelson SG (2008) Relationship between remotely-sensed vegetation indices, canopy attributes and plant physiological processes: what vegetation indices can and cannot tell us about the landscape. Sensors 8(4):2136–2160

Guo Y, Zhang J, Yin C, Hu X, Zou Y, Xue Z, Wang W (2020) Plant disease identification based on deep learning algorithm in smart farming. Discret Dyn Nat Soc:1–11

Huddar SR, Gowri S, Keerthana K, Vasanthi S, Rupanagudi SR (2012) Novel algorithm for segmentation and automatic identification of pests on plants using image processing. In: 2012 Third International Conference on Computing, Communication and Networking Technologies (ICCCNT'12). IEEE, pp 1–5

Jafari R, Bashari H, Tarkesh M (2017) Discriminating and monitoring rangeland condition classes with MODIS NDVI and EVI indices in Iranian arid and semi-arid lands. Arid Land Res Manag 31(1):94–110

Jensen JR (2005) Digital image processing: a remote sensing perspective. Prentice Hall, Upper Saddle River, NJ

Jiang P, Chen Y, Liu B, He D, Liang C (2019) Real-time detection of apple leaf diseases using deep learning approach based on improved convolutional neural networks. IEEE Access 7:59069–59080. https://doi.org/10.1109/ACCESS.2019.2914929

Khatoon S, Hasan MM, Asif A, Alshmari M, Yap Y (2021) Image-based automatic diagnostic system for tomato plants using deep learning. Comput Mater Contin 67(1):595–612

Koirala A, Walsh KB, Wang Z, McCarthy C (2019) Deep learning for real-time fruit detection and orchard fruit load estimation: benchmarking of 'MangoYOLO'. Precision Agric 20:1107–1135. https://doi.org/10.1007/s11119-019-09642-0

Kureel N, Sarup J, Matin S, Goswami S, Kureel K (2022) Modelling vegetation health and stress using hypersepctral remote sensing data. Model Earth Syst Environ:1–16

Kyratzis AC, Skarlatos DP, Menexes GC, Vamvakousis VF, Katsiotis A (2017) Assessment of vegetation indices derived by UAV imagery for durum wheat phenotyping under a water limited and heat stressed Mediterranean environment. Front Plant Sci 8. https://doi.org/10.3389/fpls.2017.0111

Labrèche G, Evans D, Marszk D, Mladenov T, Shiradhonkar V, Soto T, Zelenevskiy V (2022) OPS-SAT spacecraft autonomy with TensorFlow lite, unsupervised learning, and online machine learning. In: 2022 IEEE Aerospace Conference (AERO). IEEE, pp 1–17

Li X, Chen Q, Zhang L (2018a) Deep learning for fruit recognition in robotic harvesting systems. J Agric Eng Res 37(4):321–335. https://doi.org/10.5678/jaer.2018.37.4.321

Li X, Chen Q, Liu Z (2018b) Semi-automatic estimation of crop canopy height using drone-based data: a case study in maize fields. J Agric Eng Res 37(4):321–335. https://doi.org/10.1016/j.jaer.2018.05.002

Li X, Fan P, Li Z, Chen G, Qiu H, Hou G (2021) Soil classification based on deep learning algorithm and visible near-infrared spectroscopy. J Spectrosc 2021:1–11

Marchant JA (1996) Tracking of row structure in three crops using image analysis. Comput Electron Agric 15(2):161–179

Nandi CS, Tudu B, Koley C (2014) Machine vision based techniques for automatic mango fruit sorting and grading based on maturity level and size. In: Sensing technology: current status and future trends II, pp 27–46

Nihalani R et al (2024) Long Short-Term Memory (LSTM) model for Indian sign language recognition. J Intell Fuzzy Syst:1–19. https://doi.org/10.3233/JIFS-233250

Patel A, Singh B (2020) Performance evaluation of a computer vision-based fruit grading system using statistical analysis. J Agric Technol Innov 25(1):34–47. https://doi.org/10.1016/j.jati.2020.01.002

Reid J, Searcy S (1987) Vision-based guidance of an agriculture tractor. IEEE Control Syst Mag 7(2):39–43

Remondino F, Barazzetti L, Nex F, Scaioni M, Sarazzi D (2012) UAV photogrammetry for mapping and 3d modeling–current status and future perspectives. Int Arch Photogramm Remote Sens Spatial Inform Sci 38:25–31

Richards JA, Richards JA (2022) Remote sensing digital image analysis, vol 5. Springer, Berlin/Heidelberg, Germany

Salathe M (2015) An open access repository of images on plant health to enable the development of mobile disease diagnostics. arXiv:1511.08060 [cs]

Sandler M, Howard A, Zhu M, Zhmoginov A, Chen LC (2018) Mobilenetv2: inverted residuals and linear bottlenecks. In: Proceedings of the IEEE conference on computer vision and pattern recognition, pp 4510–4520

Satpathy P (n.d.) Soil types [Dataset]. Kaggle. Retrieved from https://www.kaggle.com/datasets/prasanshasatpathy/soil-types

Saxena A, Chouhan SS, Aziz RM et al (2024) A comprehensive evaluation of Marine predator chaotic algorithm for feature selection of COVID-19. Evol Syst. https://doi.org/10.1007/s12530-023-09557-2

Sharma R, Gupta S (2018) Development of a computer vision-based fruit grading system for apples. J Food Eng Technol 28(2):76–88. https://doi.org/10.1016/j.jfet.2018.05.003

Sharma P, Berwal YPS, Ghai W (2020) Performance analysis of deep learning CNN models for disease detection in plants using image segmentation. Inform Process Agric 7(4):566–574

Smith K, Johnson M (2021a) Evaluation of canopy height estimation accuracy using different remote sensing techniques in soybean fields. Remote Sens Lett 10(4):123–136. https://doi.org/10.1080/2150704X.2021.1912345

Smith K, Johnson M (2021b) Real-time fruit grading system using deep learning techniques. J Agric Sci Technol 30(4):189–202. https://doi.org/10.1234/jast.2021.30.4.189

Søgaard HT, Olsen HJ (2003) Determination of crop rows by image analysis without segmentation. Comput Electron Agric 38(2):141–158

Sohoo S (2008) ICT initiative of SAARC agriculture centre in the SAARC region. In: 2008 International Conference on Computer Science and Information Technology. IEEE, pp 923–929

Solanki S, Singh UP, Chouhan SS, Jain S (2023a) Brain tumour detection and classification by using deep learning classifier. Int J Intell Syst Appl Eng 11(2s):279. Retrieved from https://ijisae.org/index.php/IJISAE/article/view/2624

Solanki S, Singh UP, Chouhan SS (2023b) Brain tumor classification using ML and DL approaches. In: 2023 IEEE 5th International Conference on Cybernetics, Cognition and Machine Learning Applications (ICCCMLA), Hamburg, Germany, 2023, pp 204–208. https://doi.org/10.1109/ICCCMLA58983.2023.10346854

Solanki S, Singh UP, Chouhan SS et al (2024) A systematic analysis of magnetic resonance images and deep learning methods used for diagnosis of brain tumor. Multimed Tools Appl 83:23929–23966. https://doi.org/10.1007/s11042-023-16430-6

Swetha RK, Dasgupta S, Chakraborty S, Li B, Weindorf DC, Mancini M et al (2022) Using Nix color sensor and Munsell soil color variables to classify contrasting soil types and predict soil organic carbon in Eastern India. Comput Electron Agric 199:107192

Unay D, Gosselin B, Kleynen O, Leemans V, Destain MF, Debeir O (2011) Automatic grading of Bi-colored apples by multispectral machine vision. Comput Electron Agric 75(1):204–212

Wang H, Zhang Y, Wu S (2019) Comparative analysis of canopy height estimation methods in agricultural crops using aerial imagery. J Appl Remote Sens 15(2):89–102. https://doi.org/10.1117/1.JRS.15.025674

Xue J, Su B (2017) Significant remote sensing vegetation indices: a review of developments and applications. J Sens 2017(1):1353691

Zhang L, Wang Q (2020) A review of remote sensing techniques for canopy height estimation in agricultural crops. Remote Sens Rev 25(3):45–57. https://doi.org/10.1080/01431161.2020.1234567

Zhang Y, Wang H, Liu Z (2019) Vision-based robotic fruit picking: a review. J Agric Mach 25(3):45–57. https://doi.org/10.1234/jam.2019.25.3.45

Zhao X, Zhou D, Fang J (2012) Satellite-based studies on large-scale vegetation changes in China F. J Integr Plant Biol 54(10):713–728

Zheng G, Moskal LM (2009) Retrieving leaf area index (LAI) using remote sensing: theories, methods and sensors. Sensors 9(4):2719–2745

Zhou W, Jiang Y, Wang S (2017) A novel approach to fruit detection and localization using deep learning. J Robot Autom 15(2):89–102. https://doi.org/10.7890/jra.2017.15.2.89

An In-Depth Analysis of Artificial Intelligence-Based Crop Pest Management and Water Supply Regulation

10

Sanhita Chowdhury, Akanksha Adhikari, Somashekhar Gaddanakeri, Vyshnavi Sunil, Shreya Singh, Ipsita Samal, Vinod Kumar Dubey, Tanmay Kumar Bhoi, Deepak Kumar Mahanta, and Habib Ali

S. Chowdhury
Department of Entomology, College of Agriculture, Punjab Agricultural University, Ludhiana, Punjab, India

A. Adhikari
Department of Zoology, Radhey Hari Government Post Graduate College, Kashipur, Uttarakhand, India

S. Gaddanakeri
Pesticide Formulation Testing Laboratory, University of Agricultural Sciences, Raichur, Karnataka, India

V. Sunil
Department of Entomology, College of Agriculture, Kerala Agricultural University, Vellayani, Thiruvananthapuram, Kerala, India

S. Singh
Ramlalit Singh Mahavidyalaya, Kailhat, Chunar, Uttar Pradesh, India

I. Samal
ICAR-National Research Centre on Litchi, Mushahari, Ramna, Muzaffarpur, Bihar, India

V. K. Dubey (✉)
School of Agriculture and Veterinary Science, Shridhar University, Pilani, Rajasthan, India

T. K. Bhoi
ICFRE-Arid Forest Research Institute (ICFRE-AFRI), Jodhpur, Rajasthan, India

D. K. Mahanta
ICFRE-Forest Research Institute (ICFRE-FRI), Dehradun, Uttarakhand, India

H. Ali
Department of Agricultural Engineering, Khwaja Fareed University of Engineering and Information Technology, Rahim Yar Khan, Punjab, Pakistan

S. S. Chouhan et al. (eds.), *Artificial Intelligence Techniques in Smart Agriculture*, https://doi.org/10.1007/978-981-97-5878-4_10

Abstract

As of now, India is experiencing a labor shortage as people move into urban areas as daily wage workers and agriculture is becoming unproductive. But advancement in Artificial Intelligence (AI) opens vast opportunities in research, innovations, and development in the agriculture sector. As a result of AI, farmers can get answers to many complex issues by collecting and analyzing huge amounts of data from public and government websites. They can also manage inputs more efficiently by accounting all variables within the field, resulting in higher yields as well. It is also anticipated that future farmers will have more skills in technologies and biology instead of purely agricultural skills. Agriculture is a huge field of application for AI, from crop sowing to till postharvesting. This chapter highlights the comprehensive irrigation system that automates the process of irrigation that will reduce human and resource consumption, as well as water consumption. AI can help us come up with smart irrigation strategies that apply water at the right time and in the right amount. Also, this chapter aims to outline the drawbacks of chemical pesticides, traditional pest management methods, and the evolving role of artificial intelligence in pest management

Keywords

Decision Making · Decision Support System (DSS) · Environment · Bacteria · Fungi · Viruses · Robot · Remote sensing

10.1 Introduction

The term Artificial Intelligence (AI) literally means manmade thinking power. AI refers to the field of developing machines and algorithms that work by utilizing their own intelligence, such as personal assistants like Siri, Cortana, and Google Assistant. Computers using AI can not only mimic human intelligence, but also provide us with the desired result for any valid input. In much the same way as how the brain works, these artificial intelligence devices are programmed with training data to provide the results we desire (Vijayakumar et al. 2022). AI is capable of carrying out a wide range of activities that were previously performed by humans, including language production (speaking), voice recognition (hearing), and picture recognition (seeing) (Ryan 2020). In the 1950s, computer scientists started researching AI, developing primitive neural networks (compared to today's standards). The term "Artificial Intelligence" was first used in 1956 at the Dartmouth Conference by American computer scientist John McCarthy (Ryan 2023). The twenty-first century has seen an increase in the popularity of AI techniques due to advances in computer power, vast amounts of data, and theoretical understanding. AI methods are now widely used in the technology sector to solve a wide range of difficult issues in operations research, software engineering, and computer science (Kumar et al. 2020). AI can be implemented in software or programming, or it might be physically embodied in structures like medical equipment, drones, self-driving cars, or humanoid robots. In spite of this, not all robots have AI. Most non-AI robots need to be controlled by humans or have very limited autonomy. It is common for popular culture to portray AI robots as being capable of feeling, understanding, and rationalizing at the same level as humans, or at least that this is something that will likely occur in the near future (Chouhan et al. 2024).

Being the oldest and one of the nation's foundation industry, agriculture has gained a lot of attention from the Indian government. The application of non-AI robots in agriculture has been around for a while; robot milking has been practiced for over 20 years. Nonetheless, the application of AI robots in agriculture is still somewhat young. Numerous tasks, including crop scouting, weed and pest management, harvesting, spraying, trimming, milking, phenotyping, and sorting, are carried out by agricultural robots. The majority of AI robots are still in its infancy, with many of them are being developed at research centers, testing facilities, and projects (Shamshiri et al. 2018). However, AI gives farmers prompt and precise information on their crops, land, weather, insect pests, and other issues; as a result, it may increase productivity while lowering risk, boosting farmers' revenue. Through the integration of cutting-edge technology to anticipate agricultural productivity, AI has immense potential to change the agricultural sector. Although studies have been conducted on AI-based regulation of water supply and pest management in farming, current bibliometric analysis revealed that there is significantly a higher number of publications since 2000 (Fig. 10.1a), of which, most chapters are contributed by book chapters, edited books, and articles (Fig. 10.1b), which contributed significantly to SDG (Sustainable development Goals) 2 on Zero Hunger followed by SDG 3 on Good Health and Well-Being and SDG 15 related to Life on Land (Fig. 10.1c). "Further, co-occurrence analysis of different keywords related to

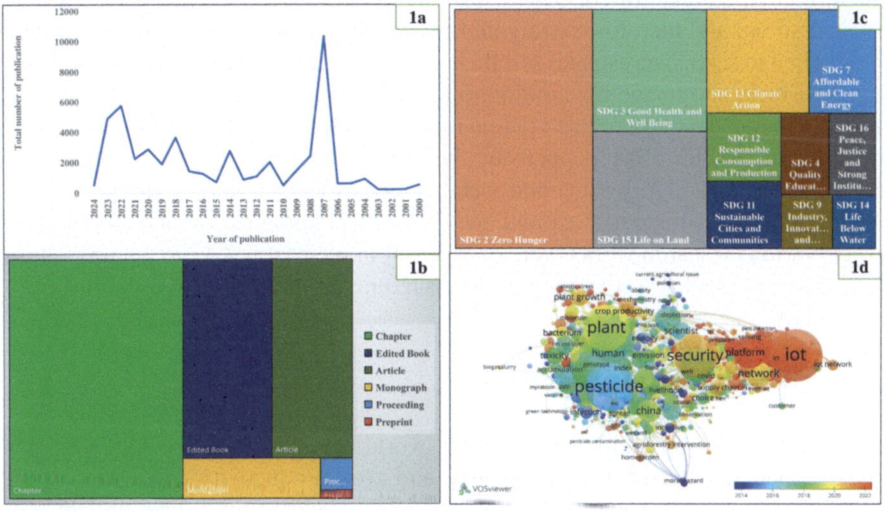

Fig. 10.1 Bibliometric analysis of the publications on AI-based regulation of water supply and pest management in farming since 2000

AI-based regulation of water supply and pest management in farming revealed that, terms like "pesticide," "security," "IoT," and "current agricultural issue" have occurred more frequently (Fig. 10.1d)." Keeping this in view, this chapter will briefly introduce the importance of AI in agriculture for solving farming challenges.

10.2 Impact of AI on Indian Agriculture

In India, agriculture is the nation's economic backbone and accounts for 18% of the country's Gross Domestic Product (GDP). In an era of growing population, demand for food rises and improving productivity as per significant population rise is not an easy task to perform. There are several challenges like land degradation, reduction in soil fertility and ground water table, pest resistance and resurgence, injudicious use of inorganic chemicals, etc. The shortage of skilled labor in agriculture is a problem that can be solved by utilizing AI techniques like automation and robots. By enhancing efficiency, dependability, and accuracy all at once and decreasing the requirement for human intervention, automation has significantly raised the output of agricultural machinery (Bechar and Vigneault 2016). Moreover, climate change becomes more sensible and unpredictable nowadays. Introduction of AI brings broader chance for agriculturists to meet with those challenges as AI can provide long-term solutions.

The first mention of computer uses in agriculture dates back to 1983 (Baker et al. 1983). McKinion and Lemmon initially suggested the use of AI techniques in crop management in their 1985 work "Expert Systems for Agriculture" (McKinion and Lemmon 1985; Bannerjee et al. 2018). Agricultural AI encompasses both AI software, which frequently offers data, insights, and recommendation, and AI robots. The former type of AI involves retrieving data from the farm, local climate, machinery, and so on, and then forecasting and prescribing recommendations for farmers about when to seed, harvest, and sell their crops, the health and behavior patterns of their livestock, and image recognition to identify plant disease, among other applications. On one hand, AI robots operate (almost) independently on farms to harvest fruit and vegetables (like peppers in the SWEEPER project), tend to gardens, remove manure (LELY Robot) and weed (Weed Wacker robot), spray pesticides, and weeds (ecorobotix), harvest and hoe (NAIO technologies' robots) (Ryan et al. 2021). Various research works were carried out in the past few years in agriculture from tillage to postharvest techniques to overcome farming challenges (Table 10.1)

10.3 AI in Irrigation Systems

Arid and semiarid agricultural lands in the world require effective irrigation methods and their management. Plant growth and crop productivity depend on smart irrigation strategies that apply water at the right time and in the right amount (Gu et al. 2017). In times of growing production needs under competitive costs while maintaining good product quality, a demand for this is most important. Consequently,

Table 10.1 Different AI techniques in agricultural practices

Agricultural practices	AI-based techniques	References
Crop and soil monitoring	**Bani rob** (agribot): Helps to monitor crops; checks soil moisture and fertilizer level; detects weed and pest infestations	Mohan et al. (2023)
Soil manipulation	**Adaptable tillage depth monitoring system**: Measures different terrain morphologies	Honglei et al. (2016)
	Laser-guided leveling equipment: Levels or grades the field precisely	Mohan et al. (2023)
Sowing and planting	**Precision seeders and planters**: Saves seeds and keeps uniform spacing	Chauhan et al. (2020)
	Electric precision seeder: Adjust row spacing and sowing depth for small size vegetable seeds with 95% sowing precision	Jin et al. (2019)
Irrigation	**Smart sensor array (real-time prototype):** Used for measuring soil moisture and soil temperature	Vellidis et al. (2007)
Inter-cultural operations	**Autonomous robotic weed control system**: Reduces labor and herbicide dependency	Slaughter et al. (2008)
	Real-time robotic weed knife control system: Eliminates hand weeding for lettuce and tomato by introducing crop signaling technology	Rekha et al. (2020)
	Laser weeding prototype robot (equipped with a dual-gimbal): Identifies weeds in the indoor environment	Xiong et al. (2017)
	Variable rate granular fertilizer applicator (VRA): Helps to set target fertilizer application	Alameen et al. (2019)
	RFID-based VRA for tree crops: Helps in precise amount of fertilizer application	Mohan et al. (2023)
Harvesting and postharvest techniques	**Robotic apple harvesting system:** Reduces labor dependency and saves time	Heng et al. (2018)
	Smart camera: A machine vision system for inspecting and grading food products	Zhonghua et al. (2020)
	Electronic nose application: Assists in quality assessment and determines the maturity of fruit	Manuela and Alphus (2015)

it has become clear that the use of meteorological and soil data for decision-making in irrigated agriculture has been a fully settled reality for decades (Pimentel et al. 2010). Nevertheless, smart methods to estimate plant water stress and automatically activate irrigation systems are crucial to saving water and energy in crop management. Plant and soil water status is currently determined by a wide variety of laboratory methods that are not automated, resulting in high costs due to both the time and expertise needed, as well as the availability of facilities near crops (Almeida et al. 2008). Based on these challenges and needs, results of current studies suggest that AI tools can be a viable alternative to increase crop production and resource efficiency, especially in the use of water (Udutalapally et al. 2020).

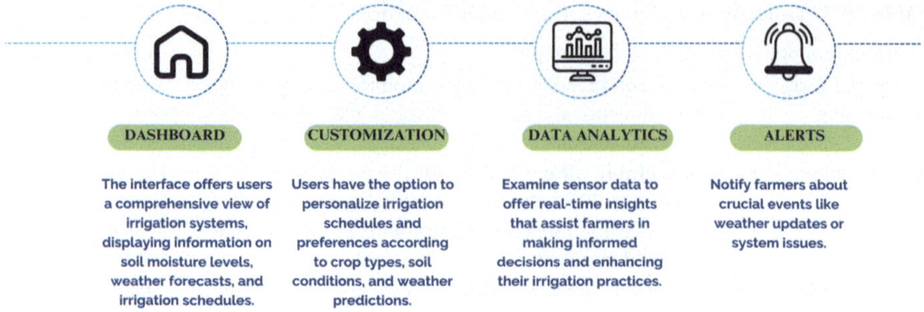

Fig. 10.2 Framework of mobile and web application in irrigation

10.4 Reviving the Irrigation System Through AI Model

Managing water is a crucial element of practicing sustainable agriculture. The field has seen the emergence of AI models, which have become effective instruments in this subject. These models provide enhanced prediction abilities and improved strategies for managing water resources as depicted in (Fig. 10.2 and Table 10.2).

10.4.1 Machine Learning Algorithms

To analyze water quality data and forecast future quality parameters, AI is applied to machine learning algorithms such as supervised and unsupervised applications. These programs can track dynamic, relatively large datasets and organize information that traditional models might overlook (Sayari et al. 2021; Saxena et al. 2024).

10.4.2 Supervised Irrigation Model

10.4.2.1 Linear Regression

A primary tactic for establishing a variable that is influenced by other variables, as well as one or more variables that have no influence, is accomplished through the process of fitting a linear equation to the collected data.

Pro	Cons
Low computational complexity	Limited flexibility
No assumptions on data distribution	Affected by multicollinearity

Table 10.2 A comprehensive comparison between web and mobile application

Criteria	Mobile application	Web application
Platform	Use app stores to obtain the application and install it on their devices	The application can be managed online and operates on a web server, so users don't need to install anything
Accessibility	Provide greater adaptability and mobility	Increased accessibility but with the demand for better internet access
User interface	Optimized for smaller screens and touch interactions, providing a more intuitive and user-friendly interface for mobile devices	Designed to work on a variety of devices and screen sizes, but may not offer the same level of optimization for mobile devices
Offline access	Certain applications include offline access, enabling users to continue using it even when they are not online	Don't offer offline access unless specifically designed to do so
Updates	Need frequent updates via app stores to get security patches and new features	Without requiring any updates from them, users always get the most recent version

10.4.2.2 Support Vector Machine (SVM)

Regression and classification tasks are performed with it. To operate, Support Vector Machine (SVM) first locates the feature space hyperplane that divides various classes (Yazici et al. 2018).

Pro	Cons
Efficient for large datasets	Intensive computational training
Versatile Kernel functions	Noise-sensitive and parameter-sensitivity
Effective in high-dimensional spaces	Limited to binary classifications

10.4.3 Unsupervised Irrigation Model

10.4.3.1 K Means Clustering

Data points are clustered or grouped into K clusters according to how similar they are. K mean is capable to handle fairly large datasets with a moderate number of features and is computationally efficient (Prabhu et al. 2014; Solanki et al. 2023a).

Pro	Cons
Easy implementation	Assumption of circular clusters
Computational efficiency	Sensitive to initial centroid selection

10.4.4 Artificial Neural Network (ANN)

It's made up of layers of interconnected nodes, or neurons. It is Influenced by the architecture and mechanisms of the brain. After applying a transformation function to the inputs, each neuron transmits the output to the layer below.

Pro	Cons
Accurate prediction	Require large data to learn a complex pattern
Suitable for dynamic irrigation management	Require careful training and design
Integrate data from multiple sources to improve irrigation decision-making	Not applicable for all irrigation applications

10.4.4.1 Deep Learning

Sustainable agriculture and effective water management directly depend on irrigation control. More intelligent and statistics irrigation systems have become the norm as a result of the development of deep learning technologies.

10.4.5 Convolutional Neural Network (CNN)

Images and other matrix data are the main types of data that CNNs process. Through convolutions, they are made to automatically and adaptively learn the spatial hierarchies of configurations from the input data. Their assigned uses of convolutional, pooling, and fully connected layers account for a considerable difference.

10.4.6 Recurrent Neural Network (RNN)

Its layout handles sequential data by processing inputs in a linear manner, so that the network's output at any given time step depends on the output of the time step before it. RNNs can now modify temporal dependencies in data because of this (Zhou et al. 2020; Solanki et al. 2023b).

10.4.7 Long Short-Term Memory (LSTM)

It eliminates the hassle of vanishing gradients that sometimes arises when training conventional RNNs on lengthy sequences. They are better at learning long-term interactions because they have a more intricate approach with a set of gates that regulate the transmission of information.

10.4.8 Reinforcement Learning (RL)

A type of machine learning in which an agent gains judgment call abilities through exposure to its surroundings. It's founded on the fact of trial-and-error learning, in which the agent attains knowledge by observing the results of its outcomes.

10.5 For Future: Optimize Agriculture with Smart Irrigation

Cutting-edge technologies like data analytics on cloud databases, e.g., Microsoft Azure, Amazon Web Services, and the IoT, are made possible by these frameworks, which increase crop yields and reduce uncertainty by providing user-friendly interfaces, real-time data analytics, and remote-control capabilities (Abioye et al. 2022 Solanki et al. 2023c).

10.6 Case Studies

10.6.1 Irrigation with NASA and AI (Application: EVAPO)

10.6.1.1 Core Value

There are irrigation action plans, monitoring, and reporting parameters in this web farm management platform according to the Penman-Monteith method. Using climate-gridded data from NASA POWER, the potential evapotranspiration of the entire planet is estimated in real time. It's updated in real situations where water is needed and is based on hydro-meteorological issues.

10.6.2 AI-Powered Crop Monitoring and Irrigation (Application: Nano Ganesh)

10.6.2.1 Core Value

Using information from soil sensors and satellite imagery, this software plans irrigation timeframes and detects soil moisture levels. An irrigation automation system allows farmers to use smartphones to oversee the irrigation pumps from a distance. However, the users' unique requirements and preferences will ultimately determine which of the two options they find most appealing. Some other major application of AI in irrigation is also depicted in Fig. 10.3.

10.7 AI in Pest Management

The insect kingdom, comprising over two-thirds of all known animal species, poses a pervasive threat to various plants, including crops, trees, medicinal plants, and weeds, causing significant damage in their quest for sustenance. A whopping

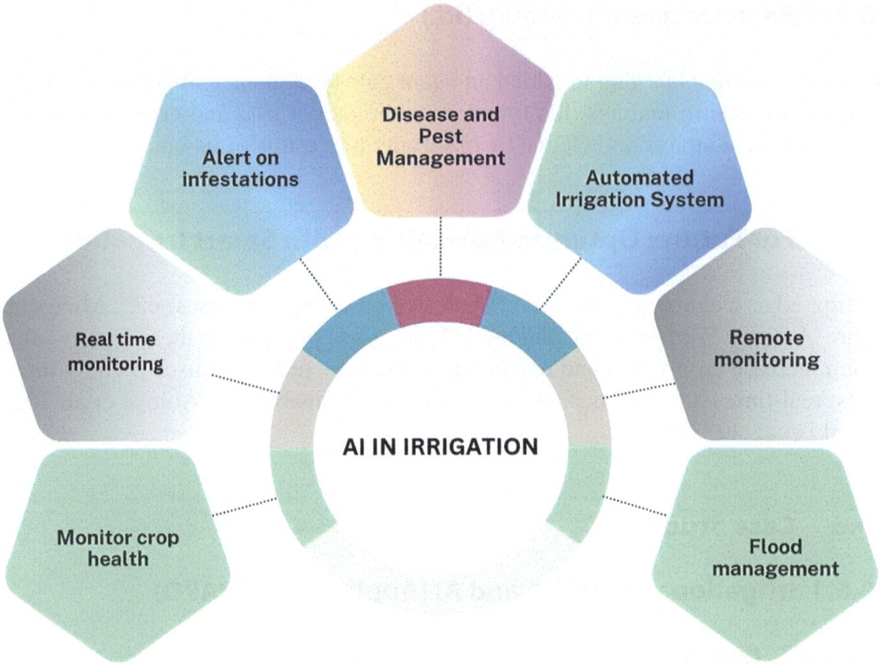

Fig. 10.3 Application of artificial intelligence in irrigation

18–20% of global crop production is destroyed by arthropods, at a cost of over US$ 470 billion a year.

Majority of losses occur before harvest (13–16%), with developing countries suffering the greatest impact. Recent studies on global food loss and waste indicate that 27–32% of all food produced worldwide is lost or wasted, as highlighted by Schuster and Torero (2016), Bali et al. (2023). However, these estimates do not encompass losses during crop production in the fields. While pesticides have traditionally been employed to control these pests, their use escalates production costs and poses environmental hazards. Integrated Pest Management (IPM) aims to mitigate these risks by minimizing pesticide impact on the environment and human and animal health (Deguine et al. 2021; Kumar et al. 2023). IPM focuses on preventing and monitoring disease and insect pests, increasingly aided by sensing techniques and AI methods. AI endeavors to develop robots capable of emulating human cognition and behavior, including problem-solving and learning.

AI is transforming agricultural systems by replacing conventional methods with more efficient approaches, thereby boosting productivity. Numerous agricultural companies are leveraging AI and machine learning to revolutionize their operations. The emergence of AI in pest identification, monitoring, and management has spawned several startups utilizing these technologies to track pest population dynamics and manage them effectively (Tables 10.3 and 10.4).

Table 10.3 Major application of AI in pests management

Sr. No.	Applications	Description	References
1.	Detection of plant diseases	Developing a Deep Neural Network to accurately diagnose and categorize diseases from images of apple plant leaves is crucial for early detection in agriculture, addressing a significant research challenge	Al-bayati and Üstündağ (2020)
2.	Detection of minute insect-pests	AI algorithms are capable of detecting small insects, triggering immediate alerts to farmers' smartphones upon detection of intrusions, enabling timely precautions to be taken	Deng et al. (2018), Roldán-Serrato et al. (2018)
3.	Pest monitoring using pheromone traps	Utilizing deep learning object detectors, automated pest counting is enabled from pheromone trap images, facilitating pest monitoring, reducing labor, and saving time	Hong et al. (2021)
4.	To predict pest spread	A bidirectional Recurrent Neural Network incorporating Long Short-Term Memory (LSTM) units is employed to forecast the occurrence of cotton pests and diseases, leveraging climatic factors for prediction. Forecasting the distribution of vine mealybug is achieved through the utilization of the Maximum Entropy Algorithm (MaxEnt)	Chen et al. (2020a, b), Ji et al. (2020)
5.	Minimizes repetitive application of pesticide	By employing machine learning algorithms like the K-nearest neighbor algorithm, the repetition of pesticide usage can be minimized by at least 20%	Baghel et al. (2022)
6.	Uniform application of pesticide	Advancements in smart robot technology enhance performance and enable efficient pesticide spraying, ensuring comprehensive coverage of all plants with optimal dosage	Chaitanya et al. (2020)

10.8 Startups Using AI and Machine Learning in Agriculture

10.8.1 Taranis

Taranis, which was established in the year 2015 is an Israeli-based intelligence platform employing sophisticated techniques like advanced computer facilities and planned algorithms to look after the fields. Utilizing high-precision aerial surveillance imagery, it can ascertain plant damages, yield losses due to insect pests, and diseases. This platform aims to improve crop quality, optimize fertilizer and crop protection usage, and safeguard soil health with minimal environmental impact.

Table 10.4 Different AI system used for pest management and their impact on pest abundance and activity

Sr. No.	AI systems used	Targeted pests	Impact	References
1.	Intelligent and Integrated Pest and Disease Management (I2PDM)	Thrips (*Frankliniella intonsa, Thrips hawaiiensis*, and *Thrips tabaci*), and whiteflies (*Bemisia argentifolii* and *Trialeurodes vaporariorum*),	Significant yearly reductions in insect pest count as high as 50.7% were observed on the farms	Rustia et al. (2022)
2.	Deep learning (DL) technology	–	DL, especially the development of CNN models, have brought many breakthroughs in the field of intelligent information processing in pest monitoring	Li et al. (2021)
3.	Internet of Things (IoT)-assisted Unmanned Aerial Vehicle (UAV) for rice pest detection image using Imagga cloud	–	It monitors the rice field at regular time intervals by capturing the rice pest images	Bhoi et al. (2021)
4.	Artificial Intelligence (AI)-based methodology to improve hand-held image acquisition of insect traps placed in vineyards	Grapevine insect pests	Providing an automated detection and counting of key vector insects, like *Lobesia botrana* (european grapevine moth), *Empoasca vitis* (smaller green leafhopper), or *Scaphoideus titanus* (American grapevine leafhopper); and improving and anticipating treatment recommendations for the detected pests	Faria et al. (2021)
5.	Faster R-CNN	Insect pests of oilseed	Users can accomplish detecting the pests of oilseed rape in the field real time by operating the intuitive and simple interface based on an Android mobile phone,	He et al. (2019)
6.	Deep learning YOLOv3 for image recognition to obtain the location of *Tessaratoma papillosa*	*Tessaratoma papillosa*	Notifies farmers of the presence of different pests before they start multiplying in large numbers	Chen et al. (2020a, b)

10.8.2 Trap View

Trap view, founded by EFOS, is a Slovenian platform focused on pest insect monitoring and prediction. Automated electronic traps enable hassle-free tracking of pest insects with minimal maintenance. Through advanced artificial intelligence analysis of trap data, Trap view delivers accurate, up-to-date information on pest population trends. Additionally, it provides forecasts on pest development stages tailored to particular regions.

10.8.3 EcoRobotix

EcoRobotix startup came into existence during 2014 in the European continent. With the help of AI, it recognizes the crops and suggests various treatments in order to improve the crop health. Their focus is on reducing the environmental impact of modern agriculture. Their leading product reduces pesticides application in fields by 70–95%, thereby reducing the input cost and increased crop yields.

10.8.4 Kishan Know

The Kishan Know device, created by Rishikesh Amit Nayak from DAV Public School, Chandrasekharpur, employs real-time data and IoT technology to detect bacteria, fungi, and viruses in crops, providing micro-level solutions. It includes a GSM communication module, a circuit chip for communication, and a thermal camera. It aims to identify causes of crop failure, particularly pest attacks.

10.8.5 Greeneye Technology

Greeneye Technology is an Israeli company focused on precision agriculture, combining expertise in computer knowledge, AI, pesticide sprayers, and mechanics. It offers need-based solutions for real-time problems faced by the farmers. Utilizing AI, Greeneye transforms agricultural pest control by transitioning from indiscriminate application of pesticides. Its specialized feature, like the selective spraying system, helps to detect weed flora and manages them pretty well thereby reducing usage of herbicides.

10.9 Pest Detection and Monitoring Systems

AI technology plays a crucial role in helping farmers enhance yields by aiding in crop selection, improving soil health and plant health, minimizing pest load, good agricultural practices, and forecasting commodity prices. Utilizing various AI tools like deep learning, robotics, IoT, artificial neural networks, Wireless Sensor

Networks (WSN), image processing, and other advanced methods, it addresses the difficulties in agriculture (Nihalani et al. 2024). In recent years, remote sensing has employed wireless and vibration sensors for plant pest detection. Acoustic sensors capable of recording specific frequencies emitted by insect-pests are also being used. However, detecting population of pest species across large areas remains a challenge. Nevertheless, these sensors can indicate the presence of certain insects in specific locations. Additionally, image processing and specialized sensor-based tools in combination with pheromone traps, enable direct pest identification (Preti et al. 2021). Accurate pest detection is crucial for implementing Integrated Pest Management (IPM) strategies. So various sensors like ultrasonic, acoustic, and optical sensors have been tailored to monitor specific pests enabling the diagnosis and density determination of pests (Das et al. 2024).

10.10 Decision Support System

In recent decades, global climate changes have resulted in rising occurrences of unexpected precipitation, rising temperatures, severe droughts, and more frequent anomalies in weather events. Alongside these global challenges, agricultural production issues such as pesticide resistance, resurgence, pest replacement and development of biotypes have compounded the problem (Weller et al. 2014). AI plays a crucial role in supporting decision-making processes in pest management, including monitoring and control. Decision Support Systems (DSSs) represent a category of computer software tools designed to assist decision-makers. Well-designed DSSs enable decision-makers to extract valuable insights from raw data, thorough knowledge, proper documentation, and models to identify and address the issues with proper solutions. These systems analyze and interpret data, providing recommendations for the most suitable actions or options. Climate change has led to increased outbreaks of existing pests or the emergence of new pests in regions where they were previously not problematic (Scherm 2004). To address these challenges, there is a critical need for sustainable and continuous productivity growth across all agricultural sectors. This necessitates the diligent and efficient use of resources such as water, energy, pesticides, and fertilizers. AI plays a pivotal role in meeting these demands.

10.11 Precision Pest Management

The application of novel technologies to boost crop yields while mitigating environmental risks is both an art and a science. These sophisticated systems serve as agricultural management tools by offering site-specific, integrated, and interpreted advice. AI plays a crucial role in identifying central plants on farms and assisting in pest control by determining optimal pesticide usage and application methods, including drone-based spraying to save time. This technology aids in understanding crop growth patterns, potential diseases, and prescribing appropriate fertilizers or pesticides based on disease patterns and plant characteristics such as leaf growth, size, or color. Utilizing various sensors like feed, weight, soil, temperature, intensity, and cameras installed on machines, such as low-flying drones or small field robots, AI is concentrated on examining faulty crops and pinpointing pests to improve crop vitality and increase yields. Through technological assistance, crop protection practices are evolving, enabling farmers to increase yield and quality while optimizing resource usage. AI applications in pest management include pest identification, counting, and predicting pest spread. In disease management, AI is utilized for disease recognition and early plant disease forecasting.

10.12 GIS and GPS Modules

In precision pest management, which relies on precise information and considers the spatial and temporal variability of pest populations, geographic information systems (GIS) play a fundamental role. GIS serves as the cornerstone for extracting valuable insights from data on pest population dynamics, acting as the central component of precision farming systems. The spatial analysis capabilities inherent in GIS are essential for precision agriculture. Through GIS, we can effectively monitor insect abundances and determine the reasons behind variations in pest populations across different areas by correlating, comparing, and analyzing information with graphical representations (Rano et al. 2022; Dubey et al. 2022). GIS modules encompass different software and hardware settings and protocols tailored to compile, store, retrieve, and analyze characteristic features and coordinates of locations for map production. These modules integrate with various other parameters, including yield data, rainfall data and pattern, crop types, nutrient composition in soils, and pest distributions. GIS facilitates the transformation of digital information into a usable format, aiding in informed decision-making processes.

Global Positioning System (GPS) modules are integral components of navigation systems reliant on a network of satellites. They enable users to accurately record positional information, including longitude, latitude, and elevation, with precision ranging from 100 to 0.01 m (Lang 1992). By harnessing GPS technology, farmers can pinpoint the precise information on different characteristics in various fields such as soil types, insect-pest infestations, different weed flora, water sources, boundaries, and obstacles. Equipped with automatic control systems, including light or Sound Guiding Panels (DGPs), antennas, and receivers, GPS modules receive signals broadcasted by GPS satellites. These signals enable users to identify their precise position.

10.13 Merits of AI

10.13.1 Reduced Use of Pesticides

With AI solutions, farmers can accurately identify the type of pest infestation and its specific location. This enables them to apply only minimal amounts of pesticides to the affected area, reducing the impact on both the soil and the environment.

10.13.2 Increased Efficiency

AI-driven pest detection systems promptly notify farmers of impending pest infestations, allowing for quick response to protect crops. Moreover, the precise and data-supported remedial actions suggested by these AI systems outperform traditional methods, which often rely on guesswork.

10.13.3 Lower Costs of Pest Control

AI-driven solutions reduce pesticide usage and minimize the necessity for frequent human intervention, resulting in time and resource savings.

10.14 Limitations of AI

While AI has shown notable advancements in agriculture, its overall impact on agricultural activities remains below average compared to its potential and influence in other sectors. Further efforts are needed to enhance agricultural practices through AI, considering the numerous limitations to its implementation.

10.14.1 Response Time and Accuracy

An essential characteristic of intelligent or expert systems is their capacity to swiftly and accurately perform tasks. However, many systems struggle with either response time, accuracy, or both, influencing users' task strategy choices.

10.14.2 Large Data Requirement

The effectiveness of an intelligent agent is also determined by the quantity of input data it can handle. Real-time AI systems must process vast amounts of data, filtering out irrelevant information while remaining responsive to critical or unforeseen events.

10.14.3 Methodology

The effectiveness of an expert system hinges on its execution methodology. Given its reliance on big data, the approach to retrieval and training must be meticulously defined to optimize both speed and accuracy.

10.14.4 High Data Cost

Governments can aid farmers by developing cost-effective web-enabled devices tailored for AI systems, expanding access to remote areas.

10.14.5 Flexibility

The versatility of a robust AI system is crucial. While advancements have been notable in applying AI to specific tasks, the forefront of AI-based robotics emphasizes integrating subsystems into a cohesive environment, demanding flexibility within these subsystems. Moreover, it should possess extensive capabilities to incorporate additional user data from field experts.

10.15 Challenges While Using AI

AI can assist humans in achieving greater accomplishments, yet is unable to substitute human discretion and judgment; making the correct choices relies heavily on having precise real-time information. Moreover, in numerous locations, the availability of reliable data can be limited, like soil moisture levels. The installation of AI-driven irrigation systems can be expensive, especially for small- and medium-scale farmers who may lack the required finances. Ensuring that AI systems can integrate effectively with existing irrigation projects and farm management software can pose challenges. Continued technological innovations and increased agricultural awareness can lower the deviation. Besides, to fully achieve the potential of AI in enhancing irrigation sustainability and efficiency, it's crucial to tackle the challenges.

10.16 Conclusion and Future Perspectives

The integration of AI in agricultural pest management represents a transformative advancement with significant potential for enhancing crop protection and yield optimization. Through the utilization of AI-driven techniques such as predictive modeling, image recognition, and autonomous decision-making systems, farmers can efficiently identify, monitor, and control pest populations while minimizing environmental impact and resource usage. Moreover, AI facilitates real-time data analysis and adaptation, enabling farmers to respond swiftly to evolving pest threats and changing environmental conditions. The use of AI in agriculture is changing the way crop monitoring and management are done, providing many advantages for farmers and the environment. AI-driven systems are improving precision irrigation, early detection of diseases, estimation of crop yields, control of weeds and pests, AI-guided monitoring of groundwater levels, as well as data collection and analysis. These advancements empower farmers to make better use of resources, increase effectiveness, and encourage sustainable farming methods. As AI technology progresses, its influence on agricultural engineering will keep reshaping how farmers handle water supply and pest control in the agriculture sector. The future promises great potential for AI-based solutions to further boost productivity, minimize environmental impact, and ensure efficient resource management in farming. As research and development in this field continue to progress, the adoption of AI technologies holds promise for revolutionizing agricultural pest management practices, ultimately contributing to sustainable food production and global food security.

References

Abioye EA, Hensel O, Esau TJ, Elijah O, Abidin MSZ, Ayobami AS, Yerima O, Nasirahmadi A (2022) Precision irrigation management using machine learning and digital farming solutions. Agricul Eng 4:70–103. https://doi.org/10.3390/agriengineering4010006

Alameen AA, Al-Gaadi KA, Tola E (2019) Development and performance evaluation of a control system for variable rate granular fertilizer application. Comp Electron Agric 160:31–39. https://doi.org/10.1016/j.compag.2019.03.011

Al-bayati JSH, Üstündağ BB (2020) Artificial intelligence in smart agriculture: modified evolutionary optimization approach for plant disease identification. In: 2020 4th International Symposium on Multidisciplinary Studies and Innovative Technologies, pp 1–6

Almeida CAB, Uchikawa R, Amazonas IB, Soares WA (2008) Modelagem computacional do erro associado à medição do fluxo de seiva. In: Encontro Regional de Matemática Aplicada e Computacional. Natal, Proceedings, pp 46–58

Baghel AS, Bhardwaj A, Ibrahim W (2022) Optimization of pesticides spray on crops in agriculture using machine learning. Comput Intell Neurosci 2022:1–10. https://doi.org/10.1155/2022/9408535

Baker DN, Lambert JR, McKinion JM (1983) GOSSYM: a simulator of cotton crop growth and yield. South Carolina Agric Exp Stn Tech Bull:1089

Bali A, Chouhan SS, Kumar G, Kumar R, Singh UP (2023) Adaptive fault-tolerant control for pure-feedback stochastic nonlinear systems with sensor and actuator faults. Circus Syst Sig Proc 42:5838–5867. https://doi.org/10.1007/s00034-023-02366-7

Bannerjee G, Sarkar U, Das S, Ghosh I (2018) Artificial intelligence in agriculture: a literature survey. Int J Scient Res Comput Sci Appl Manag Stud 7:1–6

Bechar A, Vigneault C (2016) Agricultural robots for field operations: concepts and components. Biosyst Eng 149:94–111. https://doi.org/10.1016/j.biosystemseng.2016.06.014

Bhoi SK, Jena KK, Panda SK, Long HV, Kumar R, Subbulakshmi P, Jebreen HB (2021) An Internet of Things assisted unmanned aerial vehicle based artificial intelligence model for rice pest detection. Microprocess Microsyst 1(80):103607. https://doi.org/10.1016/j.micpro.2020.103607

Chaitanya P, Kotte D, Srinath A, Kalyan KB (2020) Development of smart pesticide spraying robot. Int J Rec Technol Engg 8:2193–2202. https://doi.org/10.35940/ijrte.E6343.018520

Chauhan N, Sankhyan NK, Sharma RP, Singh J, Gourav (2020) Effect of long-term application of inorganic fertilizers, farm yard manure and lime on wheat (*Triticum aestivum* L.) productivity, quality and nutrient content in an acid alfisol. J Plant Nut 43:2569–2578. doi: https://doi.org/10.1080/01904167.2020.1783298

Chen CJ, Huang YY, Li YS, Chang CY, Huang YM (2020a) An AIoT based smart agricultural system for pests detection. IEEE 8:180750–180761. https://doi.org/10.1109/ACCESS.2020.3024891

Chen P, Xiao Q, Zhang J, Xie C, Wang B (2020b) Occurrence prediction of cotton pests and diseases by bidirectional long short-term memory networks with climate and atmosphere circulation. Comput Electron Agricul 176:105612. https://doi.org/10.1016/j.compag.2020.105612

Chouhan SS, Singh UP, Jain S (2024) Introduction to Computer Vision and Drone Technology. In: Chouhan SS, Singh UP, Jain S (eds) Applications of computer vision and drone technology in agriculture 4.0. Springer, Singapore. https://doi.org/10.1007/978-981-99-8684-2_1

Das A, Kadawala K, Nath H, Chakraborty S, Ali H, Singh S, Dubey VK (2024) Drone-based intelligent spraying of pesticides: current challenges and its future prospects. In: Chouhan SS, Singh UP, Jain S (eds) Applications of computer vision and Drone technology in agriculture 4.0. Springer, Singapore, pp 199–223. https://doi.org/10.1007/978-981-99-8684-2_12

Deguine JP, Aubertot JN, Flor RJ, Lescourret F, Wyckhuys KA, Ratnadass A (2021) Integrated pest management: good intentions, hard realities. A review. Agron Sustain Dev 41:1–38. https://doi.org/10.1007/s13593-021-00689-w

Deng L, Wang Y, Han Z, Yu R (2018) Research on insect pest image detection and recognition based on bio-inspired methods. Biosys Eng 169:139–148. https://doi.org/10.1016/j.biosystemseng.2018.02.008

Dubey VK, Sahoo SK, Sujatha B, Das A (2022) Impact of heavy metals on honey bees. Vigyan Varta 3:101–103

Faria P, Nogueira T, Ferreira A, Carlos C, Rosado L (2021) AI-powered mobile image acquisition of vineyard insect traps with automatic quality and adequacy assessment. Agronomy 11(4):731. https://doi.org/10.3390/agronomy11040731

Gu Z, Qi Z, Ma L, Gui D, Xu J, Fang Q, Yuan S, Feng G (2017) Development of an irrigation scheduling software based on model predicted crop water stress. Comput Electron Agric 143:208–221. https://doi.org/10.1016/J.COMPAG.2017.10.023

He Y, Zeng H, Fan Y, Ji S, Wu J (2019) Application of deep learning in integrated pest management: a real-time system for detection and diagnosis of oilseed rape pests. Mobile Inform Syst (1):4570808. https://doi.org/10.1155/2019/4570808

Heng W, Cameron JH, Santosh B, Manoj KC, Miller JH (2018) Simulation as a tool in designing and evaluating a robotic apple harvesting system. IFAC-PapersOnLine 51(7):135–140. https://doi.org/10.1016/j.ifacol.2018.08.076

Hong SJ, Nam I, Kim SY, Kim E, Lee CH, Ahn S, Kim G (2021) Automatic pest counting from pheromone trap images using deep learning object detectors for *Matsucoccus thunbergianae* monitoring. Insects 12:342. https://doi.org/10.3390/insects12040342

Honglei J, Guo M, Yu H, Yang L, Feng X, Zhao J, Qi J (2016) An adaptable tillage depth monitoring system for tillage machine. Biosyst Eng 151:187–199. http://www.sciencedirect.com/science/article/pii/S1537511016300642

Ji W, Han K, Lu Y, Wei J (2020) Predicting the potential distribution of the vine mealybug, *Planococcus ficus* under climate change by MaxEnt. Crop Prot 137:105268. https://doi.org/10.1016/j.cropro.2020.105268

Jin X, Li QW, Zhao KX, Zhao B, He ZT, Qiu ZM (2019) Development and test of an electric precision seeder for small-size vegetable seeds. Int J Agric Biol Eng 12:75–81. https://doi.org/10.25165/j.ijabe.20191202.4618

Kumar R, Yadav S, Kumar M, Kumar J, Kumar M (2020) Artificial intelligence: new technology to improve Indian agriculture. Int J Chem Stud 8:2999–3005. https://doi.org/10.22271/chemi.2020.v8.i2at.9208

Kumar R, Singh UP, Bali A, Chouhan SS, Tiwari AK (2023) Adaptive control of unknown fuzzy disturbance-based uncertain nonlinear systems; application to hypersonic flight dynamics. J Analysis 8:1–20. https://doi.org/10.1007/s41478-023-00687-z

Lang L (1992) GPS+ GIS+ remote sensing: an overview. Earth Observation Mag:23–26

Li W, Zheng T, Yang Z, Li M, Sun C, Yang X (2021) Classification and detection of insects from field images using deep learning for smart pest management: a systematic review. Eco Inform 1(66):101460. https://doi.org/10.1016/j.ecoinf.2021.101460

Manuela B, Alphus DW (2015) Electronic-nose applications for fruit identification, ripeness and quality grading. Sensors 15:899–931. https://doi.org/10.3390/s150100899

McKinion JM, Lemmon HE (1985) Expert systems for agriculture. Comput Electron Agric 1:31–40. https://doi.org/10.1016/0168-1699(85)90004-3

Mohan SS, Venkat R, Rahaman S, Vinayak M, Hari Babu B (2023) Role of AI in agriculture: applications, limitations and challenges: a review. Agric Rev 44:231–237. https://doi.org/10.18805/ag.R-2215

Nihalani R et al (2024) Long Short-Term Memory (LSTM) model for Indian sign language recognition. J Intell Fuzzy Syst 1-19. https://doi.org/10.3233/JIFS-233250

Pimentel JS, Silva TJA, Borges Júnior JCF, Folegatti MV, Montenegro AAA (2010) Estimativa da transpiração em cafeeiros utilizando-se sensores de dissipação térmica. Rev Brasil Engenharia Agrícola e Ambiental 14:187–195. https://doi.org/10.1590/S1415436620100002000010

Prabhu B, Sophia S, Mathew A (2014) A review of efficient information delivery and clustering for drip irrigation management using WSN. Int Comput Sci Bus Inform:1–13

Preti M, Verheggen F, Angeli S (2021) Insect pest monitoring with camera-equipped traps: strengths and limitations. J Pest Sci 94:203–217. https://doi.org/10.1007/s10340-020-01309-4

Rano SH, Afroz M, Rahman MM (2022) Application of GIS on monitoring agricultural insect pests: a review. Rev Food Agric 3:19–23. https://doi.org/10.26480/rfna.01.2022.19.23

Rekha R, Nguyen TT, Slaughter DC, Fennimore SA (2020) Real-time robotic weed knife control system for tomato and lettuce based on geometric appearance of plant labels. Biosys Eng 194:152–164. https://doi.org/10.1016/j.biosystemseng.2020.03.022

Roldán-Serrato KL, Escalante-Estrada JAS, Rodríguez-González MT (2018) Automatic pest detection on bean and potato crops by applying neural classifiers. Eng Agric Environ Food 11:245–255. https://doi.org/10.1016/j.eaef.2018.08.003

Rustia DJ, Chiu LY, Lu CY, Wu YF, Chen SK, Chung JY, Hsu JC, Lin TT (2022) Towards intelligent and integrated pest management through an AIoT-based monitoring system. Pest Manag Sci 78:4288–4302. https://doi.org/10.1002/ps.7048

Ryan M (2020) Agricultural big data analytics and the ethics of power. J Agric Environ Ethics 33:49–69. https://doi.org/10.1007/s10806-019-09812-0

Ryan M (2023) The social and ethical impacts of artificial intelligence in agriculture: mapping the agricultural AI literature. AI Soc 38:2473–2485. https://doi.org/10.1007/s00146-021-01377-9

Ryan M, van der Burg S, Bogaardt MJ (2021) Identifying key ethical debates for autonomous robots in agri-food: a research agenda. AI Ethics. https://doi.org/10.1007/s43681-021-00104-w

Saxena A, Chouhan SS, Aziz RM, Agarwal V (2024) A comprehensive evaluation of Marine predator chaotic algorithm for feature selection of COVID-19. Evol Sys 2:1–14. https://doi.org/10.1007/s12530-023-09557-2

Sayari S, Mahdavi-Meymand A, Zounemat-Kermani M (2021) Irrigation water infiltration modeling using machine learning. Comput Electron Agric 180:105921. https://doi.org/10.1016/j.compag.2020.105921

Scherm H (2004) Climate change: can we predict the impacts on plant pathology and pest management? Cana J Plant Pathol 26:267–273. https://doi.org/10.1080/07060660409507143

Schuster M, Torero M (2016) Towards a sustainable food system: reducing food loss and waste. In: 2016 Global food policy report. International Food Policy Research Institute. Available via http://ebrary.ifpri.org/cdm/ref/collection/p15738coll2/id/130211 (Accessed 28 January 2024)

Shamshiri R, Weltzien C, Hameed IA, Yule I, Grift T, Balasundram SK, Pitonakova L, Ahmad D, Chowdhary G (2018) Research and development in agricultural robotics: a perspective of digital farming. Int J Agric Biol Eng. https://doi.org/10.25165/j.ijabe.20181104.4278

Slaughter DC, Giles DK, Downey D (2008) Autonomous robotic weed control systems: a review. Comput Electron Agric 61:63–78. https://doi.org/10.1016/j.compag.2007.05.008

Solanki S, Singh UP, Chouhan SS, Jain S (2023a) A systematic analysis of magnetic resonance images and deep learning methods used for diagnosis of brain tumor. Multimed Tools Appl 83:23929–23966. https://doi.org/10.1007/s11042-023-16430-6

Solanki S, Singh UP, Chouhan SS, Jain S (2023b) Brain tumour detection and classification by using deep learning classifier. Int J Intell Sys Appl Engg 11:262–279. https://ijisae.org/index.php/IJISAE/article/view/2624

Solanki S, Singh UP, Chouhan SS (2023c) Brain tumor classification using ML and DL approaches. In: 2023 IEEE 5th International Conference on Cybernetics, Cognition and Machine Learning Applications (ICCCMLA), Hamburg, Germany, pp 204–208. https://doi.org/10.1109/ICCCMLA58983.2023.10346854

Udutalapally V, Mohanty SP, Pallagani V, Khandelwal V (2020) Crop: a novel device for sustainable automatic disease prediction, crop selection, and irrigation in Internet-of-Agro-Things for smart agriculture. IEEE Sensors J 21:17525–17538. https://doi.org/10.1109/jsen.2020.3032438

Vellidis G, Tucker M, Perry C, Kvien C, Bednarz C (2007) A real-time wireless smart sensor array for scheduling irrigation. Comput Electron Agric 61:44–50. https://doi.org/10.1016/j.compag.2007.05.009

Vijayakumar S, Kumar RM, Choudhary AK, Deiveegan M, Tuti MD, Sreedevi B, Sundaram RM (2022) Artificial intelligence (AI) and its application in agriculture. Chron Biores Manage 6:025–031

Weller SC, Van Wyk E, Simon JE (2014) Sustainable production for more resilient food production systems: case study of African indigenous vegetables in eastern Africa. In: XXIX International Horticultural Congress on Horticulture: sustaining lives, livelihoods and landscapes, pp 289–298. https://doi.org/10.17660/ActaHortic.2015.1102.35

Xiong Y, Ge Y, Liang Y, Blackmore S (2017) Development of a prototype robot and fast path-planning algorithm for static laser weeding. Comput Electron Agric 142:494–503. https://doi.org/10.1016/j.compag.2017.11.023

Yazici MT, Basurra S, Gaber MM (2018) Edge machine learning: enabling smart internet of things applications. Big Data Cogn Comput 2:1–26. https://doi.org/10.3390/bdcc2030026

Zhonghua G, Zhang M, Dah-Jye L, Taylor S (2020) Smart camera for quality inspection and grading of food products. Electronics 9:505. https://doi.org/10.3390/electronics9030505

Zhou J, Zhang Y, Tian S, Lai S (2020) Forecasting rainfall with recurrent neural network for irrigation equipment. In: IOP conference series: earth and environmental science, vol 510. IOP Publishing, p 042040. https://doi.org/10.1088/1755-1315/510/4/042040

AI for Data-Driven Decision-Making in Smart Agriculture: From Field to Farm Management

11

Harshit Mishra and Divyanshi Mishra

Abstract

The rise of global population and changing climatic conditions necessitate a paradigm shift toward smarter agricultural practices. Smart agriculture, encompassing a fusion of information and communication technologies (ICTs) with agricultural practices, offers immense potential to improve efficiency, sustainability, and productivity. Its core components include sensing and monitoring, data collection, communication technologies, and decision support systems. Data is essential to smart agriculture since it allows for data-driven decision-making in a variety of contexts. For making well-informed decisions, information on the weather, soil, crops, livestock, and growth stages and yields is essential. However, because agricultural data is real time, heterogeneous, and volumetric, processing it might be difficult. An effective tool for maximizing the potential of data in smart agriculture is artificial intelligence (AI). Artificial intelligence is able to monitor crops, identify pests and illnesses, and forecast yields with high accuracy because of advances in machine learning, deep learning, and computer vision. Farmers are empowered to improve resource allocation, decrease waste, and maximize yields through the integration of AI and data-driven decision-making. However, challenges like data security, privacy concerns, and skill gaps need to be addressed for widespread adoption. Precision agriculture, enabled by AI and smart technologies, allows for targeted interventions at the field level, optimizing inputs and minimizing environmental impact. Intelligent farm equipment and automation further enhance farm management efficiency. Data integration platforms are crucial for combining data from diverse sources, while robust data management strategies address challenges like scalability, interoperability, and security. Looking ahead, emerging technologies like

H. Mishra (✉) · D. Mishra
Department of Agricultural Economics, College of Agriculture, Acharya Narendra Deva University of Agriculture and Technology, Kumarganj, Ayodhya, Uttar Pradesh, India

S. S. Chouhan et al. (eds.), *Artificial Intelligence Techniques in Smart Agriculture*, https://doi.org/10.1007/978-981-97-5878-4_11

blockchain and Internet of Things (IoT) hold immense promise for further revolutionizing smart agriculture. Continuous innovation, coupled with a focus on sustainability and ethical considerations, will ensure that AI paves the way for a secure, efficient, and resource-conscious agricultural future.

Keywords

Artificial intelligence · Blockchain · Data integration · Data-driven decision-making · Edge Computing · Food and Agriculture Organization (FAO) · Information and Communication Technologies (ICTs) · Precision agriculture · Smart agriculture · Yield prediction

11.1 Introduction

The twenty-first century presents previously unheard-of difficulties for the agriculture industry. A large increase in food production while protecting natural resources is required due to the world's expanding population and the intensifying effects of climate change. These requirements are difficult for traditional farming methods to achieve since they frequently rely on experience and intuition (Beriya and Saroja 2019; Linaza et al. 2021). With artificial intelligence (AI) supporting it, data-driven decision-making becomes a game-changer for contemporary agriculture. With a wide range of crops, farming techniques, and environmental factors, the agricultural landscape throughout the world is enormous and varied (Saiz-Rubio and Rovira-Más 2020). But a few broad patterns present a worrying image. In order to feed the world's estimated 9.7 billion people by 2050, the Food and Agriculture Organization (FAO) projects that food production will need to rise by 70%. Simultaneously, climate change disrupts weather patterns, increases the risk of extreme weather events, and exacerbates existing challenges like water scarcity and soil degradation. These factors threaten food security and agricultural sustainability on a global scale (Huo et al. 2024; Saadane et al. 2022).

Thankfully, new developments in technology provide potential answers. The development of a network of sensors and devices that continuously gather real-time data on numerous agricultural factors has been made easier by the rise of the Internet of Things (IoT) (Chouhan et al. 2024). These include crop health, animal performance, weather, nutrient levels, and soil moisture (Rozenstein et al. 2024; Araújo et al. 2023). Big data, the term used to describe this deluge of data, has the potential to greatly improve decision-making throughout the whole agricultural value chain. However, advanced analytical tools are needed to derive relevant insights from this massive volume of data (Bhat and Huang 2021). AI, which includes deep learning, machine learning, and other cutting-edge methods, is essential to this process. By analyzing historical data, identifying patterns and making predictions, AI algorithms can empower farmers to optimize their practices in several key areas (Paul et al. 2021).

One of the most significant applications of AI in agriculture lies in precision farming. This approach leverages data-driven insights to tailor resource allocation and management practices to specific field conditions and individual crop needs. For instance, AI models can analyze soil sensor data to determine the optimal application of fertilizers and irrigation water, minimizing waste and maximizing resource efficiency (Liang and Shah 2023; Rao et al. 2021). Similar to this, early crop disease and pest detection made possible by AI-powered picture identification can reduce the need for broad-spectrum pesticides by permitting customized interventions. AI is also essential for optimizing farm management. Artificial intelligence (AI) algorithms can help farmers make informed decisions regarding planting dates, crop selection, and resource allocation by analyzing data on variables like yield history, market pricing, and weather forecasts. AI may also automate monotonous processes like data collecting, analysis, and reporting, giving farmers back vital time to concentrate on other crucial areas of farm management and strategic decision-making (Shaikh et al. 2022a, b).

The integration of AI into agricultural practices extends beyond individual farms. AI-powered platforms can connect farmers with agricultural experts, providing them with real-time advice and support. Additionally, AI can facilitate the creation of digital marketplaces, enabling farmers to connect directly with consumers and access fair market prices for their produce. Artificial intelligence has a wide range of possible applications in agriculture. Among the encouraging results are higher crop yields, better resource management, more efficiency, and less environmental impact (Narra et al. 2020). However, overcoming a number of obstacles is necessary for the effective application of AI in agriculture. Ensuring data accessibility and quality is a significant task. It's possible that farmers, especially those in poorer nations, lack the resources and knowledge needed to efficiently gather and handle data. Concerns around data security and privacy must also be addressed in order to guarantee the ethical use of agricultural data (Spanaki et al. 2021; Chen et al. 2022).

To further ensure that all farmers have equitable access to technology and close the digital divide, the use of AI-powered solutions is necessary. Initiatives to increase capacity are required, as is the creation of user-friendly interfaces that accommodate a range of technical skill levels. All things considered, artificial intelligence (AI) offers a potent instrument for converting agriculture into a more data-driven, sustainable, and effective industry (El-Gayar and Ofori 2020; Javaid et al. 2022; Koshariya et al. 2024). Farmers may successfully traverse the difficult problems of the twenty-first century and help to ensure global food security for future generations by utilizing AI. But in order to fully utilize AI in agriculture, it will be necessary to solve data-related issues, close the digital gap, and promote equitable access to technology for all parties involved (Tiwari et al. 2023). As we move forward, collaborative efforts between researchers, policymakers, and agricultural communities will be crucial in shaping the responsible and sustainable development of AI-powered solutions for the future of agriculture (Tomar and Kaur 2021).

11.2 Overview of Smart Agriculture

The term "smart agriculture" describes the fusion of cutting-edge technologies with data-driven methodologies to improve the productivity, sustainability, and efficiency of agricultural activities. This paradigm shift turns traditional farming practices into more intelligent and responsive systems by utilizing cutting-edge technologies like the Internet of Things (IoT), artificial intelligence (AI), and data analytics (Mandapuram et al. 2019).

Using digital technologies at every stage of the agricultural value chain—from planting and cultivation to harvesting and farm management—is known as "smart agriculture." Optimizing resource use, raising crop output, lessening environmental effect, and making defensible judgments based on real-time data are the main objectives (Mitra et al. 2022).

11.2.1 Definition and Scope

11.2.1.1 Definition

The term "smart agriculture" describes the use of sensors, automation, data analytics, and information and communication technologies (ICT) in the agricultural industry. In order to maximize crop productivity, resource use, and overall farm management, farmers can make timely decisions by utilizing connected devices and systems for data collection, processing, and analysis (Kumar et al. 2022; Issa et al. 2024).

11.2.1.2 Scope

The scope of smart agriculture encompasses a wide range of technologies and practices that aim to revolutionize traditional farming methods. Some key components of the scope include:

- *IoT Devices and Sensors*: The field installation of Internet of Things (IoT) devices and sensors is a critical component of smart agriculture. These gadgets gather data in real time on a number of variables, including crop health, temperature, humidity, and soil moisture. After then, the data is sent to a centralized system for examination.
- *AI and data analytics*: These two areas of study are at the core of smart agriculture. To deliver insights that may be put to use, sophisticated algorithms analyze the massive volume of data produced by IoT devices. Based on past and present data, machine learning algorithms can optimize irrigation schedules, forecast agricultural illnesses, and suggest appropriate crop varieties.
- *Precision Farming Techniques*: The exact control of resources like water, fertilizer, and pesticides is a component of precision agriculture. With the use of smart agriculture technologies, farmers can deploy these resources more strategically, increasing productivity and lowering waste and environmental effect.

- *Farm Management Software*: Smart agriculture frequently incorporates farm management software to help with decision-making. These platforms assist farmers with the planning and supervision of a range of tasks, such as pest management, planting, and harvesting. They might also provide tools for inventory tracking, compliance reporting, and financial management.
- *Automation and Robotics*: Smart agriculture uses robotics and automation to accomplish activities like weed control, planting, and harvesting. Drones and autonomous cars with cameras and sensors can swiftly survey big regions and deliver insightful data for decision-making.
- *Remote Monitoring and Control*: Farmers may monitor and operate a range of farm operations remotely with the use of connected technologies. Large-scale farms or those with scattered land holdings will especially benefit from this, as it enables real-time decision-making without the requirement for physical presence.

As highlighted in Table 11.1, the definition and scope of smart agriculture encompass a broad spectrum of technologies, including AI, which are crucial for data-driven decision-making and optimizing farming practices. By integrating data-driven decision-making processes, smart agriculture aims to increase productivity, sustainability, and profitability in the agricultural sector. The following sections will delve deeper into specific aspects of smart agriculture, focusing on data-driven decision-making from the field to overall farm management (Mishra and Mishra 2023).

Table 11.1 Importance of AI for data-driven decision-making in smart agriculture

Area of impact	Traditional approach	AI-driven approach	Benefits
Crop monitoring	Manual scouting, subjective assessments	Real-time data from sensors, automated analysis using computer vision	Early detection of stress, disease, and nutrient deficiencies
Pest and disease management	Reactive spraying based on experience or calendar schedules	Predictive models based on historical data and weather patterns	Targeted interventions, reduced use of pesticides and fungicides
Yield prediction	Historical averages, expert estimates	Machine learning models considering weather, soil conditions, and crop health data	Improved resource allocation, informed decisions about planting and harvesting
Irrigation Management	Manual adjustments based on visual cues	Sensor-based data analysis for soil moisture and plant water needs	Optimized water usage, reduced waste, and environmental impact
Fertilization	Uniform application based on soil type	Precision application based on real-time nutrient deficiencies	Increased crop yield, reduced fertilizer costs, and environmental pollution

11.2.2 Components of Smart Agriculture

11.2.2.1 Sensing and Monitoring

A smart agriculture's core is sensing and monitoring. To collect data in real time, a variety of sensors are used in the field, including temperature sensors, drones with cameras, and sensors for measuring soil moisture. These sensors gather data on weather patterns, crop health, soil conditions, and other pertinent factors. The information gathered is necessary to comprehend the farm's current situation and make wise decisions (Nyakuri et al. 2022; Holzinger et al. 2022).

11.2.2.2 Data Collection

The process of gathering data entails compiling information from several sources, such as farm equipment, satellite images, and sensors. Numerous factors are included in this data, including crop growth phases, soil quality, weather, and mechanical performance. Modern techniques for gathering data provide a thorough and accurate depiction of the dynamics of the farm, giving farmers a comprehensive understanding of their activities.

11.2.2.3 Communication Technologies

In order to provide smooth data flow between the many elements of the smart agriculture ecosystem, communication technologies are essential. Connectivity solutions, IoT devices, and communication protocols make it easier to transfer data from field sensors to cloud platforms or centralized databases. By ensuring that all parties involved have access to the most recent information, this real-time communication promotes prompt decision-making (Sharma et al. 2023).

11.2.2.4 Decision Support Systems (DSS)

The brains driving smart agriculture are DSS. To deliver useful insights, these systems examine the massive volume of data gathered from sensors and other sources. After processing data, AI algorithms and machine learning models provide forecasts, suggestions, and optimization techniques for farm management. DSS give farmers the ability to make knowledgeable decisions about crop management in general as well as irrigation, fertilization, and pest control.

Components of smart agriculture work synergistically to create a data-driven ecosystem. Sensing and monitoring provide real-time information, data collection ensures comprehensive data sets, communication technologies enable seamless connectivity, and decision support systems leverage AI to turn data into actionable insights. This integrated approach enhances the overall efficiency and sustainability of modern agricultural practices (Shaikh et al. 2022a, b).

11.3 Role of Data in Smart Agriculture

Data is a key component of smart agriculture, which uses it to improve and optimize a number of farming activities. Traditional farming practices have undergone a revolution with the incorporation of new technologies and data-driven decision-making processes, resulting in enhanced production, resource utilization, and efficiency.

11.3.1 Importance of Data-Driven Decision-Making

Making decisions based on data has been essential to the development of contemporary agriculture. With the use of historical and current data, it enables farmers and farm managers to make well-informed decisions that maximize yields, minimize risks, and allocate resources optimally. Farmers may take proactive measures to improve results by using data analytics to obtain insightful knowledge about crop health, soil conditions, and climate patterns (Keswani et al. 2020; Ouafiq et al. 2022).

Data-driven decision-making in smart agriculture involves the use of sophisticated algorithms and AI models that analyze vast datasets. This not only enhances precision in farming but also contributes to sustainable agricultural practices by minimizing environmental impact.

11.3.2 Types of Data in Agriculture

11.3.2.1 Environmental Data

Environmental data encompasses information about weather patterns, soil conditions, and other external factors affecting the farm. Weather data provides insights into temperature, humidity, precipitation, and wind speed, aiding farmers in predicting optimal planting times, irrigation scheduling, and pest management.

11.3.2.2 Crop Data

Crop data includes information related to the health, growth, and yield of crops. Sensors, satellite imagery, and drones are used to collect data on factors such as plant health, crop density, and nutrient levels. This data is crucial for monitoring crop development, identifying potential diseases, and optimizing fertilizer and water usage (Akkem et al. 2023).

11.3.2.3 Livestock Data

Data on animal health, nutrition, and behavior are crucial for managing livestock. Real-time data on an animal's health can be obtained by sensors and tracking devices, which helps farmers identify symptoms of disease, plan feeding regimens optimally, and enhance animal welfare in general.

11.3.3 Challenges in Handling Agricultural Data

While the benefits of utilizing data in smart agriculture are substantial, there are also challenges associated with handling agricultural data effectively.

11.3.3.1 Data Security and Privacy

Data security and privacy are issues because agricultural data, which includes sensitive information like farm locations, crop varieties, and production numbers, is sensitive. Strong systems are necessary for farmers and agricultural groups to safeguard their data from illegal access and guarantee that privacy laws are followed (Kose et al. 2022; Qazi et al. 2022).

11.3.3.2 Data Integration

Data about agriculture is frequently produced by a variety of sources, such as satellite photography, machines, and sensors. It can be difficult to integrate these disparate datasets; uniform formats and protocols are needed to guarantee smooth compatibility.

11.3.3.3 Connectivity Issues

Rural areas may face connectivity challenges, limiting the real-time transfer of data from remote farms. This can hinder the effectiveness of data-driven decision-making processes, emphasizing the need for improved infrastructure and connectivity solutions in agriculture.

In order to promote sustainable and effective agricultural methods and realize the full potential of data-driven decision making in smart agriculture, it is imperative that these problems be addressed (Javaid et al. 2023; Kumar et al. 2023).

11.4 AI in Agriculture

Agriculture is not an exception to the way AI has emerged as a transformational force in different industries. Farmers may increase productivity by optimizing resource allocation, making better decisions, and utilizing AI technologies in their operations. The introduction to artificial intelligence (AI) in agriculture is covered in this section, with a particular emphasis on AI methods like computer vision, deep learning, and machine learning that are essential to the revolution of smart agriculture.

11.4.1 Introduction to AI

AI is the term for machines that have been taught to think and learn like humans, simulating human intelligence. AI in agriculture seeks to use cutting-edge technologies for data analysis and interpretation, offering insightful information for decision-making. Artificial Intelligence (AI) is a broad field that includes several methods. In the agricultural field, AI is essential for converting conventional farming practices into data-driven and precision agriculture (Addas et al. 2023).

11.4.2 AI Techniques in Agriculture

AI encompasses various techniques and in the agricultural sector, three major ones stand out—Machine Learning, Deep Learning, and Computer Vision.

11.4.2.1 Machine Learning

A type of artificial intelligence called machine learning (ML) enables computers to learn from data and gradually enhance their performance without the need for explicit programming. Large-scale datasets in agriculture can be analyzed by ML algorithms to find trends, correlations, and patterns (Muniasamy 2020). For instance, ML can be used to forecast pest outbreaks, improve irrigation plans, and estimate crop yields based on past data. ML models are useful tools for improving decision-making in smart agriculture because of their capacity to adjust to changing circumstances (Thilakarathne et al. 2021).

11.4.2.2 Deep Learning

Deep Learning (DL) is a subset of ML that focuses on neural networks with multiple layers (deep neural networks). DL algorithms excel in processing and analyzing complex data, such as images and sensor data (Nihalani et al. 2024). In smart agriculture, deep learning can be applied to tasks like image recognition, disease detection in crops, and even predicting optimal planting times. The deep neural networks' ability to automatically extract features from raw data enhances the accuracy and efficiency of decision-making in farm management.

11.4.2.3 Computer Vision

Teaching machines to analyze and comprehend visual data from the outside world is known as computer vision (CV). In the realm of agriculture, CV can be used to examine photos taken by sensors or drones in order to track crop health, identify weeds, and evaluate the general state of the fields. Through the utilization of CV, farmers can obtain up-to-date knowledge about the condition of their crops, allowing them to take immediate and proactive action to resolve problems. In smart agriculture, this visual intelligence plays a major role in data-driven decision-making and is essential for precision farming (Doshi and Varghese 2022; Barriguinha and Moysiadis 2021).

By giving farmers access to cutting-edge tools for better decision-making, resource optimization, and overall farm management, the integration of artificial intelligence (AI), including machine learning, deep learning, and computer vision, is transforming the agricultural industry.

11.4.3 Applications of AI in Smart Agriculture

11.4.3.1 Crop Monitoring

AI is essential for managing and keeping an eye on crops at every stage of growth. AI algorithms can analyze enormous volumes of data, including crop health indicators, soil conditions, and weather patterns, by using remote sensing technology like satellites and drones. This helps farmers to make well-informed choices about fertilization, irrigation, and pest management. Real-time data from AI-driven crop monitoring systems enable farmers to identify irregularities and take immediate action to resolve problems.

11.4.3.2 Pest and Disease Detection

Early pest and disease detection and management is one of agriculture's biggest concerns. Early crop threat identification is made easier by AI applications like computer vision and machine learning techniques. These technologies can identify minute indicators of pest infestations or diseases that are invisible to the human eye by analyzing photos of fields and plants. Farmers may carry out focused treatments thanks to this early detection, which minimizes the use of pesticides and lowers crop losses (De Alwis et al. 2022).

11.4.3.3 Yield Prediction

In order to estimate projected harvests, AI-powered yield prediction models combine historical data, weather trends, and crop health information. The various factors that affect agricultural yields are analyzed by machine learning algorithms, which then provide precise forecasts to help farmers plan their harvest, use resources wisely, and improve logistics. Better decision-making about crop selection, planting techniques, and market planning is made possible by this predictive capacity, which eventually helps farmers become more profitable.

The use of artificial intelligence (AI) into smart agriculture improves decision-making by offering predictive analytics and real-time insights. Artificial intelligence (AI) applications provide farmers with the means to maximize resource allocation, minimize environmental effect, and boost output in a sustainable way. These tools range from monitoring crop health to identifying possible risks and forecasting harvests (Rabhi et al. 2021). As technology continues to advance, the role of AI in agriculture is likely to expand, further transforming the industry into a more data-driven and efficient ecosystem.

11.5 Data-Driven Decision Making

A key component of smart agriculture is data-driven decision-making, which makes use of information to improve farming operations' productivity, sustainability, and efficiency. This section examines the idea, tenets, and application of AI to decision-making processes. It also looks at the advantages and disadvantages of this strategy.

11.5.1 Concept and Principles

Large-scale data created within the agricultural ecosystem is systematically collected, analyzed, and interpreted in order to facilitate data-driven decision-making in smart agriculture. To make well-informed and optimal decisions, this approach uses real-time data from multiple sources, including sensors, satellites, and on-farm devices (Sharma et al. 2024; Panchal et al. 2024).

The requirement for timely and accurate data, the integration of many data sets, and the use of advanced analytics to extract actionable insights are the guiding principles of data-driven decision-making (Sabrina et al. 2022). These insights are used by farmers and other agricultural stakeholders to help them make decisions that maximize crop management, resource allocation, and overall farm performance.

11.5.2 Integration of AI in Decision-Making

The incorporation of data into decision-making processes is greatly aided by artificial intelligence. Complex datasets are analyzed using machine learning algorithms, predictive analytics, and other AI technologies. Farmers will be able to make better decisions thanks to AI's ability to spot patterns, forecast trends, and offer advice based on both historical and current data.

Several facets of smart agriculture, such as yield forecasting, insect management, weather forecasting, and crop monitoring, are connected with AI systems. Over time, AI systems improve their accuracy through ongoing learning and adaptation, giving farmers more useful decision support (Ashraf and Akanbi 2023).

11.5.3 Benefits and Challenges

11.5.3.1 Benefits

- *Precision Farming*: Data-driven decision-making enables precision farming, allowing farmers to optimize the use of resources such as water, fertilizers, and pesticides. This results in increased efficiency and reduced environmental impact.
- *Increased Yield and Quality*: AI-driven insights contribute to better crop management, leading to increased yield and improved crop quality. Farmers can take proactive measures based on data to mitigate risks and enhance overall production.
- *Cost Reduction*: By optimizing resource allocation and minimizing waste, data-driven decision-making helps in reducing operational costs. This is particularly crucial in the context of volatile market conditions and fluctuating input prices.

11.5.3.2 Challenges

- *Data Quality and Security*: The effectiveness of data-driven decision-making relies on the quality and security of the data. Ensuring the accuracy, reliability, and security of the data collected from various sources can be challenging.

- *Implementation Costs*: The initial costs associated with implementing AI technologies, such as sensors and data analytics tools, can be a barrier for some farmers. Additionally, there may be ongoing costs related to system maintenance and updates.
- *Skill Gaps*: Successfully implementing AI in agriculture requires farmers and agricultural professionals to have a certain level of technical expertise. Bridging the skill gap through training and education is a critical challenge.

Data-driven decision-making in smart agriculture, facilitated by the integration of AI, holds immense potential for transforming traditional farming practices. While there are notable benefits in terms of efficiency and productivity, addressing challenges related to data quality, security, implementation costs, and skill gaps is essential for widespread adoption and sustainable success (Pradhan et al. 2022).

11.6 From Field to Farm Management

The use of AI technologies in smart agriculture has transformed conventional methods and made data-driven decision-making possible for all types of farming operations. The main facets of this transition are explored in this part, which includes automation in farm management, intelligent farm equipment, and precision agriculture.

11.6.1 Precision Agriculture

Using cutting-edge technologies, such as artificial intelligence (AI), precision agriculture maximizes field-level control in crop production. Ensuring sustainability, profitability, and environmental preservation are the goals.

- *Data Collection and Analysis*: An important component of precision agriculture is data. Artificial intelligence (AI) algorithms collect data from multiple sources, including sensors, satellites, and drones, to offer comprehensive insights on crop health, weather patterns, and soil conditions. Machine learning algorithms are capable of analyzing past data to forecast trends and provide suggestions for the future.
- *Variable Rate Technology (VRT)*: AI facilitates the implementation of VRT, where inputs like water, fertilizers, and pesticides are applied at variable rates across a field. This targeted approach, guided by AI analysis, optimizes resource utilization and enhances crop yields.
- *Decision Support Systems*: AI-driven decision support systems assist farmers in making informed choices. These systems consider a multitude of factors, including weather forecasts, market conditions, and historical data, to recommend optimal planting times, irrigation schedules, and harvesting windows.

11.6.2 Intelligent Farm Equipment

Intelligent Farm Equipment refers to the incorporation of AI into machinery and devices used in agriculture, making them smarter, more efficient, and capable of autonomous decision-making.

- *Autonomous Vehicles*: AI enables the development of autonomous tractors and drones. These machines can navigate fields, plant seeds, apply fertilizers, and harvest crops with minimal human intervention. Machine learning algorithms allow these vehicles to adapt to changing conditions in real time.
- *IoT Integration*: Intelligent sensors and IoT devices embedded in farm equipment collect and transmit data. AI algorithms analyze this data to optimize equipment performance, detect potential issues, and schedule predictive maintenance, reducing downtime and improving overall efficiency.
- *Robotics in Agriculture*: Robots with AI capabilities carry out jobs including packing, gathering fruit, and weeding. By identifying and managing crops using computer vision, these robots increase accuracy and decrease the need for manual labor.

11.6.3 Automation in Farm Management

Automation in farm management involves the application of AI to streamline and optimize the administrative and operational aspects of running a farm.

- *Crop Monitoring and Management*: AI-powered systems continuously monitor crops, detecting diseases, pests, and nutrient deficiencies. Automated alerts and recommendations help farmers take timely action, preventing potential losses.
- *Supply Chain Optimization*: AI streamlines every step of the supply chain, from distribution to harvest. Farmers may better manage their production and cut waste by using machine learning algorithms and predictive analytics to predict market demand.
- *Financial Management*: AI assists in financial planning and budgeting. It analyses data on crop yields, market prices, and input costs to provide insights that aid in making financially sound decisions.

Integration of AI in smart agriculture, from precision agriculture to intelligent farm equipment and automation in farm management, offers a holistic approach to farming that enhances productivity, sustainability, and profitability. The synergy between data-driven insights and advanced technologies is reshaping the agricultural landscape, paving the way for a more efficient and resilient future (Musanase et al. 2023).

11.7 Data Integration and Management

Data integration and management plays a pivotal role in leveraging AI for data-driven decision-making in smart agriculture. This section delves into the key aspects of data integration platforms, challenges associated with data management, and the critical considerations regarding security and privacy.

11.7.1 Data Integration Platforms

In the context of smart agriculture, data integration platforms act as a linchpin in collecting, aggregating, and harmonizing diverse datasets from various sources across the agricultural ecosystem. These platforms serve to bridge the gap between different types of data, such as weather patterns, soil conditions, crop health, and machinery telemetry (Dineva and Atanasova 2022). Here are some points to consider:

- *Interoperability*: Data integration platforms must ensure interoperability among disparate data sources, such as sensors, drones, satellite imagery, and farm machinery. This interoperability facilitates a seamless flow of information, allowing farmers to have a comprehensive view of their operations.
- *Real-time Data Processing*: With the dynamic nature of agriculture, real-time data processing is crucial. Integration platforms should be capable of handling and processing data in real time, enabling farmers to make timely decisions based on the most recent information.
- *Scalability*: The scalability of data integration platforms is essential as agricultural operations can vary significantly in scale. Whether it's a small family farm or a large agribusiness, the platform should be able to scale to accommodate different levels of data volume and complexity.

11.7.2 Challenges in Data Management

Though there are some advantages, there are drawbacks to using efficient data management in smart agriculture. These issues must be resolved in order to guarantee the validity and precision of data-driven insights:

- *Data Quality and Accuracy*: Ensuring the quality and accuracy of data is a significant challenge. Factors such as sensor errors, calibration issues, or data transmission problems can compromise the integrity of the information, leading to erroneous decision-making.
- *Data Standardization*: The lack of standardized formats for agricultural data poses a hurdle. Different devices and systems may generate data in varying formats, making it challenging to integrate and analyze information cohesively.

- *Data Ownership and Sharing*: Farmers may have concerns about data ownership and sharing. Establishing clear protocols for data ownership and incentivizing data sharing among stakeholders are critical considerations.

11.7.3 Security and Privacy Concerns

Given the sensitive nature of agricultural data, security and privacy concerns demand careful attention. Farmers need assurance that their data is handled with the utmost confidentiality and integrity:

- *Data Encryption*: Implementing robust encryption mechanisms ensures that sensitive data remains secure during transmission and storage. This is particularly crucial when dealing with financial records, crop yield projections, or proprietary farming practices.
- *Access Control*: Strict access control measures must be in place to regulate who can access and modify the data. This helps prevent unauthorized access and ensures that only authorized personnel can make decisions based on the information (Mishra and Singh 2023).
- *Compliance with Regulations*: Adherence to data protection regulations is essential. Compliance with regional and international data protection laws helps build trust among farmers and stakeholders, demonstrating a commitment to ethical data handling practices.

An important prerequisite for the successful application of AI for data-driven decision-making in smart agriculture is efficient data management and integration. Building a solid basis for precision agriculture and farm management requires overcoming obstacles and resolving security and privacy issues (Steup et al. 2019).

11.8 Future Trends in AI for Smart Agriculture

In the rapidly evolving landscape of smart agriculture, the integration of AI plays a pivotal role in shaping the future of farming practices. This section explores the anticipated trends that are expected to redefine the way agriculture is approached, from leveraging emerging technologies to addressing critical considerations such as sustainability and ethics.

11.8.1 Emerging Technologies

11.8.1.1 Internet of Things (IoT) Integration

The seamless connectivity provided by IoT is set to become a cornerstone in smart agriculture. By interconnecting devices, sensors, and machinery across the agricultural landscape, farmers can collect real-time data on various parameters such as soil moisture, crop health, and weather conditions. AI algorithms can then analyze this data to provide actionable insights, enabling more informed decision-making (Katyayan et al. 2021; Goel et al. 2021).

11.8.1.2 Robotics and Automation

Robotics integration in agriculture is growing in popularity, with the promise of lower labor costs and more productivity. Precision operations such as planting, harvesting, and weeding can be carried out by autonomous vehicles, drones, and robotic arms that are outfitted with AI algorithms. This lowers operating expenses and resource consumption while simultaneously increasing production.

11.8.1.3 Edge Computing

Processing nearer to the source is essential as the amount of data collected in agriculture keeps growing. Data processing can take place locally, close to the devices that are producing the data, thanks to edge computing. This not only lowers latency but also takes bandwidth and data privacy concerns into account (Karunathilake et al. 2023). Real-time insights can be obtained from AI models that are installed at the edge, independent of centralized cloud processing.

11.8.2 Potential Innovations

11.8.2.1 Predictive Analytics for Crop Management

AI-powered predictive analytics can revolutionize crop management by forecasting disease outbreaks, pest infestations, and optimal harvest times. Machine learning algorithms trained on historical data can identify patterns and correlations, enabling farmers to implement proactive measures and optimize yield.

11.8.2.2 Precision Agriculture at Scale

AI has the potential to enable the expansion of precision agriculture in the future. Artificial intelligence (AI) systems can produce extremely accurate resource allocation prescriptions by combining data from several sources, such as satellite imaging, weather forecasts, and soil sensors. This reduces waste and the negative effects on the environment by ensuring that water, herbicides, and fertilizers are sprayed exactly where and when they are needed (Hassan et al. 2022; Fishman et al. 2020).

11.8.2.3 Blockchain for Supply Chain Transparency

AI and blockchain technologies together can improve the agricultural supply chain's traceability and transparency. Every stage of the production and distribution process can be tracked and verified with the use of distributed ledgers and smart contracts (Nishad et al. 2023). Customers will be more trusting as a result, and problems like contamination or fraud will be easier to spot and address fast.

11.8.3 Sustainability and Ethical Considerations

11.8.3.1 Resource Optimization and Environmental Impact

AI in smart agriculture must prioritize sustainability by optimizing resource usage. This includes minimizing water consumption, reducing chemical inputs, and optimizing energy use. Smart algorithms can help farmers make data-driven decisions to achieve a balance between productivity and environmental conservation (Maduranga and Abeysekera 2020).

11.8.3.2 Ethical Use of Data

As AI relies heavily on data, ethical considerations become paramount. Farmers and technology developers must establish transparent practices regarding data ownership, usage, and privacy (Jain and Chouhan 2014). Ensuring that data is collected and utilized responsibly will be crucial in building and maintaining trust within the agricultural community.

11.8.3.3 Addressing Socioeconomic Impact

The adoption of AI in agriculture should be mindful of its socioeconomic impact. While enhancing efficiency is crucial, it should not lead to job displacement in rural communities. Training programs and initiatives should be put in place to empower farmers and workers with the skills needed to navigate the evolving agricultural landscape (Routis et al. 2022; Ferehan et al. 2022).

The future trends in AI for smart agriculture promise a paradigm shift in the industry. By embracing emerging technologies, fostering innovations, and addressing sustainability and ethical considerations, the agriculture sector can embark on a transformative journey toward data-driven, efficient, and responsible practices from the field to farm management (Chergui and Kechadi 2022).

11.9 Conclusion

AI-powered smart agriculture and data-driven decision-making have the potential to completely transform the agriculture industry. Farmers can obtain real-time insights into their fields and animals by utilizing a network of sensors, data gathering instruments, and communication technology. This allows farmers to maximize resource allocation, enhance yields, and reduce their environmental effect. A plethora of opportunities arises from the integration of AI techniques such as computer vision, deep learning, and machine learning. AI enables farmers to make more accurate and efficient decisions by enabling automated pest and disease detection, yield prediction, and intelligent farm equipment.

But there are still difficulties. Steady infrastructure and standards are necessary for data management, integration, and security. For inclusive and sustainable growth, privacy concerns must be addressed and equal access to these technologies must be guaranteed. With so much potential, smart agriculture has a bright future ahead of it. Robotics and blockchain are two emerging technologies that could lead to even more automation and transparency. Future AI-driven agriculture will support food security, environmental sustainability, and universal economic success through ongoing research and development and ethical use.

References

Addas A, Tahir M, Ismat N (2023) Enhancing precision of crop farming towards smart cities: an application of artificial intelligence. Sustain For 16(1):355

Akkem Y, Biswas SK, Varanasi A (2023) Smart farming using artificial intelligence: a review. Eng Appl Artif Intel 120:105899

Araújo SO, Peres RS, Filipe L, Manta-Costa A, Lidon F, Ramalho JC, Barata J (2023) Intelligent data-driven decision support for agricultural systems-ID3SAS. IEEE Access

Ashraf H, Akanbi MT (2023) Sustainable agriculture in the digital age: crop management and yield forecasting with IoT, Cloud, and AI. Tensorgate J Sustain Technol Infrastruct Dev Countries 6(1):64–71

Barriguinha A, Moysiadis T (2021) Data-driven artificial intelligence applications for sustainable precision agriculture. Agronomy 11:1227

Beriya A, Saroja VN (2019) Data-driven decision making for smart agriculture. ICT India Working Paper

Bhat SA, Huang NF (2021) Big data and ai revolution in precision agriculture: survey and challenges. IEEE Access 9:110209–110222

Chen Q, Li L, Chong C, Wang X (2022) AI-enhanced soil management and smart farming. Soil Use Manage 38(1):7–13

Chergui N, Kechadi MT (2022) Data analytics for crop management: a big data view. J Big Data 9(1):1–37

Chouhan SS, Singh UP, Jain S (2024) Introduction to computer vision and Drone technology. In: Chouhan SS, Singh UP, Jain S (eds) Applications of computer vision and drone technology in agriculture 4.0. Springer, Singapore. https://doi.org/10.1007/978-981-99-8684-2_1

De Alwis S, Hou Z, Zhang Y, Na MH, Ofoghi B, Sajjanhar A (2022) A survey on smart farming data, applications and techniques. Comput Ind 138:103624

Dineva K, Atanasova T (2022) Cloud data-driven intelligent monitoring system for interactive smart farming. Sensors 22(17):6566

Doshi M, Varghese A (2022) Smart agriculture using renewable energy and AI-powered IoT. In: AI, edge and IoT-based smart agriculture. Academic Press, pp 205–225

El-Gayar OF, Ofori MQ (2020) Disrupting agriculture: the status and prospects for AI and Big Data in smart agriculture. In: AI and Big Data's potential for disruptive innovation. IGI Global, pp 174–215

Ferehan N, Haqiq A, Ahmad MW (2022) Smart farming system based on intelligent internet of things and predictive analytics. J Food Qual 2022:1–8

Fishman R, Ghosh M, Mishra A, Shomrat S, Laks M, Mayer R, Jog A, Ben-Dor E, Shacham-Diamand Y (2020) Digital villages: a data-driven approach to precision agriculture in small farms. In: SENSORNETS, pp 161–166

Goel RK, Yadav CS, Vishnoi S, Rastogi R (2021) Smart agriculture–urgent need of the day in developing countries. Sustain Comput Inform Syst 30:100512

Hassan M, Malhotra K, Firdaus M (2022) Application of artificial intelligence in IoT security for crop yield prediction. ResearchBerg Rev Sci Technol 2(1):136–157

Holzinger A, Saranti A, Angerschmid A, Retzlaff CO, Gronauer A, Pejakovic V, Medel-Jimenez F, Krexner T, Gollob C, Stampfer K (2022) Digital transformation in smart farm and forest operations needs human-centered AI: challenges and future directions. Sensors 22(8):3043

Huo D, Malik AW, Ravana SD, Rahman AU, Ahmedy I (2024) Mapping smart farming: addressing agricultural challenges in data-driven era. Renew Sustain Energy Rev 189:113858

Issa AA, Majed S, Ameer SA, Al-Jawahry HM (2024) Farming in the digital age: smart agriculture with AI and IoT. In: E3S Web of conferences, vol 477. EDP Sciences, p 00081
Jain S, Chouhan SS (2014) Cyclic redundancy codes: study and implementation. Int J Emerg Technol Adv Eng 4(4):213–217
Javaid M, Haleem A, Singh RP, Suman R (2022) Enhancing smart farming through the applications of Agriculture 4.0 technologies. Int J Intell Netw 3:150–164
Javaid M, Haleem A, Khan IH, Suman R (2023) Understanding the potential applications of artificial intelligence in agriculture sector. Adv Agrochem 2(1):15–30
Karunathilake EM, Le AT, Heo S, Chung YS, Mansoor S (2023) The path to smart farming: innovations and opportunities in precision agriculture. Agriculture 13(8):1593
Katyayan A, Mashelkar S, DC AG, Morajkar S (2021) Design of smart agriculture systems using artificial intelligence and big data analytics. In: 2021 3rd International Conference on Advances in Computing, Communication Control and Networking (ICAC3N). IEEE, pp 592–595
Keswani B, Mohapatra AG, Keswani P, Khanna A, Gupta D, Rodrigues J (2020) Improving weather dependent zone specific irrigation control scheme in IoT and big data enabled self driven precision agriculture mechanism. Enterp Inf Syst 14(9–10):1494–1515
Kose U, Prasath VS, Mondal MR, Podder P, Bharati S (2022) Artificial intelligence and smart agriculture technology. CRC Press
Koshariya AK, Rameshkumar PM, Balaji P, Cavaliere LP, Dornadula VH, Singh B (2024) Data-driven insights for agricultural management: leveraging industry 4.0 technologies for improved crop yields and resource optimization. In: Robotics and automation in industry 4.0. CRC Press, pp 260–274
Kumar P, Singh A, Rajput VD, Yadav AK, Kumar P, Singh AK, Minkina T (2022) Role of artificial intelligence, sensor technology, big data in agriculture: next-generation farming. In: Bioinformatics in agriculture. Academic Press, pp 625–639
Kumar N, Kushwaha RR, Meena NR, Mishra H, Yadav APS (2023) A study on costs and returns of paddy cultivation in Ambedkar Nagar district of Uttar Pradesh. Int J Statist Appl Math 8(3):107–111
Liang C, Shah T (2023) IoT in agriculture: The future of precision monitoring and data-driven farming. Eigenpub Rev Sci Technol 7(1):85–104
Linaza MT, Posada J, Bund J, Eisert P, Quartulli M, Döllner J, Pagani AG, Olaizola I, Barriguinha A, Moysiadis T, Lucat L (2021) Data-driven artificial intelligence applications for sustainable precision agriculture. Agronomy 11(6):1227
Maduranga MW, Abeysekera R (2020) Machine learning applications in IoT based agriculture and smart farming: a review. Int J Eng Appl Sci Technol 4(12):24–27
Mandapuram M, Mahadasa R, Surarapu P (2019) Evolution of smart farming: integrating IoT and AI in agricultural engineering. Glob Discl Econ Bus 8(2):165–178
Mishra H, Mishra D (2023) Artificial intelligence and machine learning in agriculture: transforming farming systems. Res Trends Agric Sci 1:1–6
Mishra H, Singh M (2023) Market liberalization and agricultural sector transformation: paving the path to a dynamic future. Agric Mag 2(7):230–234
Mitra A, Vangipuram SL, Bapatla AK, Bathalapalli VK, Mohanty SP, Kougianos E, Ray C (2022) Everything you wanted to know about smart agriculture arXiv preprint arXiv, 2201.04754
Muniasamy A (2020) Machine learning for smart farming: a focus on desert agriculture. In: 2020 International Conference on Computing and Information Technology (ICCIT-1441). IEEE, pp 1–5
Musanase C, Vodacek A, Hanyurwimfura D, Uwitonze A, Kabandana I (2023) Data-driven analysis and machine learning-based crop and fertilizer recommendation system for revolutionizing farming practices. Agriculture 13(11):2141
Narra N, Nevavuori P, Linna P, Lipping T, Dahanayake A, Huiskonen J, Kiyoki Y, Thalheim B, Jaakkola H (2020) A data driven approach to decision support in farming. In: Information modelling and knowledge bases XXXI, vol 321. IOS Press, Amsterdam, The Netherlands, pp 175–185

Nihalani R et al (2024) Long Short-Term Memory (LSTM) model for Indian sign language recognition. J Intell Fuzzy Syst:1–19. https://doi.org/10.3233/JIFS-233250
Nishad DC, Mishra H, Tiwari AK, Pandey A (2023) Towards sustainable agriculture: mitigating the adverse effects of stubble burning in India. Res Trends Environ Sci I:42–48
Nyakuri JP, Bizimana J, Bigirabagabo A, Kalisa JB, Gafirita J, Munyaneza MA, Nzemerimana JP (2022) IoT and AI based smart soil quality assessment for data-driven irrigation and fertilization. Am J Comput Eng 5(2):1–4
Ouafiq EM, Saadane R, Chehri A (2022) Data management and integration of low power consumption embedded devices IoT for transforming smart agriculture into actionable knowledge. Agriculture 12(3):329
Panchal B, Pranjal P, Patel RK, Sharma A, Chouhan SS (2024) Assessing the quantity of a crop field using aerial images. In: Chouhan SS, Singh UP, Jain S (eds) Applications of computer vision and Drone technology in agriculture 4.0. Springer, Singapore. https://doi.org/10.1007/978-981-99-8684-2_11
Paul PB, Biswas S, Bairagi AK, Masud M (2021) Data-driven decision making for smart cultivation. In: 2021 IEEE International Symposium on Smart Electronic Systems (iSES). IEEE, pp 249–254
Pradhan AK, Swain S, Kumar Rout J (2022) Role of machine learning and cloud-driven platform in IoT-based smart farming. In: Machine learning and internet of things for societal issues. Springer Nature Singapore, Singapore, pp 43–54
Qazi S, Khawaja BA, Farooq QU (2022) IoT-equipped and AI-enabled next generation smart agriculture: a critical review, current challenges and future trends. IEEE Access 10:21219–21235
Rabhi L, Falih N, Afraites L, Bouikhalene B (2021) Digital agriculture based on big data analytics: a focus on predictive irrigation for smart farming in Morocco. Indonesian J Electr Eng Comput Sci 24(1):581–589
Rao EP, Rakesh V, Ramesh K (2021) Big Data analytics and artificial intelligence methods for decision making in agriculture. Indian J Agron 66(5):279–287
Routis G, Paraskevopoulos M, Vetsikas IA, Roussaki I, Stavrakoudis D, Katsantonis D (2022) Data-driven and interoperable smart agriculture: an IoT-based use-case for arable crops. In: 2022 IEEE International Conference on Omni-layer Intelligent Systems (COINS). IEEE, pp 1–8
Rozenstein O, Cohen Y, Alchanatis V, Behrendt K, Bonfil DJ, Eshel G, Harari A, Harris WE, Klapp I, Laor Y, Linker R (2024) Data-driven agriculture and sustainable farming: friends or foes? Precis Agric 25(1):520–531
Saadane R, Chehri A, Jeon S (2022) AI-based modeling and data-driven evaluation for smart farming-oriented big data architecture using IoT with energy harvesting capabilities. Sustain Energy Technol Assess 52:102093
Sabrina F, Sohail S, Farid F, Jahan S, Ahamed F, Gordon S (2022) An interpretable artificial intelligence based smart agriculture system. Comput Mater Contin:3777–3797
Saiz-Rubio V, Rovira-Más F (2020) From smart farming towards agriculture 5.0: a review on crop data management. Agronomy 10(2):207
Shaikh TA, Mir WA, Rasool T, Sofi S (2022a) Machine learning for smart agriculture and precision farming: towards making the fields talk. Arch Comput Methods Eng 29(7):4557–4597
Shaikh TA, Rasool T, Lone FR (2022b) Towards leveraging the role of machine learning and artificial intelligence in precision agriculture and smart farming. Comput Electron Agric 198:107119
Sharma A, Sharma A, Tselykh A, Bozhenyuk A, Choudhury T, Alomar MA, Sánchez-Chero M (2023) Artificial intelligence and internet of things oriented sustainable precision farming: towards modern agriculture. Open Life Sci 18(1):20220713
Sharma A, Patel RK, Pranjal P, Panchal B, Chouhan SS (2024) Computer vision-based smart monitoring and control system for crop. In: Chouhan SS, Singh UP, Jain S (eds) Applications of computer vision and Drone technology in agriculture 4.0. Springer, Singapore. https://doi.org/10.1007/978-981-99-8684-2_5
Spanaki K, Karafili E, Despoudi S (2021) AI applications of data sharing in agriculture 4.0: a framework for role-based data access control. Int J Inf Manag 59:102350

Steup R, Dombrowski L, Su NM (2019) Feeding the world with data: visions of data-driven farming. In: Proceedings of the 2019 on designing interactive systems conference, pp 1503–1515

Thilakarathne NN, Yassin H, Bakar MS, Abas PE (2021) Internet of things in smart agriculture: challenges, opportunities and future directions. In: 2021 IEEE Asia-Pacific Conference on Computer Science and Data Engineering (CSDE). IEEE, pp 1–9

Tiwari AK, Mishra H, Nishad DC, Pandey A (2023) Sustainable water management in agriculture: irrigation techniques and water conservation. In: Research trends in agriculture science, vol II. Bhumi Publishing

Tomar P, Kaur G (2021) Artificial intelligence and IoT-based technologies for sustainable farming and smart agriculture. IGI Global

AI-Based Regulation of Water Supply and Pest Management in Farming

12

Murugasridevi Kalirajan, V R. Mageshen, K. Aswitha, and M. Saranya

Abstract

With the growing global population, effectively managing water demands for farming and undertaking appropriate pest management are challenging tasks to attain sustainable crop production. Since water shortages and losses are intolerable, efficient management calls for sound leak detection procedures and proactive decision-making. On the other hand, Current Agriculture is being confronted with climate change, unexpected pest outbreaks, frequent invasions of transboundary pests, and pest management problems. At this juncture, Industry 4.0 has plenty of technical answers, but artificial intelligence (AI) is an unexplored potential tool in smart farming. AI's core competence is scalable, cost-effective personalized services, which the agriculture business needs. By understanding farmers' needs, identifying the right systemic mechanisms to meet them, and assessing their potential impact pathways, we can use AI to create resilient and feasible solutions that surpass current technological advances. Water supply and pest management are the domains that possess significant potential to enhance

M. Kalirajan (✉)
Department of Agricultural Entomology, Amrita School of Agricultural Sciences, Amrita Vishwa Vidyapeetham, Coimbatore, Tamil Nadu, India

V R. Mageshen
Department of Soil Science and Agricultural Chemistry, Amrita School of Agricultural Sciences, Amrita Vishwa Vidyapeetham, Coimbatore, Tamil Nadu, India

K. Aswitha
Department of Soil Science and Agricultural Chemistry, JKK Munirajah College of Agricultural Science, Erode, Tamil Nadu, India

M. Saranya
Department of Agricultural Entomology, Tamil Nadu Agricultural University, Coimbatore, Tami Nadu, India

S. S. Chouhan et al. (eds.), *Artificial Intelligence Techniques in Smart Agriculture*, https://doi.org/10.1007/978-981-97-5878-4_12

the livelihoods of subsistence farmers. In this chapter, we want to highlight promising AI approaches that can help farmers overcome water supply and pest control challenges. This chapter discusses the deployment of AI-powered pest detection, identification, classification, prediction, decision making, and pesticide spraying in real time.

Keywords

Decision Trees (DT) · Naïve Bayes (NB) · Random Forests (RF) · Genetic Algorithm (GA) · Soft Computing (SC) · Evolutionary Computing (EC)

12.1 Introduction

According to predictions from the United Nations (2019), the World's population is expected to reach 9.7 billion in 2050, and 10.9 billion in 2100. In this scenario, agriculture faces a great challenge to meet out the food demand of this rising population amidst the changing climate, frequent invasion of unwelcomed insect pests, and issues related to irrigation. In these circumstances, smallholding producers encounter several significant obstacles in the realm of water management, encompassing concerns such as availability to water, efficacy of water use, and renewal of water resources. The underlying mentalities and actions of producers regarding decision-making make it so difficult to resolve these issues. Optimizing water distribution from an energy-saving and a quality-of-service standpoint is challenging because it involves a complicated system with numerous nodes, hidden factors, and operational restrictions. Water distribution systems must therefore manage a careful trade-off between the solution's computational time and effectiveness. Artificial intelligence (AI), a subfield of computer science is concerned with simulating intelligent behaviour in computers. Artificial intelligence (AI) and machine learning (ML) are primarily used in decision-making tasks related to provide an efficient water supply (Chouhan et al. 2024). These tasks include figuring out how water utilities can optimize data and information to make better decisions while improving service delivery, optimizing capital expenditure (CAPEX), and lowering operating costs (OPEX), which includes externalities related to society and the environment (Zouizza et al. 2023). The decision-making capacity of artificial intelligence depends on its three principal components is indicated in Fig. 12.1. In this connection, advanced AI engines have the capability to efficiently interpret and analyze data flows obtained from satellite, aircraft, or drone photography. Machine learning, specifically deep-learning algorithms, may assist in analyzing visual data and detecting patterns that highlight irrigation problems, as well as other concerns like pests (Solanki et al. 2024). By integrating imaging with soil- and plant-based sensors, we may obtain very precise data on watering requirements in real time, while also receiving notifications about any problems (Sharma et al. 2023).

On the other hand, the globe is much challenged to produce sufficient food for the growing population amidst the risk of pest and disease which causes 20–40% yield loss (IPPC Secretariat 2021). Prevailing climate change, tremendous increase

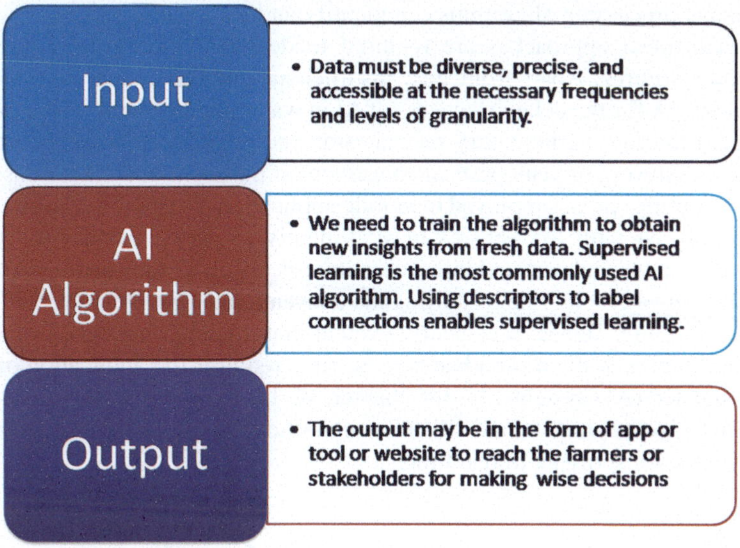

Fig. 12.1 Components of AI

in exchange of planting materials across the globe, breakage of host plant resistance by pests, sudden pest outbreak, development of insecticide resistance by insect pests, pest resurgence, and secondary pest outbreak are important points urging the growers to take up necessary cautious steps called Integrated Pest Management (IPM). In the modern era, IPM must be facilitated by employing emerging techniques to smoothen the observation and monitoring of pest at real time. As monitoring the population dynamics of the pest is a crucial element in IPM, which remains time consuming and labour-intensive, this can be automated with the help of AI. Additionally, pest detection, identification, and classification involve knowledge and expertise; these conventional methods can be replaced with the help of AI. Targeted pesticide application with precision spraying is one of the emerging trends in minimizing the pesticide hazards.

Holistically, AI provides golden opportunities in farming, viz., provides surveillance systems to monitor crop health using AI-powered drones, detects pest and disease infestation, tracks soil nutrition, schedules irrigation for a farm, predicts the weather parameters, identifies weed infestation and recommends herbicides to spray based on nature of weeds, and also helps the growers to sow seeds at optimal time frame to earn maximum yields. These AI-powered drones provide the images captured in the field and presents the crop health status of the entire farm, from which growers can make appropriate decisions. Recently, time-consuming and strenuous agricultural tasks are done by employing agricultural robots in the farm level. These robots drastically reduce the volume of work done manually and saves money for the growers by lowering the labour charges (Kumar et al. 2020). Yet, a major problem is to acquire diverse pest datasets on multisource temporal and spatial resolution for training AI to eliminate blurs and noisy background in the images.

While image processing algorithms can identify and isolate affected areas within a picture, additional approaches are required to deal with background noise and superfluous features. Apart from this, insufficient, unreliable data on water is a major drawback for the delay in success of AI in water management. These data can be gathered through participatory mechanisms and crowded sources. In addition, real-time monitoring of plant pests, management of the pests, and sensor-based irrigation scheduling can be employed to mobile applications or drones. Another option of integrating AI with IoT sensors may virulently enhance in data collection of groundwater status and pest and infestation images acquisition. Emerging AI trends potentially play a crucial role in shifting the conventional agricultural sector to precise smart farming. Hence, it is clearly evident that tracking, managing, and using agricultural inputs at the most advantageous time requires the application of artificial intelligence (AI) systems. In this chapter, we have discussed the utilization of AI in terms of pest detection, identification, classification, prediction, decision making, and pesticide spraying in real time.

12.1.1 The Rise of AI and Its Application in Agriculture Water Management

Water management is a crucial obstacle in the pursuit of sustainable growth across the vast expanses of the world, where agriculture is the lifeblood of many countries. Artificial Intelligence (AI) is a ray of hope that appears at the nexus of technology and agriculture, promising to transform the way we address one of the most critical problems. AI offers a transformative solution through precision agriculture. Different techniques exploited under AI is depicted in Fig. 12.2. Through the integration of IoT devices, machine learning algorithms, and smart sensors, farmers may improve water use by utilizing real-time data. Artificial intelligence (AI)-powered smart irrigation systems allow for accurate water delivery to crops, reducing waste and increasing efficiency. This results in increased crop productivity and water conservation, strengthening the agricultural sector's sustainability, and resilience (Sharma et al. 2021). A water-secure and prosperous agricultural future can be achieved by addressing the problems of water shortage, unequal distribution, and wasteful use through the combination of technology innovation, cross-disciplinary collaboration, and community participation. The implementation of AI-powered solutions, as illustrated in Fig. 12.3, holds promise for mitigating existing challenges and advancing a sustainable and resilient water management paradigm. AI's planned exponential scaling is its power. Numerous standard algorithms would be required to account for all agricultural pest images. However, an AI computing could identify farm pests from a fresh photograph by finding similarities across images. Standard algorithms can't provide hyper-local weather forecasts since they must account for each location's circumstances. However, AI algorithms may find patterns to anticipate at any level. For these reasons, AI in agriculture is revolutionary (Vijayakumar et al. 2022).

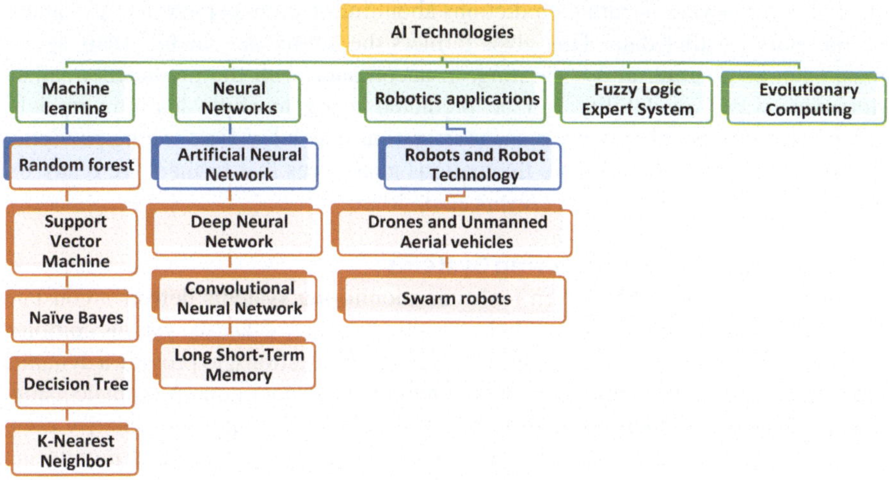

Fig. 12.2 Different technologies under AI

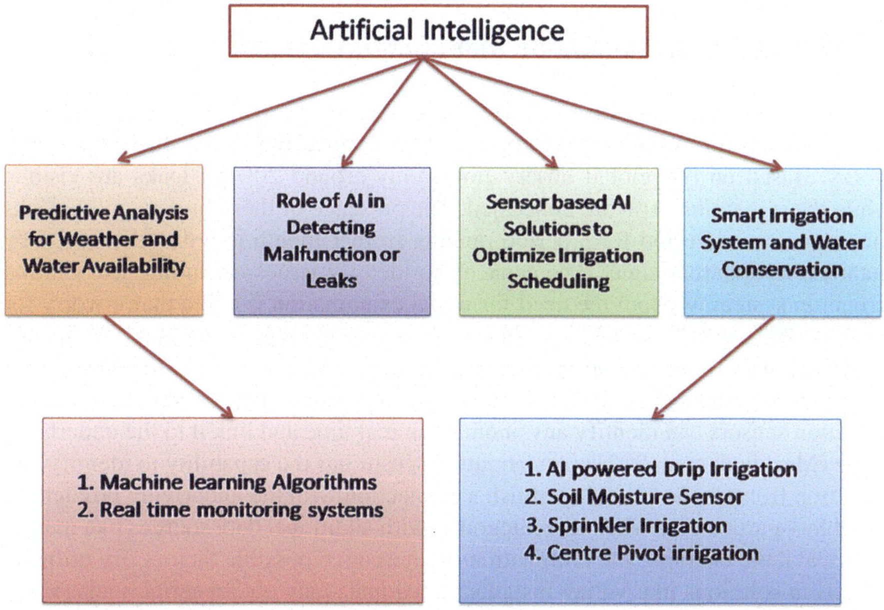

Fig. 12.3 Role of AI in regulation of water supply

12.1.2 Predictive Analytics for Weather and Water Availability

12.1.2.1 Machine Learning Algorithms

Predictive analytics, which makes use of machine learning, is becoming indispensable for predicting weather patterns and water supply. Machine learning algorithms are able to make very accurate predictions about future climate patterns by examining previous weather data. This gives farmers the ability to schedule their operations, such as planting and harvesting, in accordance with the predicted weather. Moreover, water availability can be predicted using machine learning models, which helps farmers effectively manage irrigation schedules. Forecasting precipitation, droughts, and other climate fluctuations guarantees that farmers are ready for any challenge that nature poses (Awais et al. 2023).

12.1.2.2 Real-time Monitoring Systems

Artificial intelligence (AI)-driven real-time monitoring systems deliver a constant flow of information about the weather and water resources. These systems combine data from several sources, such as sensors and weather stations, to provide a dynamic and realistic viewpoint. This instantaneous nature of data facilitates prompt decision-making, enabling farmers to modify irrigation schedules or put preventative measures in place in reaction to unforeseen circumstances. Real-time monitoring combined with AI-powered analytics forms a strong ecosystem that can adjust to agriculture's unpredictability.

12.1.3 AI to Detect Irrigation Malfunctions or Leaks

Leakage is typically the primary factor contributing to the inefficiency of UFW (Unaccounted for Water) and refers to the losses incurred in the distribution networks. Based on our global observations, only around 20% of leaks are visible, while the remaining 80% are concealed. The presence of these imperceptible leaks may remain undetected for a period ranging from 1 month to a decade, until they manifest as apparent signs. The capacity to identify issues, including leaks in the irrigation system, without the need for visual examination is still a major worry for farmers. The use of linked devices in irrigation systems is growing as AI's influence in agriculture expands, enabling software to send an alert when something seems suspicious (Balaska et al. 2023). Farmers are thus able to take prompt action since irrigation sensors can identify any anomaly in real time and link it to the underlying cause (Mashhadi et al. 2021). An irrigation sensor has the capability to identify any deviation from the norm and establish a connection with the underlying problem or variable—particularly when it is integrated with additional data sources like meteorological data, enabling the elimination of alternative possible factors. By utilizing AI-driven solutions like Valley Insights, individuals may obtain aerial imagery and additional field data, like thermal imaging of individual plants. The collected images and AI field analysis may deliver precise warnings that identify the specific trouble areas, including irrigation concerns that want quick action. This indicates their capability to address challenges like pivot-related leaks, which are challenging to identify by visual observation. The capabilities of artificial intelligence extend beyond merely highlighting a problem. It offers valuable information on how to correct an irrigation imbalance.

12.1.4 Sensor-based AI Solutions to Optimize Irrigation Scheduling

Farmers strive to develop an ideal irrigation timetable for their crops that will maximize both productivity and quality, while effectively managing expenses. Evapotranspiration is a crucial measure used to design an irrigation system that is specifically suited to the requirements of a plant (Vathsala et al. 2023). Aiming to optimize output and quality while controlling water use, farmers are working to improve agricultural irrigation schedules in response to the growing need to address issues with water efficiency in agriculture. IoT sensors can be utilized in AI-based irrigation systems that can provide strategic insights into the best times to water crops, apply fertilizer or pesticides, etc., depending on the type of data to be recorded, such as soil, light, weather, and plant. Soil-based sensors are strategically placed across the field to gather pertinent data on salinity, volumetric water content, and other essential characteristics. This information is then used to quickly assess the needs of the soil and make real-time irrigation predictions (Sinwar et al. 2020). Contemporary satellite imaging and weather forecasts assist farmers in enhancing the evaluation of evapotranspiration. Advancements in Internet of Things (IoT) sensor technology have enabled more precise irrigation judgements by monitoring the plants' activity along with the soil and weather conditions.

Plant-based sensors attached to a plant's fruit or stalks are still relatively new, but they are already showing great potential for monitoring the plant's water status. These sensors have the ability to detect and monitor minute changes, such as swelling or shrinking, and can notify farmers of yield-reducing behaviours or inadequate water content. In this sense, artificial intelligence (AI) in agriculture has the potential to be very helpful in determining field water levels and properly scheduling irrigation operations to maximize favourable outcomes (Saggi and Jain 2022).

12.1.5 Smart Irrigation Systems and Water Conservation

Smart irrigation technology is designed to enhance production efficiency by monitoring water levels, soil temperature, fertilizer content, and weather forecasts, hence reducing the need for extensive human labour. The operation of the irrigator pump is controlled by the microprocessor, which turns it on or off. The M2M, or Machine to Machine, technology is being created to facilitate interaction and transfer of information between various nodes inside the agricultural field, as well as with the server or cloud, using the primary network. The soil moisture sensor and rain drop sensor were employed to identify the requirement for water delivery to the fields. These sensors were directed over a wireless broadband network and supplied by solar panels. The raindrop sensor and soil moisture sensor provide the farmer with information about the soil's moisture content via SMS on their mobile phone using a module that supports GSM. The farmer may control the water supply by sending SMS orders to activate or deactivate it. Therefore, this technology has the capability to identify certain sections or regions in the fields that require additional water and can prevent the farmer from irrigating during rainfall (Navinkumar et al. 2021).

12.1.5.1 AI-Powered Drip Irrigation

Water is frequently wasted in traditional irrigation techniques because of over-irrigation or ineffective distribution. Here come AI-powered drip irrigation systems, in which several sources of data are analyzed by algorithms to determine each plant's exact water requirements. These systems maximize crop output, conserve resources, and optimize water use by supplying water straight to the roots based on current needs. This not only helps with sustainable farming methods but also tackles the pressing problem of water scarcity (Krishnan et al. 2022). These irrigation techniques assure equal distribution, lower evaporation and runoff, and increase irrigation efficiency by applying water directly to the plant's roots. Artificial intelligence (AI) algorithms can be used to precisely control how much water is applied to a given crop (Yousaf et al. 2023).

12.1.5.2 Soil Moisture Sensors

Soil moisture sensors are used at the core of smart irrigation. These ground-based sensors continuously assess the amount of moisture present at various depths. AI systems analyze this information to produce a sophisticated picture of the soil conditions. Consequently, farmers can adjust their irrigation tactics to make sure crops get the proper quantity of water at the right time. This degree of accuracy reduces the negative effects that excessive water use has on the environment while simultaneously optimizing water efficiency. The use of AI in agricultural water management is a complex strategy that transforms how farmers work with their land. AI enables the agricultural community to traverse the complexity of water management through proactive decision-making and precise data collecting, ultimately promoting resilient and sustainable growth in agriculture (Talaviya et al. 2020).

12.1.6 AI in Different Types of Irrigation Systems

Automation in agriculture is increasingly based on AI. Simple and routine chores like compiling field data, labelling and analyzing data, creating reports, and sending notifications can all be automated with its assistance. In addition to cost savings, smart irrigation reduces water use, minimizes human labour, unifies the understanding of soil properties, and improves long-term landscape health. These days, a variety of water irrigation systems heavily rely on AI systems and IoT sensors to accomplish these benefits.

12.1.6.1 Sprinkler Irrigation

In order to plan the next irrigation, an AI-based irrigation system gathers information from thermal and acoustic rain sensors that gauge the amount of rainfall. After this data is analyzed and computed, the system automatically notifies sprinklers to avoid overwatering or excessive water usage (Mohammed et al. 2023).

12.1.6.2 Centre Pivot Irrigation

The AI-based system that regulates circle irrigation sprinklers and modifies the stream or angle of water flow receives data insights from in-field sensors. This helps reach plants that are far from the water supply instead of overwatering the closest ones.

12.1.7 Key Areas for Policy Action

The utilization of artificial intelligence (AI), both in broad terms and specifically in the context of water supply, carries many repercussions for policy aimed at enhancing the efficiency of water systems and the calibre of service provision. Ethics and governance encompass the safeguarding of consumers' confidential and economic information, as well as the security of scientific and accounting information about water utilities. Regulation involves the process of comparing and evaluating performance, since unaccounted-for water (UFW) is a crucial factor in assessing the effectiveness of a water utility in decreasing both physical losses (such as leaks in the water system and pipe breaks) and business losses (such as illegal connections and metering problems). Two technical policies collaborate with branch ministries and water organizations to revise the national and water "Code of Practice" and guide water utilities in their digitized modernization. Financial policies focus on meeting immediate capital expenditure (CAPEX) needs for funding effective water utilities through the use of innovative funding mechanisms.

12.1.8 Advancing: Effective Partnerships Between Government and Private Sector

The functioning of AI is a remarkable phenomenon in its own right. The fundamental power of AI rests in its capacity to convert real time, accurate data into practical insights through the skill of forecasting and prediction. India aims to transform the handling of water via the utilization of AI optimization. This ambitious goal requires an effective alliance between the government and the commercial sector. The G20 and similar platforms have the potential to stimulate international cooperation. A notable instance of this partnership is the ANIC-ARISE initiative, which is a component of the Aatmanirbhar Bharat Abhiyan. This project facilitates collaborations among Indian start-ups and Micro, Small, and Medium Enterprises (MSMEs) to use the potential of AI in diverse industries, such as water management. Furthermore, the DST-Intel Collaborative Research on real-time river water and air quality monitoring serves as another notable illustration. The cooperation between the Department of Science and Technology (DST) and Intel seeks to utilize sophisticated technologies such as AI to enhance the effectiveness of the handling of water resources by enabling accurate and real-time monitoring of water quality, which is a crucial component. The journey progresses by creating strong data infrastructures, protecting data privacy, and fostering a group of highly qualified AI specialists, all driven by collaborative efforts.

12.2 Significance of Artificial Intelligence in Pest Management

Precise recognition of the insect pest paves the way for the proper decision-making on management of the pest. The conventional method of insect identification relies on observing morphological characters of the insect that takes notable time frame, biased and unsuccessful. Also, explicit taxonomic identification of pests can only be done by the taxonomic expertise, highly skilled farmer, and experienced field staffs. On the other hand, routine pest monitoring is a significant strategy in Integrated Pest Management programme. Pest monitoring involves constant watch kept on population dynamics of the pest to gather information update about pest density, associated natural enemies, and pest hotspots through regular field surveillance. Based on these monitoring data, appropriate management decision is chosen to combat the pest. Hence, pest recognition and pest surveillance data strengthen the future predictions of pest occurrence (Arnaudov et al. 2012). In recent years, AI is a hot topic dealt to solve a range of applied problems in smart farming. This emerging technology includes precise pest recognition, monitoring, and early warning and facilitates timely decision-making for better management (Fig. 12.4). The advantages of AI-based pest management are represented in Fig. 12.5.

12.2.1 Pest Detection, Identification, and Classification

Accurate identification of pest is a crucial element in Integrated Pest management strategy. However, pest recognition and classification can only be done with the help of taxonomic expertise and highly skilled personnel. To overcome conventional insect detection and classification techniques, AI-powered machine learning

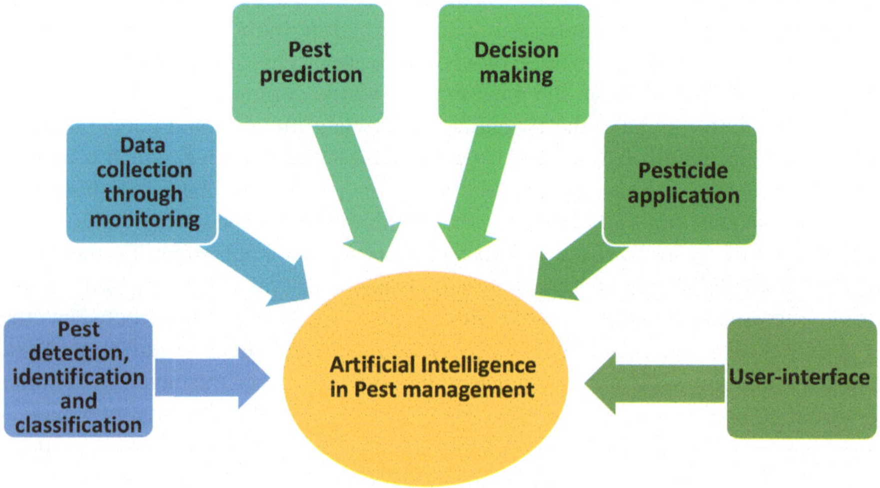

Fig. 12.4 Successful pest monitoring and management achieved through AI

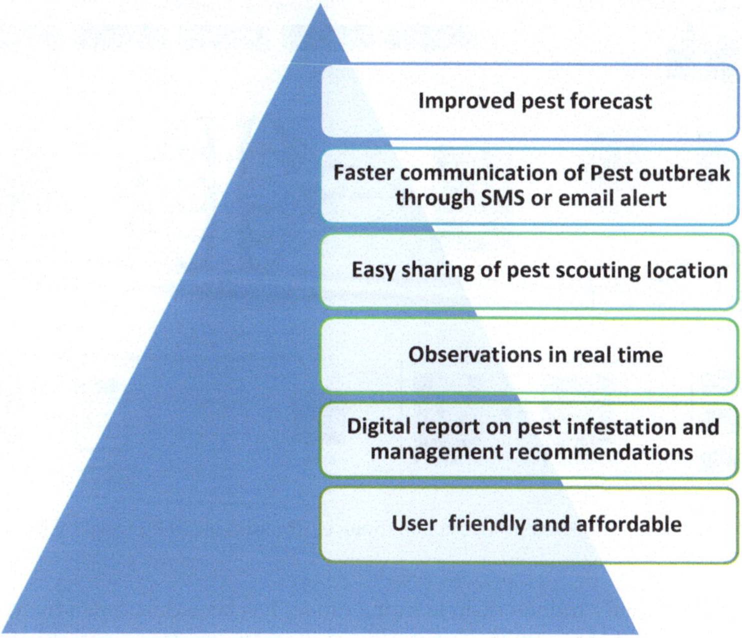

Fig. 12.5 Advantages of AI-based pest management

employs artificial neural networks (ANN), support vector machine (SVM), k-nearest neighbours (KNN), naive bayes (NB), and convolutional neural network (CNN) model for identifying and classifying the pests (Solanki et al. 2023; Jain and Chouhan 2014; Nihalani et al. 2024). Insect classification deploys shape features isolated from images of the insects. The initial RGB form of insect image is transformed to a greyscale image for feature extraction and those images are classified into various classes by applying classifier model. This way of image rescaling, preprocessing, and augmenting the pest dataset is done to enhance the accuracy of the images. CNN model is found to be the best in presenting insect classification with high accuracy amidst the practical difficulties such as shade, dirt, leaves, branches, twigs, and flower buds in field environment. Hence, AI utilizes massive datasets such as noises, photographs, and environmental data in a farmer's field to attain accuracy of detection (Kasinathan et al. 2021). Larger pest dataset can be acquired from the field through massive gathering of high-resolution images under certain environmental conditions. This dataset has to be marked with coordinates in a bounding box by using annotation tool like yolo mark. After this, the data is trained using deep learning models, validated and tested for its performance in real time under field condition (Fig. 12.6). Many researches have documented the performance metrics of deep learning models for pest detection, identification, and classification (Table 12.1). Adult cigarette beetles and warehouse beetle's images are shot by the camera, further processed by the computer, and detected by the deep learning model. Insect detection setup comprising a camera, a Jetson Nano

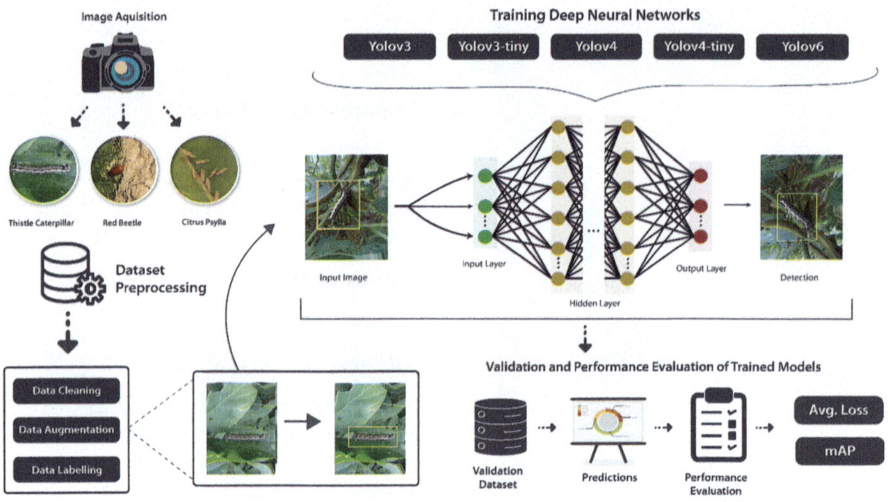

Fig. 12.6 Deep learning-based pest detection (Source: Khalid et al. 2023) (used by permission)

computer with highly trained deep learning model has been developed by Mendoza et al. (2023). Rimal et al. (2023) utilized Tensorflow lite model to classify 102 common agricultural pests online and offline. More and Nighot (2016) evolved a web-based search engine in the name of AgroSearch to detect and classify the pests of pomegranate. The input images are processed with multi-class support vector machine and the final results showed 84% accuracy. Ebrahimi et al. (2017), by integrating the SVM approach with image processing approaches, successfully classified the thrips in strawberry field. Furthermore, multiple task sparse representation and multiple-kernel learning (MKL) techniques was employed by Xie et al. (2015) to classify 24 common field pests. Kasinathan and Uyyala (2023) proposed Fall Armyworm (FAW), *Spodoptera frugiperda* (J.E. Smith), moth detection system based on Mask Region Convolutional Neural Network (Mask R-CNN). To train the Mask R-CNN model, the FAW insect dataset is pre-processed, enhanced, and annotated. The Mask R-CNN detection model's results were compared to those of RetinaNet, Faster R-CNN, Region CNN (R-CNN), and Single Shot Multi-box Detector (SSD). The results revealed that Mask R-CNN model with the Resnet-101 backbone had a mean average precision of 94.21%. Molina-Rotger et al. (2023) successfully detected olive fly through electronic trap by utilizing ML techniques such as Random Forest and Support Vector Machine algorithms. Integration of these techniques to detect olive fly yielded an accuracy of 89.1%. IoT technology uses many affordable sensor devices to collect real-time data on pest-related agricultural development characteristics. Ali et al. (2023) developed an AI-enabled real-time, IoT-based automatic pest detection system using sound pest analytics. They have collected 500 pest sounds and trained, validated, and evaluated them using the CNN, LSTM, Bi-LSTM, and CNN-Bi-LSTM network models. Among these, the

Table 12.1 Deep learning models employed for pest detection

Dataset used	Target pests	AI technique	Model	Performance metrics			Purpose	References
				Accuracy (%)	mAP (%)	Recall (%)		
IP102 dataset	102 different insect pests	CNN	YOLOv7	88.50	93.30	88.80	Pest detection	Nayar et al. (2022)
IP102 dataset	50 different insect pests	Transformer-like architecture	Vision Transformer (ViT)	92.8	–	–	Pest species recognition	Agarwal et al. (2023)
IP102 dataset	102 different insect pests	CNN	Efficient NetB4	95	95	95	Pest identification	Dewari et al. (2022)
Ortho images obtained by UAV	*Spodoptera frugiperda*	CNN	Pest Region-CNN (Pest R-CNN)	43.60	–	–	Pest monitoring	Du et al. (2022)
Zengcheng District Experimental database and Gaoyao District test base	*Cnaphalocrocis medinalis*, *Chilo suppressalis*	NCNN	Improved YOLOv5s	93.70	92.60	93.90	Pest detection	Deng et al. (2023)
Images collected from field	Thistle caterpillars (*Vanessa cardui*), red beetles (*Aulacophora foveicollis*), and citrus psylld (*Diaphorina citri*)	CNN	Yolov8	–	84	–	Pest detection	Khalid et al. (2023)
IP102 dataset	102 different insect pests	CNN	MobileNet	82.04	–	–	Pest recognition	Rimal et al. (2023)
Turkey-Plant Dataset	*Eriophyes erineus*, *Eriosomalanigerum*, *Aphis* spp.	CNN	SVM	97.56	–	–	Pest classification	Turkoglu et al. (2022)
Images collected from field	Banana corm weevil	DCNN	ResNet50 and InceptionV2	90.00	98	–	Pest detection	Selvaraj et al. (2019)
Maryland Biodiversity dataset	Brown marmorated stink bug, *Halyomorpha halys*	R-CNN	YOLOv5-m	95.3	99.20	98.40	Pest detection and classification	Sava et al. (2022)
Images collected from field	*Tessaratoma papillosa*	LSTM (long short-term memory)	YOLOv3	90.00	–	–	Pest identification, classification, and prediction	Chen et al. (2020)
Images collected from field	*Liriomyza sativae*	CNN	U-Net	91.58	87.840	87.59	Pest detection	Martins Crispi et al. (2023)

CNN-Bi-LSTM model achieved 98.91% accuracy, 96.8%sensitivity, 97.96% specificity, 98.54% recall, 98.63% precision, and 98.58% F1 score.

12.2.2 Pest Monitoring

A key component of Integrated Pest Management (IPM) systems is a precise estimation of pest populations. However, current practices necessitate a large amount of human effort to physically verify traps set in probable insect hotspots, leaving space for automated operations. Ciampi et al. (2023) developed a program using artificial intelligence (AI) to count whiteflies automatically from images of sticky chromotropic traps shot with digital cameras or smartphones. Chiwamba et al. (2019) constructed an improved FAW trap to track *S. frugiperda* by combining a Raspberry Pi 3 Model B+ micro-computer, Atmel 8-bit AVR microcontroller, 3G cellular modem, and multiple sensors operated with an off-grid solar photovoltaic system to gather real-time FAW moth images and agrometeorological data. The photos that are taken, along with the agrometeorological data, are transmitted to the cloud server. There, machine learning is used to instantaneously classify the images, identify and confirm the presence of FAW moth. Thus, status of real-time occurrence of FAW pests is indicated to the users through an intuitive web application platform.

Tannous et al. (2023) developed an automated pest monitoring method to detect two tephritid fruit flies, namely, Mediterranean fruit fly *Ceratitis capitata*, and the olive fruit fly, *Bactrocera oleae* by employing convolutional neural network (CNN), which showed 93% accuracy in detection in real time. Song et al. (2023) utilized image recognition and AI algorithm YOLOV4+SENet model to identify and count the number of pests. Thomas et al. (2023) deployed IR sensors and acoustic sensors to detect and monitor the presence of Berry Borer, Stem Borer, and Thrips by training the ANN model. Rustia and Lin (2017), by utilizing image processing and machine learning algorithm, successfully counted white flies and fruit flies with an accuracy of 98%. Qiao et al. (2008) developed a straightforward image processing method that determined the whitefly density on sticky traps automatically. Nagasaka et al. (2004) developed a "dog" robot for the purpose of cotton farming, with integrated controller area network (CAN) bus, laser system, and camera to scout fields. It visits the farms twice a week at a nominal cost of roughly $22/hectare and gathers data regarding field issues like pest infestations that are above economic threshold levels or diseases of plants symptoms that necessitate the placement of pesticides.

12.2.3 Pest Prediction and Decision-Making by Employing AI

By incorporating AI-based technology into the system of surveillance, real-time data may be used to boost farm efficiency and enable automated pest management. AI also helps in predicting the pest outbreak and its dispersal pattern. Growers can use a straightforward website or smartphone app to access system information.

Cutting-edge agricultural monitoring systems include tools for making decisions and visualizing data that are easy to use. Farmers may manage their agronomic operations, resources, and harvests with the use of charts, maps, and dashboards. Drones with sensors and high-resolution cameras might be used to survey pests and crops from the air. Multispectral or hyperspectral imaging can be used by farmers to detect pest infestations and water stress. Thanks to sophisticated agricultural monitoring technologies, farmers can check and modify farm data using their phones and tablets. Based on sensor data, smartphone apps alert and counsel farmers. Due to its significance, smart agriculture monitoring systems may incorporate weather and pest monitoring (Ali et al. 2023). The FL-based decision-making method was created by Karthik et al. (2020) to assess the detrimental effects of several pesticides on human health. In addition to providing weather forecasts, fuzzy expert models assist farmers in making timely and informed decisions and in taking preventative action to minimize losses due to insect pests.

Halbig et al. (2024) evolved a phenology model named, "PHENTHAUproc" to predict the occurrence of oak processionary moth (OPM), *Thaumetopoea processionea* in Europe. This Decision Support System helps in early warning of the pest by monitoring and hazard assessment. With the help of IR sensors and acoustic sensors, Thomas et al. (2023) detected, monitored the pest and thereby alerted the farmer's community through their mobile phones using Wi- Fi and IoT. By eliminating human labour and utilizing everyday image analysis, smart pest surveillance traps with sensors and cameras have transformed the accuracy of pest predictions (Marković et al. 2021). The hybrid AI-based strategy proved to be effective in forecasting environmental factors that contribute to insect outbreaks in sugarcane with sufficient accuracy. In order to promptly adopt an appropriate management strategy and lower the pest load, it therefore functions as an early warning tool (Figueredo et al. 2021). Crop monitoring is no longer a hardship thanks to automated robotic technology, which has improved pest monitoring and prediction efficiency and decreased unwarranted application of dangerous chemicals.

12.2.4 Pesticide Application Using AI

One technique for minimizing pesticide application losses and eliminating chemical residues in the soil is precision spraying. When combined with machine learning, optical remote sensors are widely used for targeted pesticide application, protecting beneficial arthropods that are not intended targets (Kirkeby et al. 2021). AI-based smart machines are currently being developed to minimize pesticide amounts, lower usage, and achieve outstanding crop coverage all at the same time, to minimize the adverse impacts of pesticides on applicators, customers, and the environment. Although UAV, Drones and Self-propelled sprayers used for pesticide spraying has the advantage of minimizing the negative effects of pesticides on farmers and spraying a vast area quickly (Kanase et al. 2018), these sprayers would still not be a viable choice. Its main drawback is that it sprays pesticide at a height that causes an excessive amount of chemical to disperse into the air and unevenly cover non-target areas, resulting in pollution of surrounding region (Martin et al. 2019). To overcome these ill effects, targeted pesticide spraying with mobile robot is utilized. Meshram et al. (2022) developed an autonomous robot for pesticide spraying by training suitable navigation and trajectory planning algorithm for short distance movement within plants. This aided the robots to spray the pesticide precisely over the crops and avoid wastage of pesticides.

12.3 AI in Agriculture: Opportunities

Despite the obstacles that prevented artificial intelligence from being widely adopted in the agriculture industry, AI-driven projects have recently been welcomed by the sector at an accelerating rate. The widespread integration of AI into agricultural techniques is primarily motivated by four factors:

- *Global Food Demand*: In order to meet the need for more food, agriculture businesses need to implement technologies that increase productivity in environmentally favourable and sustainable ways. Precision farming should be widely used as it is the most efficient way to increase yield, reduce expenses, and leave the least possible environmental impact.
- •*Availability and Affordability of Infrastructures for Data Collection*: The amount of data collected in the agriculture sector has increased dramatically during the last 20 years. Twenty years ago, farmers and producers could not have imagined having access to the massive amounts of granular and comprehensive data that they have access to today.
- *Competitiveness and Increasing Cost of Labour and Materials*: Data-driven farming methods increase yield while cutting costs. Early on in its development, technology is expensive and used in very limited ways, but as it gains traction, it advances exponentially. This tendency is especially true for small-scale farming, where the substantial yield gap caused by inadequate technology could be reduced by innovative innovations.
- *International/Governmental Initiatives for AI in Agriculture*: Data-driven agriculture initiatives are receiving attention from an increasing number of governments, non-profits, NGOs, and international organizations. Their initiatives take the shape of grants and funds, facilities, instruction, and—above all—open data and infrastructures for data collection.

12.4 AI in Agriculture: Challenges

Any significant effort to use AI in agriculture must assess the real-world obstacles that have prevented this sector from changing as quickly as certain other industries. The upstream agriculture sector faces four primary obstacles to the adoption of suitable AI-based solutions:

- *Data Collection*: Despite the fact that precision agriculture has grown over the last three decades, two major issues have prevented machine learning predictions from being widely used in the field. First, a sizable percentage of growers and farmers have not embraced precision-Ag′s utility. Second, utilizing precision-Ag technologies does not include gathering all relevant data via IoT or other channels in order to make predictions.
- *Data Ownership and Privacy*: Farmers, greenhouse managers, breeders, agronomists, and other producers are the owners and users of data, but that is not the whole picture. Owners have access to certain vital data, but unless they pay for access and receive training, they don't have instant access to many other sources of pertinent data, such as satellite imaging or historical environmental data for the area.
- *Agricultural Data Has Been Scattered*: Data gathering among farms is not centralized, even in cases where extensive data collection is made and a farmer (or advisor) has access to all pertinent information from the farm. This implies that a body of information or insight obtained by AI-based methods does not fully develop.
- *Cost of Change Towards AI in Agriculture, Cost of Mistakes*: In the case of the farming industry, this high cost of change is especially evident. Compared to certain other industries that welcomed artificial intelligence with open arms, this cost structure is significantly different.

12.5 Conclusions

Presently, artificial intelligence has grown into a demanding tool because of its rigorous learning skills, which are needed to address the issues in the agricultural area. AI guarantees not just progressive enhancements, but also ensures effectiveness, environmental sustainability, endurance, and dependability in water delivery. Entering the next phase of water governance involves the utilization of highly sophisticated algorithms and real-time data analysis to protect India's future water profile. The use of artificial intelligence (AI) in India's water sector signifies a significant and revolutionary advancement, effectively complementing government endeavours such as the Jal Jeevan Mission and the achievement of Sustainable Development Goal 6 (SDG 6). The innovative advancements in artificial intelligence pave the way for a future in which water is distributed effectively, sustainably, and fairly, guaranteeing universal access to pure water for all individuals. Nevertheless, to completely achieve this goal, partnership remains the crucial element. India can accelerate its progress towards water security by establishing collaborations with foreign bodies, business sectors, and research institutes. This cooperative strategy not only ensures everyone access to clean water, but also aligns with the wider global objective of attaining water-related sustainability goals—a significant tribute to India's dedication to a more promising and water-protected future. Like many technological advances, making an AI-based product practical and approachable may be difficult. Smartphones are still out of reach for many farmers. Especially in outlying rural regions, connectivity is a persistent problem. Risks of bias and lack of transparency are inherent in using AI-based recommendation systems to make important decisions in agriculture, especially when those suggestions go against established norms and practices. In other ways, AI helps in automated insect pest monitoring accurate pest identification and classification, aids in pest prediction and timely optimal pesticide application in the field. Subsequent studies may concentrate on increasing the performance of AI through more useful datasets with a variety of field locations from different agro-ecological zones. As the majority of farmers are not acquainted with AI applications, they are unable to completely understand the fundamental ideas behind the various AL algorithms. To tackle these problems, user-friendly programmes ought to be released. Further investigation could be carried out to enhance the efficacy and precision of pest management techniques in agriculture by incorporating the suggested pest detection model into precision spraying technology.

References

Agarwal N, Kalita T, Dubey AK, Om S, Dogra A (2023) An improved deep learning model implementation for pest species detection. In: International conference on artificial intelligence: towards sustainable intelligence. Cham, Springer Nature Switzerland, pp 119–131

Ali MA, Dhanaraj RK, Nayyar A (2023) A high performance-oriented AI-enabled IoT-based pest detection system using sound analytics in large agricultural field. Microprocess Microsyst 103:104946. https://doi.org/10.1016/j.micpro.2023.104946

Arnaudov V, Raykov S, Davidova R, Hristov H, Vasilev V, Petkov P (2012) Monitoring of pest populations-an important element of integrated pest management of field crops. Agric Sci Technol 4(1):77–80

Awais M, Naqvi SM, Zhang H, Li L, Zhang W, Awwad FA, Ismail EA, Khan MI, Raghavan V, Hu J (2023) AI and machine learning for soil analysis: an assessment of sustainable agricultural practices. Bioresour Bioprocess 10(1):90. https://doi.org/10.1186/s40643-023-00710-y

Balaska V, Adamidou Z, Vryzas Z, Gasteratos A (2023) Sustainable crop protection via robotics and artificial intelligence solutions. Mach Des 11(8):774. https://doi.org/10.3390/machines11080774

Chen CJ, Huang YY, Li YS, Chang CY, Huang YM (2020) An AIoT based smart agricultural system for pests detection. IEEE Access 8:180750–180761. https://doi.org/10.1109/ACCESS.2020.3024891

Chiwamba SH, Phiri J, Nkunika PO, Sikasote C, Kabemba MM, Moonga MN (2019) Automated fall armyworm (*Spodoptera frugiperda*) pheromone trap based on machine learning. J Comput Sci 15(12):1759–1779. https://doi.org/10.3844/jcssp.2019.1759.1779

Chouhan SS, Singh UP, Jain S (2024) Introduction to Computer Vision and Drone Technology. In: Chouhan SS, Singh UP, Jain S (eds) Applications of computer vision and drone technology in agriculture 4.0. Springer, Singapore. https://doi.org/10.1007/978-981-99-8684-2_1

Ciampi L, Zeni V, Incrocci L, Canale A, Benelli G, Falchi F, Amato G, Chessa S (2023) A deep learning-based pipeline for whitefly pest abundance estimation on chromotropic sticky traps. Eco Inform 78:102384. https://doi.org/10.1016/j.ecoinf.2023.102384

Deng J, Yang C, Huang K, Lei L, Ye J, Zeng W, Zhang J, Lan Y, Zhang Y (2023) Deep-learning-based rice disease and insect pest detection on a mobile phone. Agronomy 13(8):2139. https://doi.org/10.3390/agronomy13082139

Dewari S, Gupta M, Kumar R (2022) Agricultural insect pest's recognition system using deep learning model. In: Congress on intelligent systems. Springer Nature Singapore, Singapore, pp 287–299. https://doi.org/10.1007/978-981-19-9225-4_22

Du L, Sun Y, Chen S, Feng J, Zhao Y, Yan Z, Zhang X, Bian Y (2022) A novel object detection model based on faster R-CNN for spodoptera frugiperda according to feeding trace of corn leaves. Agriculture 12(2):248. https://doi.org/10.3390/agriculture12020248

Ebrahimi MA, Khoshtaghaza MH, Minaei S, Jamshidi B (2017) Vision-based pest detection based on SVM classification method. Comput Electron Agric 137:52–58. https://doi.org/10.1016/j.compag.2017.03.016

Figueredo L, Villa-Murillo A, Colmenarez Y, Vásquez C (2021) A hybrid artificial intelligence model for *Aeneolamia varia* (Hemiptera: Cercopidae) populations in sugarcane crops. J Insect Sci 21(2):11. https://doi.org/10.1093/jisesa/ieab017

Halbig P, Stelzer AS, Baier P, Pennerstorfer J, Delb H, Schopf A (2024) PHENTHAUproc—an early warning and decision support system for hazard assessment and control of oak processionary moth (*Thaumetopoea processionea*). For Ecol Manage 552:121525. https://doi.org/10.1016/j.foreco.2023.121525

Jain S, Chouhan SS (2014) Cyclic redundancy codes: study and implementation. Int J Emerg Technol Adv Eng:213–217

Kanase S, Patwegar S, Patil P, Pore A, Kadam Y (2018) Agriculture Drone sprayer. Int J Recent Trends Eng Res 4(3):181–185

Karthik S, Dash SK, Punithavelan N (2020) A fuzzy decision-making system for the impact of pesticides applied in agricultural fields on human health. Int J Fuzzy Syst Appl 9(3):42–62. https://doi.org/10.4018/IJFSA.2020070103

Kasinathan T, Uyyala SR (2023) Detection of fall armyworm (*Spodoptera frugiperda*) in field crops based on mask R-CNN. Signal Image Video Process:1–7. https://doi.org/10.1007/s11760-023-02485-3

Kasinathan T, Singaraju D, Uyyala SR (2021) Insect classification and detection in field crops using modern machine learning techniques. Inform Process Agric 8(3):446–457. https://doi.org/10.1016/j.inpa.2020.09.006

Khalid S, Oqaibi HM, Aqib M, Hafeez Y (2023) Small pests detection in field crops using deep learning object detection. Sustain For 15(8):6815. https://doi.org/10.3390/su15086815

Kirkeby C, Rydhmer K, Cook SM, Strand A, Torrance MT, Swain JL, Prangsma J, Johnen A, Jensen M, Brydegaard M, Græsbøll K (2021) Advances in automatic identification of flying insects using optical sensors and machine learning. Sci Rep 11(1):1555. https://doi.org/10.1038/s41598-021-81005-0

Krishnan SR, Nallakaruppan MK, Chengoden R, Koppu S, Iyapparaja M, Sadhasivam J, Sethuraman S (2022) Smart water resource management using Artificial Intelligence—a review. Sustain For 14(20):13384. https://doi.org/10.3390/su142013384

Kumar R, Yadav S, Kumar M, Kumar J, Kumar M (2020) Artificial intelligence: new technology to improve Indian agriculture. Int J Chem Stud 8(2):2999–3005. https://doi.org/10.22271/chemi.2020.v8.i2at.9208

Marković D, Vujičić D, Tanasković S, Đorđević B, Ranđić S, Stamenković Z (2021) Prediction of pest insect appearance using sensors and machine learning. Sensors 21(14):4846. https://doi.org/10.22271/chemi.2020.v8.i2at.9208

Martin DE, Woldt WE, Latheef MA (2019) Effect of application height and ground speed on spray pattern and droplet spectra from remotely piloted aerial application systems. Drones 3(4):83. https://doi.org/10.3390/drones3040083

Martins Crispi G, Valente DS, Queiroz DM, Momin A, Fernandes-Filho EI, Picanço MC (2023) Using deep neural networks to evaluate leafminer fly attacks on tomato plants. Agri Eng 5(1):273–286. https://doi.org/10.3390/agriengineering5010018

Mashhadi N, Shahrour I, Attoue N, El Khattabi J, Aljer A (2021) Use of machine learning for leak detection and localization in water distribution systems. Smart Cities 4(4):1293–1315. https://doi.org/10.3390/smartcities4040069

Mendoza QA, Pordesimo L, Neilsen M, Armstrong P, Campbell J, Mendoza PT (2023) Application of machine learning for insect monitoring in grain facilities. AI 4(1):348–60. https://doi.org/10.3390/ai4010017

Meshram AT, Vanalkar AV, Kalambe KB, Badar AM (2022) Pesticide spraying robot for precision agriculture: a categorical literature review and future trends. J Field Robot 39(2):153–171. https://doi.org/10.1002/rob.22043

Mohammed M, Hamdoun H, Sagheer A (2023) Toward sustainable farming: implementing artificial intelligence to predict optimum water and energy requirements for sensor-based micro irrigation systems powered by Solar PV. Agronomy 13(4):1081. https://doi.org/10.3390/agronomy13041081

Molina-Rotger M, Morán A, Miranda MA, Alorda-Ladaria B (2023) Remote fruit fly detection using computer vision and machine learning-based electronic trap. Front Plant Sci 14:1241576. https://doi.org/10.3389/fpls.2023.1241576

More S, Nighot M (2016) AgroSearch: a web based search tool for pomegranate diseases and pests detection using image processing. In: Proceedings of the second international conference on information and communication technology for competitive strategies, pp 1–6

Nagasaka Y, Zhang Q, Grift TE, Knetani Y, Umeda N, Kokuryu T (2004) Control system design for an autonomous field watching-dog robot. In: International conference on automation technology for off-road equipment, ATOE 2004, pp 298–304

Navinkumar TM, Kumar RR, Gokila PV (2021) Application of artificial intelligence techniques in irrigation and crop health management for crop yield enhancement. Mater Today: Proc 45:2248–2253. https://doi.org/10.1016/j.matpr.2020.10.227

Nayar P, Chhibber S, Dubey AK (2022) Implementation of YOLOv7 for pest detection. In: International conference on applied machine learning and data analytics. Springer Nature Switzerland, Cham, pp 156–165. https://doi.org/10.1007/978-3-031-34222-6_13

Nihalani R et al (2024) Long Short-Term Memory (LSTM) model for indian sign language recognition. J Intell Fuzzy Syst:1–19. https://doi.org/10.3233/JIFS-233250

Qiao M, Lim J, Ji CW, Chung BK, Kim HY, Uhm KB, Myung CS, Cho J, Chon TS (2008) Density estimation of Bemisia tabaci (Hemiptera: Aleyrodidae) in a greenhouse using sticky traps in conjunction with an image processing system. J Asia Pac Entomol 11(1):25–29. https://doi.org/10.1016/j.aspen.2008.03.002

Rimal K, Shah KB, Jha AK (2023) Advanced multi-class deep learning convolution neural network approach for insect pest classification using TensorFlow. Int J Environ Sci Technol 20(4):4003–4016. https://doi.org/10.1007/s13762-022-04277-7

Rustia DJ, Lin TT (2017) An IoT-based wireless imaging and sensor node system for remote greenhouse pest monitoring. Chem Eng Trans 58:601–606. https://doi.org/10.3303/CET1758101

Saggi MK, Jain S (2022) A survey towards decision support system on smart irrigation scheduling using machine learning approaches. Arch Comput Methods Eng 29(6):4455–4478. https://doi.org/10.1007/s11831-022-09746-3

Sava A, Ichim L, Popescu D (2022) Detection of Halyomorpha halys using neural networks. In: 2022 8th International Conference on Control, Decision and Information Technologies (CoDIT), vol 1, pp 437–442

IPPC Secretariat (2021) Scientific review of the impact of climate change on plant pests—a global challenge to prevent and mitigate plant pest risks in agriculture, forestry and ecosystems. FAO, Rome

Selvaraj MG, Vergara A, Ruiz H, Safari N, Elayabalan S, Ocimati W, Blomme G (2019) AI-powered banana diseases and pest detection. Plant Methods 15:1–11. https://doi.org/10.1186/s13007-019-0475-z

Sharma T, Singh J, Singh A, Chauhan G (2021) Artificial intelligence in water management. RASSA J Sci Soc 3(3):186–189

Sharma S, Verma K, Hardaha P (2023) Implementation of artificial intelligence in agriculture. J Comput Cogn Eng 2(2):155–162. https://doi.org/10.47852/bonviewJCCE2202174

Sinwar D, Dhaka VS, Sharma MK, Rani G (2020) AI-based yield prediction and smart irrigation. Internet Things Anal Agric 2:155–180. https://doi.org/10.1007/978-981-15-0663-5_8

Solanki S, Singh UP, Chouhan SS (2024) A systematic analysis of magnetic resonance images and deep learning methods used for diagnosis of brain tumor. Multimed Tools Appl 83:23929–23966. https://doi.org/10.1007/s11042-023-16430-6

Solanki S, Singh UP, Chouhan SS (2023) Brain tumor classification using ML and DL approaches. In: 2023 IEEE 5th International Conference on Cybernetics, Cognition and Machine Learning Applications (ICCCMLA). IEEE, pp 204–208. https://doi.org/10.1109/ICCCMLA58983.2023.10346854

Song CC, Chen WZ, Chen HY, Chen YK (2023) Development of a pest automatic diagnosis system for intelligent agriculture using image recognition. In: 2023 International Technical Conference on Circuits/Systems, Computers, and Communications (ITC-CSCC). IEEE, pp 1–5

Talaviya T, Shah D, Patel N, Yagnik H, Shah M (2020) Implementation of artificial intelligence in agriculture for optimization of irrigation and application of pesticides and herbicides. Artif Intell Agric 4:58–73. https://doi.org/10.1016/j.aiia.2020.04.002

Tannous M, Stefanini C, Romano D (2023) A deep-learning-based detection approach for the identification of insect species of economic importance. Insects 14(2):148. https://doi.org/10.3390/insects14020148

Thomas JC, Manikandarajan S, Subha TK (2023) AI based pest detection and alert system for farmers using IoT. In: E3S web of conferences, vol 387, p 05003

Turkoglu M, Yanikoğlu B, Hanbay D (2022) PlantDiseaseNet: convolutional neural network ensemble for plant disease and pest detection. SIViP 16(2):301–309. https://doi.org/10.1007/s11760-021-01909-2

United Nations (2019) World Population Prospects 2019: Highlights (ST/ESA/SER. A/423). Department of Economic and Social Affairs. Available at: https://www.un.org/development/desa/publications/world-population-prospects-2019-highlights (accessed: 05 January 2023)

Vathsala AV, Lakshmi HN, Yedukondalu G, Rao CR, Kotha M, Changala R (2023) Optimization of Irrigation and Herbicides Using Artificial Intelligence in Agriculture. Int J Intell Syst Appl Eng 11(3):503–518

Vijayakumar S, Gobinath R, Kumar SA, Bandeppa S, Gompa SJ, Basavaraj K, Prakasam V, Manasa V (2022) Artificial intelligence (AI) and its application in agriculture. Chronicle Bioresource Manage 6:025–031

Xie C, Zhang J, Li R, Li J, Hong P, Xia J, Chen P (2015) Automatic classification for field crop insects via multiple-task sparse representation and multiple-kernel learning. Computers and Electronics in Agriculture 1(119):123–32. https://doi.org/10.1016/j.compag.2015.10.015

Yousaf A, Kayvanfar V, Mazzoni A, Elomri A (2023) Artificial intelligence-based decision support systems in smart agriculture: Bibliometric analysis for operational insights and future directions. Front Sustain Food Syst 6:1053921. https://doi.org/10.3389/fsufs.2022.1053921

Zouizza M, Lachgar M, Zouani Y, Hrimech H, Kartit A (2023) AIDSII: an AI-based digital system for intelligent irrigation. Softw Impacts 17:100574. https://doi.org/10.1016/j.simpa.2023.100574

Advancement and Challenges of Implementing Artificial Intelligence of Things in Precision Agriculture

13

Shashank Shekhar, Maheshwar Durgam, Suyog Balasaheb Khose, Chwadaka Pohshna, and Dattatray G. Bhalekar

Abstract

Artificial Intelligence (AI) integration has ushered in a transformative era in agriculture known as Smart Agriculture. This chapter comprehensively explores the AI concept within the context of smart agriculture. It conducts a systematic literature review, revealing a growing trend in AI-related publications and highlighting its applications and advantages. The evolution of agriculture from traditional methods to the current Agriculture 4.0 era is outlined, showcasing the pivotal role of advanced technologies such as Wireless Sensor Networks (WSN), Internet of Things (IoT), AI, etc. The ongoing revolution enables intelligent agriculture systems with real-time decision-making capabilities, automation, and data-driven approaches, addressing critical challenges like climate change, water scarcity, and food security. A detailed examination of the technological integration in smart agriculture emphasizes the role of machine learning, deep learning, WSN, and robotics. These technologies facilitate data collection, analysis, decision-making, and the optimization of various aspects of farming, such as soil and water management, weed control, disease and pest management, crop prediction, harvesting, weather forecasting, and supply chain management. Despite the promising advancements, this chapter identifies challenges in smart agriculture, ranging from data quality issues and network connectivity challenges to

S. Shekhar · M. Durgam
Krishi Vigyan Kendra Ankushpur Ghazipur, Acharya Narendra Deva University of Agriculture and Technology, Kumarganj, Ayodhya, Uttar Pradesh, India

S. B. Khose · C. Pohshna (✉)
Agricultural and Food Engineering Department, Indian Institute of Technology Kharagpur, Kharagpur, West Bengal, India

D. G. Bhalekar
Center for Precision and Automated Agricultural Systems, Department of Biological Systems Engineering, Washington State University, Prosser, WA, USA

S. S. Chouhan et al. (eds.), *Artificial Intelligence Techniques in Smart Agriculture*, https://doi.org/10.1007/978-981-97-5878-4_13

storage and processing complexities. Furthermore, the chapter explores the cybersecurity challenges within the smart farming ecosystem, categorizing potential cyber-attacks into data attacks, networking and equipment attacks, supply chain attacks, compliance and regulation challenges, cyber terrorism, and cloud computing attacks. The discussion underscores the need for a holistic approach to cybersecurity involving stakeholders, adherence to best practices, and continuous innovation to ensure the sustainability and security of smart agriculture.

Keywords

Artificial intelligence · Smart agriculture · Agricultural intelligence · Challenges · Security

13.1 Introduction

In recent decades, the global population has experienced exponential growth, leading to an unprecedented demand for food resources. Agriculture, one of the oldest human activities, remains indispensable for human survival, providing essential commodities such as food, oils, and clothing (Ahmed et al. 2018). The urgent need to address the escalating demand for food has become a global imperative, and it was documented that the food demand needs to be increased by 70% by 2050 (Shekhar et al. 2018, 2021a, b; Shekhar 2022; Khose et al. 2022; Chouhan et al. 2024). Achieving this ambitious goal requires a collective effort on a global scale, involving advancements in traditional farming practices, dietary changes, and reductions in food wastage (Godfray and Garnett 2014; Shekhar et al. 2021c, 2022, 2023a, b).

Despite innovative approaches utilizing available croplands to meet the growing demand for crops, challenges persist (Godfray et al. 2010). Many farmers still practice traditional farming methods, which may affect crop health, soil fertility, water management, and grain yield (Somov et al. 2018; Shekhar et al. 2020; Verma et al. 2023a, b, c, d). Farmers often resort to trial-and-error methods with fertilizer, irrigation, pesticides, and crop variety, which results in suboptimal yields and unhealthy production. The agriculture sector is susceptible to climate changes, pests, weeds, and diseases, collectively impacting global crop yields. The emergence of Artificial Intelligence (AI) in agriculture marks a transformative force, with implications across various sectors, including mechanical, medical, and computer sciences (Kumar et al. 2023; Solanki et al. 2023a, b, 2024). In contrast to traditional methods, AI provides a more accurate, efficient, and safer means of performing tasks without direct human intervention. The application of AI in agriculture encompasses a spectrum of technologies, including computer vision, machine learning, robotics, intelligent systems, and the Internet of Things (Mojumdar and Chakraborty 2020; Puranik et al. 2019; Abrishambaf et al. 2019; Li et al. 2020; Khose et al. 2023).

These AI-driven technologies have proven instrumental in aiding farmers in various tasks, such as analyzing land suitability, water management, fertilizer management, site-specific chemical application (Bhalekar et al. 2023), and selecting the most suitable crops for specific regions to maximize yields. Water management in agriculture, coupled with AI applications, enables accurate evapotranspiration estimation on a daily, monthly, or need basis, enhancing the effectiveness of irrigation practices (Waleed et al. 2020). AI also facilitates better measurement and prediction of daily dew point temperatures, recognizing expected weather patterns and forecasting evapotranspiration. Similarly, integrating AI into smart farming management systems allows farmers to optimize other parameters, such as chemical input, based on location, time, and safety considerations, contributing to improved food quality and reduced waste. Farmers are increasingly adopting AI models that leverage sensors and robots to monitor crop conditions, collecting valuable data to enhance their understanding of crops, potential diseases, and the genetic basis of farming practices. AI-driven machines with data-driven insights unveil hidden traits that significantly impact crop quality.

However, the widespread adoption of AI in agriculture is not without challenges. Spatial data, crucial for decision-making and the operation of precision machinery, directly impacts crop variability. Using advanced devices and sensors on farming vehicles and machinery introduces challenges related to network infrastructure, uncertainty in environmental conditions, have problems with parts like actuators and sensors and weather changes (Zude-Sasse et al. 2016; Bali et al. 2023). These limitations compel AI systems to navigate infrequent data and manage increased processing complexity and decision-making intricacies. Moreover, the agricultural sector faced significant economic repercussions due to the COVID-19 pandemic, grappling with challenges such as pest attacks and bacterial infections that led to widespread crop diseases (Deutsch et al. 2018; FAO (Food and Agriculture Organization) 2024; Saxena et al. 2024). Traditional approaches demonstrated limited effectiveness, prompting a paradigm shift toward adopting Industry 4.0 (IR4.0) technology to revitalize and transform the agricultural industry. Post-pandemic, smart agriculture emerges as a compelling solution to address labor shortages and ensure an uninterrupted food supply chain (Javaid et al. 2022; Pant et al. 2021). The vital technological paradigms such as AI and the Internet of Things (IoT) played a pivotal role during the pandemic and are recognized as the most influential set of technologies for the twenty-first century (Khan et al. 2022; Said et al. 2021; Zhang et al. 2021; Chen et al. 2021; Khose et al. 2023).

In the realm of agriculture, the integration of AI presents both limitations and promising opportunities. While AI holds immense potential to revolutionize farming practices, several challenges impede its widespread adoption. Spatial data complexities pose a significant hurdle, impacting decision-making processes and the operation of precision machinery. Additionally, the infrastructure required to support AI applications faces uncertainties related to network connectivity and environmental factors, further complicating implementation. Moreover, the processing complexity inherent in handling infrequent data and navigating decision-making intricacies presents challenges for AI systems. Economic repercussions, such as those experienced during the COVID-19 pandemic, exacerbate existing limitations, leading to pest attacks, crop diseases, and labor shortages. Despite these challenges,

AI offers compelling opportunities for the agricultural sector. Precision agriculture stands out as a key opportunity, enabling farmers to make informed decisions and optimize operational efficiency. Post-pandemic, AI solutions present a viable strategy for addressing labor shortages and ensuring the resilience of the food supply chain. Furthermore, the paradigm shift toward Industry 4.0 technologies, including AI and the IoT, holds immense promise for transforming agricultural practices. Through synergistic integration, AI and IoT technologies offer opportunities for enhancing operational efficiency, robust risk management, and scalability in agriculture. In navigating these limitations and embracing opportunities, the agricultural industry stands to benefit significantly from the transformative potential of AI.

The symbiotic relationship between AI and IoT extends to the application, implementation, infrastructure, and management of IoT technology, enhancing operational efficiency, providing robust risk management, and augmenting scalability (Bhat and Huang 2021; Fraga-Lamas et al. 2021). This study delves into the AI concept in smart agriculture, offering a focused exploration of the intersection of AI and IoT with an overview of their adoption. The systematic literature review sheds light on the growing trend in article publications related to AI across various applications, emphasizing the evolving landscape of AI and IoT (Zhang et al. 2021). Furthermore, this book chapter explores the potential of AI and IoT technologies in smart agriculture, highlighting their promising applications and challenges for adoption. It aims to contribute to the broader discourse on these transformative technologies.

13.2 Traditional Agriculture to Smart Agriculture

Agriculture plays a crucial role in economic and social stability, and its evolution is illustrated in Fig. 13.1, depicting the schematic view of agricultural evolution from the past centuries to the future (Javaid et al. 2022). Distinct practices and challenges characterize each era. Agriculture 1.0 (1784–1870 BC) featured traditional farming methods employing indigenous tools and animals, demanding substantial human resources. The subsequent era, Agriculture 2.0, marked the mechanized phase with the introduction of gas and oil, replacing steam energy. The twentieth century saw the advent of assembly lines for mass production, enhancing efficiency but grappling with challenges related to low resource utilization. Agriculture 3.0 witnessed the integration of communication and software engineering technologies, introducing automation to make precision agriculture more sustainable. However, concerns persisted regarding intelligence levels. The current era, Agriculture 4.0 (2017 onward), embraces advanced technologies like WSN, IoT, AI, big data, and blockchain technology. This ongoing revolution enables intelligent agriculture systems with real-time decision-making capabilities, leveraging high automation and data-driven approaches to enhance production, supply chain management, and food security. Despite these advancements, security challenges pose a significant bottleneck. Smart agriculture in Revolution 4.0 encompasses three development modes (Yang et al. 2021): Precision agriculture, Facility agriculture, and Order agriculture.

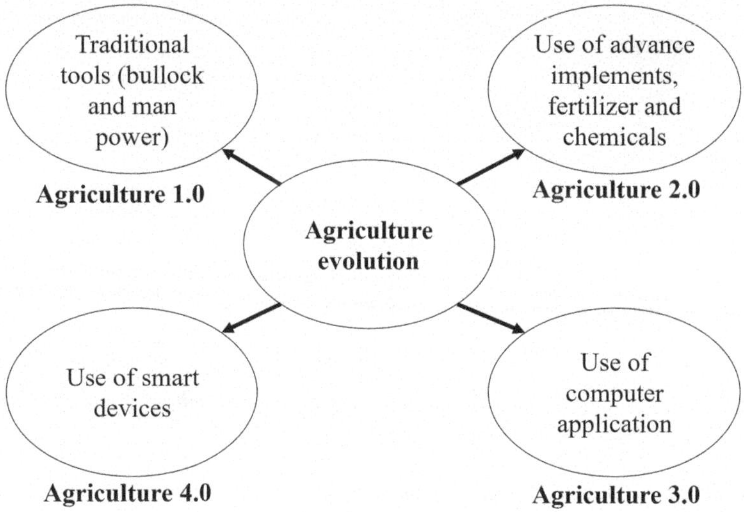

Fig. 13.1 Schematic view of agriculture evolutions

The adoption of smart agriculture empowers farmers to make informed decisions, optimize resource usage (fertilizer, water, pesticides, etc.), reduce costs, and enhance crop grain yields. Furthermore, it addresses critical challenges such as climate change, water scarcity, and food security within the agricultural industry.

13.2.1 Artificial Intelligence of Things (AIoT)

The Artificial Intelligence of Things (AIoT) paradigm results from the rapid developments of 5G infrastructures, AI, and IoT sensors into reliable technologies (Zhang et al. 2021). Even though AIoT is still in its infancy, the trends and applications it covers are rapidly changing the landscape with huge commercial possibilities. The AIoT utilizes field data for informed decision-making, and farm management systems incorporate various data acquisition methods, including IoT sensors, remote sensing, and historical farm records. The collected data spans soil conditions, disease and pest occurrences, historical yields, and weather information obtained through IoT sensors. This data is stored in local and cloud-based systems and undergoes analysis using specific algorithms to train models. These processes are repeated iteratively throughout the harvesting period to sustain farms with minimal adverse effects and optimize overall production (Mkrttchian 2021). Figure 13.2 illustrates the workflow of an AI-based smart agriculture system. The decision system employs the processed information to execute advanced machinery commands or provide recommendations. The model yields valuable insights and facilitates the processing of real-time data.

In smart agriculture, data handling, involving data acquisition and storage, is pivotal. For data acquisition, a range of internal and external devices, including

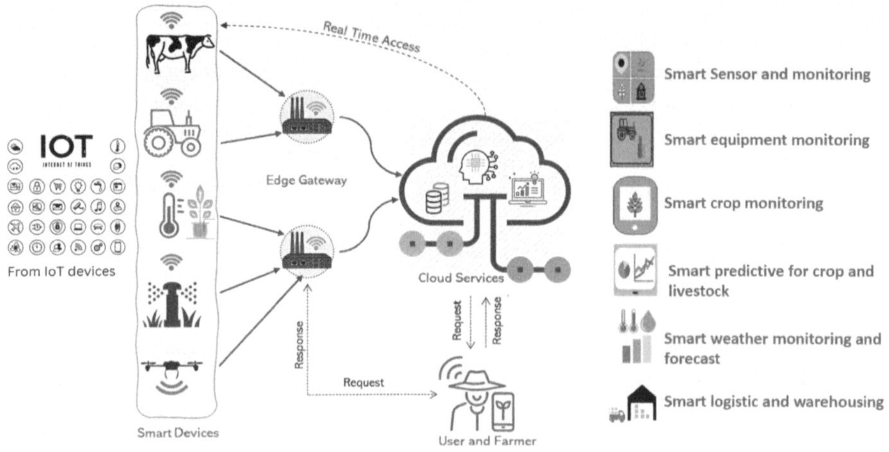

Fig. 13.2 A model of a smart agriculture system

sensors, actuators, network resources, and historical data storage tools, are employed to gather comprehensive information. Sensors play a crucial role by sensing the environment and transmitting corresponding values to storage devices. These sensors, categorized by functionality, encompass location sensors providing crop location details, electrochemical sensors detecting soil chemical reactions, light-dependent resistors (LDR sensors) assessing light conditions, and nanostructured sensors analyzing soil humidity (Reynolds et al. 2019). This diverse set of sensors collects information on soil, plant ecosystems, location, and more. Satellites contribute to data generation through remote sensing technology, aiding in monitoring ecological conditions, estimating areas, and tracking growth (Goetz and Dubayah 2011).

Smart agriculture leverages historical data related to factors such as crop yield, weather conditions, and evapotranspiration to gain valuable insights. The data is transmitted from various sources and securely stored for future use. Storage environments can adopt either a local or cloud-based approach. Local storage involves direct connections to the system without centralized management, offering data access privileges solely to the host system. On the other hand, cloud-based storage is a network-driven system, which facilitates distributed data storage, ensuring scalability for seamless access.

On data analysis and implementation, the surge in agricultural data volume has spurred the application of AI to analyze data, extract valuable features, and enhance overall productivity. Data analysis involves leveraging appropriate methods to extract useful insights into the data, fostering informed decision-making and recommendations. This process, along with associated actions, encompasses autonomous robotics, monitoring systems, and predictive analysis. Agricultural robots, integrating technologies such as computer vision and AI, prove instrumental in tasks like seed sowing, spraying, and harvesting (Cheng et al. 2023). For instance, a smart weeder system utilizing computer vision can differentiate between weeds and crops, effectively handling weeds without causing crop loss (Xie et al. 2020). Farm-picking robots utilize computer vision to classify ready-to-harvest fruits and vegetables, determine their location, and efficiently collect them in a box (Kokane et al. 2022). Monitoring systems employ specific sensors for continuous farm monitoring, with data analysis conducted by a processing unit leading to decision-making outputs, actions, and recommendations. Predictive analysis employs machine learning models to predict crop yield and forecast the impact of weather changes and other factors on farming (Crane-Droesch 2018)

13.3 Technological Integration in Smart Agriculture

The integration of AI in agriculture brings efficiency and intelligence to farming through technologies like machine learning, deep learning, WSN, and robotics. These innovations enable data collection from sensors, remote sensing, and unmanned aerial surveys, offering farmers insights for better decision-making. Machine learning and deep learning algorithms analyze data for tasks such as regression, classification, and association analysis, contributing to crop health predictions and issue detection (Khose et al. 2022). Digital image processing enhances disease control and reduces farmers' workload.

Sensors and robotics are essential in smart farm management, monitoring soil, irrigation, weeds, and diseases, while unmanned vehicles aid in pest detection and crop yield estimation. IoT and WSN are crucial in livestock monitoring, precision farming, weather tracking, and supply chain management, reducing human intervention. Cloud computing, big data analytics, and edge computing contribute to storing, processing, and visualizing large datasets and optimizing agricultural practices. Machine learning and deep learning applications, including random forest and deep neural networks, offer precise predictions and real-time insights (Nihalani et al. 2024). Decision-making systems like Agriculture Decision Support Systems (DSS) assist farmers in making informed decisions while considering complex data. Ongoing research focuses on refining these technologies for sustainable and profitable smart agriculture.

13.4 Application of Smart Agriculture

The section presents a detailed analysis of various application domains that are part of smart agriculture. It highlights the numerous opportunities and innovations that smart agriculture presents. The potential of smart agriculture is enormous, and it has the potential to revolutionize traditional farming processes. As illustrated in Fig. 13.2, the farming workflow follows a specific sequence of steps that occur at specific time intervals.

13.4.1 Soil Management

The physical and chemical properties of soil are a crucial factor in agriculture, directly impacting the yield capacity of crops. Deviations from desired soil quality standards can act as bottlenecks in plant growth. Various methods for regularly assessing soil quality have been developed, and AI can play a significant role in investigating soil nutrient deficits and potential flaws (Shaikh et al. 2022; Folorunso et al. 2023; Jain and Sethia 2023). AI-based applications can analyze crop growth patterns in agriculture, helping farmers understand different nutrient values and soil imperfections.

Zhang et al. (2017) developed a real-time IoT-based system for monitoring moisture and nutrient levels in a citrus crop. The system comprises four layers: the transmission layer (responsible for data transmission), the perception layer (responsible for data acquisition), the information layer (providing insights from the data), and the application layer. This model expanded the detection range and improved accuracy. In a study on soil management, Chen et al. (2021) demonstrated the adoption of Decision Tree (DT), Support Vector Machine (SVM), and Artificial Neural Network (ANN) techniques. These techniques assist in adjusting soil parameters, leading to positive agricultural growth. However, the authors discussed certain barriers related to data analysis, identifying them as the primary bottlenecks for improving agricultural systems.

13.4.2 Water Management

The agricultural sector, being water-intensive, requires the minimization of water loss from farming (Shekhar et al. 2018, 2020; Shekhar 2022; Khose et al. 2022; Chawla and Balasaheb 2023). Integrating modern technologies into agriculture has significantly enhanced the efficiency compared to traditional approaches. Various techniques, such as decision support systems, embedded systems, IoT-based systems, and other approaches, have been crucial in advancing irrigation systems.

Aggarwal and Kumar (2019) proposed a system that collects data on soil moisture, humidity, pH level, and temperature using specific sensors. The ANN model then analyzes the collected data to generate highly accurate results. Subsequently, based on the obtained results, the system triggers and controls the flow of water to the required portion of the farm. In another study, Mohanraj et al. (2017) employed WSNs to automate an irrigation system. The model incorporated real-time monitoring of weather conditions and soil, aiding farmers in reducing overall irrigation costs and improving farming practices intelligently. The proposed system demonstrated successful operation for both surface and subsurface drip irrigation. Khose and Mailapalli (Khose et al. 2023) use Unmanned Aerial Vehicle (UAV)-based multispectral image analytics and machine learning for surface soil moisture estimation.

13.4.3 Weed Management

Unwanted plants, known as weeds, that grow spontaneously in specific environments, pose a significant challenge to agricultural systems by competing for water, light, and essential nutrients. This competition impacts a plant's microclimate and leads to crop yield degradation (Nojavan 2001). Traditional methods of weed control require substantial effort and resources. However, advancements in knowledge and technology have revolutionized weed management, enabling faster and more extensive control with minimal resources through the integration of AI.

Alam et al. (2020) introduced a real-time vision-based sprayer system, and it was deployed in a real environment to perform weed detection tasks and calculate plant pixel area. The processed results were then transmitted to the microcontroller, which controlled the on/off switch of the sprayer, achieving an impressive 95% prediction accuracy. Brilhador et al. (2019) proposed a model employing a deep neural network for real-time weed detection, achieving an accuracy of 83.44%. The researchers also investigated the impact of data augmentation on pixel-wise classification of weeds and crops. Tejeda and Castro (2019) presented an algorithm for successful weed detection in crops utilizing image processing. The algorithm demonstrated high prediction accuracy in classifying weeds in real-time environments.

13.4.4 Diseases and Pest Management

The spread of crop diseases, whether due to nutrient deficiency or pests, poses a significant challenge to food production and quality (Chakraborty and Newton 2011; Gupta et al. 2017). Identifying the cause of such problems can be challenging for farmers lacking expertise. In this context, AI can detect diseases and nutrient deficiencies at early stages, aid in disease prevention, suggest precise pesticide quantities (Liu et al. 2016), and control other factors contributing to plant degradation. Singh (2018) introduced a cost-effective, automated, and user-friendly solution that empowers farmers to diagnose plant diseases accurately without needing a human expert. The proposed model achieves a 95% detection accuracy for diseases in four groups (bacteria, mildew, phoma, and red rust). Bahtiar et al. (2020) proposed a model employing the region-based Convolutional Neural Network (R-CNN) architecture for detecting three types of mineral deficiencies in chili plants.

13.4.5 Crop Yield Prediction

Accurate prediction of crop yield is instrumental in empowering farmers to make critical decisions related to crop harvesting, providing valuable insights into the overall damage caused by climate change, assisting in optimal crop selection, determining the best planting times, and enhancing overall crop production (Priyanka et al. 2018). Crop yield predictions also play a pivotal role in influencing decisions regarding storage space requirements, addressing potential food shortages, pricing strategies, and managing import and export activities. To make farming smarter, there is a growing need for predictive models that can deliver accurate and timely yield predictions using minimal resources (Gupta and Nahar 2023). Factors affecting crop yield predictions include climate conditions, soil quality, seed varieties, and fertilizer usage (Xu et al. 2019).

Medar et al. (2019) introduced a crop selection method designed to help farmers achieve maximum yield. The proposed method involved predicting crop yield rates in three phases: dataset management, testing, and analysis. The Naive Bayes algorithm demonstrated the highest prediction accuracy, achieving 91.11%. Oikonomidis et al. (2023) explored various deep-learning approaches for crop prediction, identifying convolutional neural networks as the most prevalent and effective algorithm. Challenges identified included the lack of access to large and accurate datasets, the need for improved parameter tuning, and other obstacles that future research must address. Ikram et al. (2022) developed a Smart Crop Selection (SCS) system using IoT and ML algorithms for the selection of crops for maximal yield. The developed model is trained for 11 different crop predictions, with the highest prediction accuracy of 97–98%.

13.4.6 Harvesting

Harvesting crops post-ripening is essential but time-consuming, labor-intensive, and prone to errors. The harvesting frequency varies based on the crop type, with some requiring a single cycle and others necessitating multiple cycles, either periodically or daily. The timing of harvesting is crucial, as both late and early harvesting can significantly impact the overall crop yield (Abbas et al. 2019). Furthermore, this process requires substantial resources. In response to these challenges, agricultural experts advocate for integrating robotics into agriculture to introduce flexibility and efficiency into harvesting. Over recent decades, advancements in robotic technology have led to more accurate and automated harvesting methods. Researchers focus on refining techniques to measure fruits' shape, color, and size, often utilizing locally available facilities. Specialized tools distinguish characteristics among targeted fruits (Gongal et al. 2015; Bhargava and Bansal 2021).

De-An et al. (2011) developed a robot for apple harvesting in challenging environments. The equipment included an image manipulator module, an end-effector with a spoon-like shape and a pneumatic gripper for apple harvesting. The developed robot demonstrated autonomous harvesting with a success rate of 77%, achieving an average harvesting time of 15 s. Li et al. (2020) designed a litchi harvesting robot with a camera and picking arm. The system featured a 6-Degree of Freedom (DoF) liftable manipulator, utilizing Deep Lab V3 image segmentation. Tested on real crops, the robot achieved an impressive accuracy ratio of 83.33%.

13.4.7 Weather Forecast

Deep learning proves pivotal in anticipating meteorological events, aiding farmers with precise weather forecasts for optimal crop management (Mohan and Patil 2018; El Hachimi et al. 2021). Trained on diverse data sources, deep learning algorithms enable rainfall predictions, guide improved irrigation schedules, and mitigate water-related risks. These algorithms also forecast catastrophic weather events, influencing strategic decisions in agriculture. Long-term weather pattern predictions empower farmers to plan crop choices, resource management, and adaptation strategies. An optimized GRU neural network model, integrating field-transportable rain forecasting equipment and cloud image prediction, notably demonstrates accurate precipitation forecasting. A deep Q-learning irrigation decision-making system utilizes short-term weather forecasts, minimizing risks associated with unpredictable weather conditions (Chen et al. 2021; Vivas et al. 2023; Ravi and Sai 2022; Ambildhuke and Banik 2022; Fahad et al. 2023).

13.4.8 Supply Chain Management

The agri-food industry consists of multiple transactional steps, each with distinct conditions, forming a supply chain that encompasses food processing, transportation, food storage, and distribution of information. Managing and maintaining the vast amount of data associated with these steps can be challenging and tedious if done manually. Therefore, integrating smart technologies, including smart contracts, can streamline supply chain management.

Awan et al. (2021) introduced a framework that utilized front-end and base technologies for smart manufacturing processes. Front-end technologies involve interactions between manufacturing, smart products, supply chain management, and working processes. In contrast, base technologies encompass IoT and cloud services. Xiong et al. (2020) demonstrated that blockchain technology can establish a trustworthy relationship between consumers and producers. This data-driven approach incorporates smart contracts, enabling stakeholders to execute timely payments recorded in the blockchain database.

13.5 Challenges in Smart Agriculture

Smart agriculture, integrating advanced technologies such as AI, faces challenges across various domains, each impacting its successful implementation. Figure 13.3 shows the roadmap of challenges in the smart agriculture system. This section delves into the complexities surrounding data quality, network connectivity, storage and processing, data analysis, decision-making, and general issues, shedding light on potential hindrances and proposing solutions for a robust smart agriculture framework.

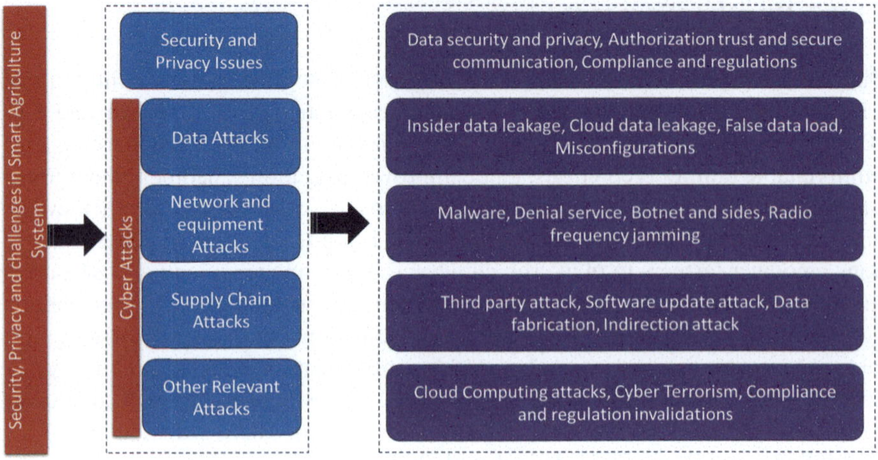

Fig. 13.3 A roadmap of cybersecurity, privacy, and challenge in smart agriculture systems

13.5.1 Data Level Challenges

The effectiveness of smart agriculture heavily relies on high-quality data from diverse sources. Issues such as data privacy, integrity, and standardization pose significant challenges. Precision agriculture employs sensors and GPS for data collection, but modifications or malfunctions can compromise data quality, impacting farm management and crop production. Future considerations should enhance data quality dimensions to address this, ensuring trustworthiness and effectiveness in smart agriculture systems (Maddikunta et al. 2021; Dhanya et al. 2022).

13.5.2 Network Layer Challenges

Wireless connectivity is fundamental in smart agriculture, providing flexibility, cost-effectiveness, and scalability. However, challenges arise from network size, interference, low latency, and transmission range (Avşar and Mowla 2022). The network layer's crucial role in signal transmission, especially in large areas, demands optimization for data confidentiality. Selecting appropriate wireless communication technologies and addressing issues like latency and interference contribute to efficient smart agriculture systems (Hu et al. 2019).

13.5.3 Storage and Processing Challenges

Devices generating satellite imagery and sensor data in agriculture present challenges in integrating diverse data from different sources. Data interoperability is crucial, but security and privacy concerns limit cloud computing adoption.

Transitioning between cloud service providers and meeting security standards poses additional constraints. Overcoming these challenges requires careful consideration of smart agriculture data integration and security measures (Cravero et al. 2022; Sharma et al. 2022).

13.5.4 Data Analysis Challenges

Deep learning and machine learning are effective for data analysis in agriculture, and challenges include high processing power demands and data quality requirements (Bhat and Huang 2021). Suggestions include adopting cloud-based platforms for flexible infrastructure and load balancing. Concerns about the diversity of algorithms for data analysis, such as SVM, RF, fuzzy logic, and Decision Trees, add complexity to choosing the most suitable one. Addressing these challenges involves leveraging cloud computing and ensuring data quality for efficient data-driven solutions (Cravero et al. 2022).

13.5.5 Decision-Making and Recommendation Challenges

Transforming data into actionable insights is crucial for enhancing decision-making in smart agriculture. Real-time processing challenges include accessibility, scalability, interoperability, and uncertainty. Improving accessibility through graphical user interfaces and addressing scalability concerns are essential. Interoperability ensures the integration of heterogeneous knowledge, while managing uncertainty in decision and recommendation systems requires addressing variable factors like climate change and environmental conditions (Chu et al. 2016; Zhai et al. 2020).

13.5.6 General Issues Challenges

Implementing tasks like irrigation and pest control involves specific hardware, contributing to setup and processing costs. Quality sensors are essential but costly, posing challenges in agricultural monitoring. Unclear data ownership and farmers' educational levels are significant constraints. Establishing corporate standards, laws, and initiatives to incentivize farmers, addressing economic conditions, and enhancing education on advanced technologies are crucial for overcoming these challenges (Yang et al. 2021; Nicolas et al. 2022). Smart agriculture requires collaboration among researchers, policymakers, and industry stakeholders. We can create a sustainable and efficient ecosystem by implementing advanced technologies and robust strategies.

13.6 Cybersecurity Challenges in the Smart Farming Ecosystem

Smart farming, integrating IoT and advanced technologies, brings unprecedented benefits to agriculture. However, this technological evolution also introduces a myriad of cybersecurity challenges. Figure 13.3 shows the roadmap of cybersecurity and privacy in smart agriculture systems. This section explores potential cyberattacks categorized into data attacks, networking and equipment attacks, supply chain attacks, compliance and regulation challenges, cyber terrorism, and cloud computing attacks within the smart farming ecosystem.

13.6.1 Data Attacks

Farmers are concerned about insider threats, where disgruntled employees may leak confidential data, affecting commodity markets. Instances, such as a former engineer accessing valuable data at Allen and Hoshall, highlight the vulnerability to intentional data leaks (Shirazi et al. 2017; Gupta and Nahar 2023). Recent research by Kaspersky (ITPro 2024) reveals that a notable portion of cyber incidents can be attributed to employees bypassing their organization's security protocols. Sensitive smart farming data stored in cloud centers globally poses risks. Legislation like China's cybersecurity law requires domestic data storage. Data stored abroad may face less stringent security, leading to potential data leaks. Laws and regulations are crucial in addressing such challenges (Yi et al. 2015).

Attackers may inject false data to influence real-time decisions, disrupting crucial processes. For example, altering soil moisture data could lead to overwatering and crop damage. Ensuring the integrity of real-time data is vital for preventing such attacks. Threats to data integrity involve attackers releasing false information about diseases or issues in crops or livestock. Disseminating false reports mimicking authentic sources poses challenges in verifying and rectifying the misinformation. This demands robust mechanisms to ensure data accuracy.

13.6.2 Networking and Equipment Attacks

Smart farming equipment relying on radio frequency communication, including GNSS for precision agriculture, is susceptible to jamming. Attackers may disrupt GNSS using low-power jammers, impacting path planning, auto steering, and other critical processes. Malware injection is prevalent, especially with the widespread adoption of precision agriculture. Malicious software can steal information, recruit devices into botnets, and hinder the functions of smart equipment, affecting crop harvests and farm areas.

With the increasing connectivity of IoT devices, smart farms have become potential targets for forming botnets. Compromised farm IoT devices controlled by a central system can be used to infect other networks, making smart farms vulnerable to cybercriminal activities. Side channel attacks exploit information gained from system implementation rather than inherent weaknesses. In the context of smart farming, attacks exploiting timing channels, power consumption patterns, and electromagnetic leaks pose risks, emphasizing the need for comprehensive security measures (Ukil et al. 2011).

13.6.3 Supply Chain Attacks

The interconnected supply chain in agriculture introduces vulnerabilities at various stages. Companies providing critical machinery can disable connected equipment, manipulate nutrient levels in fertilizers, and cause substantial damage to crops. Uploading false information related to calibration can affect critical decisions, potentially leading to suboptimal agricultural practices. Compromised water treatment facilities used for irrigation can lead to widespread crop contamination. Attacks on smart grids may cause blackouts, spoiling stored produce in large storage houses. Livestock-related attacks can spread false information about disease outbreaks or unauthorized genetic modifications. Manipulating sensors in livestock buildings can harm animals and affect livestock feed, potentially reaching humans (Halachmi et al. 2019).

13.7 Conclusions

Smart agriculture powered by AI offers an innovative solution to meet the growing global demand for food. This chapter examines various applications of AI technologies, including soil and water management, crop prediction, disease detection, and supply chain optimization. The study proposes a comprehensive and efficient approach to farm management that utilizes real-time data to make informed decisions. However, the widespread adoption of smart agriculture faces challenges related to data quality, network connectivity, storage, and processing complexities that must be addressed to ensure the seamless operation of AI-driven systems. Moreover, cybersecurity challenges, including data attacks, networking vulnerabilities, supply chain risks, and compliance issues, highlight the need for robust security measures to safeguard the integrity and functionality of smart farming ecosystems. To overcome these challenges, interdisciplinary collaboration among researchers, policymakers, and industry stakeholders is crucial. Implementing strategies to enhance data quality, optimize network connectivity, and strengthen cybersecurity measures will contribute to developing a sustainable and efficient smart agriculture ecosystem. As technology advances, continuous innovation and the integration of best practices will play a pivotal role in realizing the full potential of AI and IoT in shaping the future of agriculture.

References

Abbas G, Younis H, Naz S, Fatima Z, Hussain S, Ahmed M, Ahmad S (2019) Effect of planting dates on agronomic crop production. Agronomic crops. Prod Technol 1:131–147

Abrishambaf O, Faria P, Vale Z (2019) Energy resource scheduling in an agriculture system using a decision tree approach. In: 2019 20th International Conference on Intelligent System Application to Power Systems (ISAP), vol 1–5. IEEE

Aggarwal S, Kumar A (2019) A smart irrigation system to automate irrigation process using IOT and artificial neural network. In: 2019 2nd International Conference on Signal Processing and Communication (ICSPC). IEEE, pp 310–314

Ahmed N, De D, Hussain I (2018) Internet of Things (IoT) for smart precision agriculture and farming in rural areas. IEEE Internet Things J 5(6):4890–4899

Alam M, Alam MS, Roman M, Tufail M, Khan MU, Khan MT (2020) Real-time machine-learning based crop/weed detection and classification for variable-rate spraying in precision agriculture. In: 2020 7th International Conference on Electrical and Electronics Engineering (ICEEE). IEEE, pp 273–280

Ambildhuke G, Banik BG (2022) IoT based Portable weather station for irrigation management using real-time parameters. IJACSA 13(5)

Avşar E, Mowla MN (2022) Wireless communication protocols in smart agriculture: a review on applications, challenges and future trends. Ad Hoc Netw 136:102982

Awan S, Ahmed S, Ullah F, Nawaz A, Khan A, Uddin MI et al (2021) IoT with blockchain: a futuristic approach in agriculture and food supply chain. Wirel Commun Mob Comput 2021:1–17

Bahtiar AR, Pranowo AJ, Santoso AJ, Juhariah J (2020) Deep learning detected nutrient deficiency in chili plant. In: 2020 8th international conference on information and communication technology (ICoICT). IEEE, pp 1–4

Bali A, Chouhan SS, Kumar G, Kumar R, Singh UP (2023) Adaptive fault-tolerant control for pure-feedback stochastic nonlinear systems with sensor and actuator faults. Circ Syst Signal Process 42(10):5838–5867

Bhalekar DG, Parray RA, Mani I, Kushwaha H, Khura TK, Sarkar SK et al (2023) Ultrasonic sensor-based automatic control volume sprayer for pesticides and growth regulators application in vineyards. Smart Agric Technol 4:100232

Bhargava A, Bansal A (2021) Fruits and vegetables quality evaluation using computer vision: a review. J King Saud Univ Comput Inf Sci 33(3):243–257

Bhat SA, Huang NF (2021) Big data and AI revolution in precision agriculture: survey and challenges. IEEE Access 9:110209–110222

Brilhador A, Gutoski M, Hattori LT, de Souza IA, Lazzaretti AE, Lopes HS (2019) Classification of weeds and crops at the pixel-level using convolutional neural networks and data augmentation. In: 2019 IEEE Latin American Conference on Computational Intelligence (LA-CCI). IEEE, pp 1–6

Chakraborty S, Newton AC (2011) Climate change, plant diseases and food security: an overview. Plant Pathol 60(1):2–14

Chawla R, Balasaheb KS (2023) Optimizing water use efficiency and yield of wheat crops through integrated irrigation and nitrogen management: a comprehensive review. Int J Environ Clim Change 13(11):4059–4067

Chen M, Cui Y, Wang X, Xie H, Liu F, Luo T et al (2021) A reinforcement learning approach to irrigation decision-making for rice using weather forecasts. Agric Water Manag 250:106838

Cheng C, Fu J, Su H, Ren L (2023) Recent advancements in agriculture robots: benefits and challenges. Machines 11(1):48

Chouhan SS, Singh UP, Jain S (2024) Introduction to computer vision and Drone technology. In: Chouhan SS, Singh UP, Jain S (eds) Applications of computer vision and Drone technology in agriculture 4.0. Springer, Singapore. https://doi.org/10.1007/978-981-99-8684-2_1

Chu H, Yang Y, Li Q, Xu Y, Wei H (2016) A scalable clinical intelligent decision support system. Inclusive smart cities and digital health. In: Proceedings of the 14th International Conference on Smart Homes and Health Telematics, ICOST 2016, Wuhan, China, 25–27 May, 2016. Springer International Publishing, pp 159–165

Crane-Droesch A (2018) Machine learning methods for crop yield prediction and climate change impact assessment in agriculture. Environ Res Lett 13(11):114003

Cravero A, Pardo S, Sepúlveda S, Muñoz L (2022) Challenges to use machine learning in agricultural big data: A systematic literature review. Agronomy 12:748

De-An Z, Jidong L, Wei J, Ying Z, Yu C (2011) Design and control of an apple harvesting robot. Biosyst Eng 110(2):112–122

Deutsch CA, Tewksbury JJ, Tigchelaar M, Battisti DS, Merrill SC, Huey RB, Naylor RL (2018) Increase in crop losses to insect pests in a warming climate. Science 361(6405):916–919

Dhanya VG, Subeesh A, Kushwaha NL, Vishwakarma DK, Nagesh Kumar TN, Ritika G, Singh AN (2022) Deep learning based computer vision approaches for smart agricultural applications. Artif Intell Agric 6:211–229

El Hachimi C, Belaqziz S, Khabba S, Chehbouni A (2021) Towards precision agriculture in Morocco: a machine learning approach for recommending crops and forecasting weather. In: 2021 International Conference on Digital Age & Technological Advances for Sustainable Development (ICDATA). IEEE, pp 88–95

Fahad S, Su F, Khan SU, Naeem MR, Wei K (2023) Implementing a novel deep learning technique for rainfall forecasting via climatic variables: an approach via hierarchical clustering analysis. Sci Total Environ 854:158760

FAO (Food and Agriculture Organization) (2024) Food and Agriculture Organization of the United Nations FAOSTAT pesticides use. https://www.fao.org/faostat/en/#data/RP/visualize (Accessed 10 Jan 2024)

Folorunso O, Ojo O, Busari M, Adebayo M, Joshua A, Folorunso D et al (2023) Exploring machine learning models for soil nutrient properties prediction: a systematic review. Big Data Cogn Comput 7(2):113

Fraga-Lamas P, Lopes SI, Fernández-Caramés TM (2021) Green IoT and edge AI as key technological enablers for a sustainable digital transition towards a smart circular economy: an industry 5.0 use case. Sensors (Basel) 21(17):5745

Godfray HCJ, Garnett T (2014) Food security and sustainable intensification. Philos Trans R Soc Lond B Biol Sc 369(1639):20120273

Godfray HCJ, Beddington JR, Crute IR, Haddad L, Lawrence D, Muir JF et al (2010) Food security: the challenge of feeding 9 billion people. Science 327(5967):812–818

Goetz S, Dubayah R (2011) Advances in remote sensing technology and implications for measuring and monitoring forest carbon stocks and change. Carbon Manag 2(3):231–244

Gongal A, Amatya S, Karkee M, Zhang Q, Lewis K (2015) Sensors and systems for fruit detection and localization: a review. Comput Electron Agric 116:8–19

Gupta A, Nahar P (2023) Classification and yield prediction in smart agriculture system using IoT. J Ambient Intell Hum Comput 14(8):10235–10244

Gupta N, Debnath S, Sharma S, Sharma P, Purohit J (2017) Role of nutrients in controlling the plant diseases in sustainable agriculture. Agriculturally important microbes for sustainable agriculture. Appl Crop Prod Prot 2:217–262

Halachmi I, Guarino M, Bewley J, Pastell M (2019) Smart animal agriculture: application of real-time sensors to improve animal well-being and production. Annu Rev Anim Biosci 7:403–425

Hu Z, Xu L, Cao L, Liu S, Luo Z, Wang J et al (2019) Application of non-orthogonal multiple access in wireless sensor networks for smart agriculture. IEEE Access 7:87582–87592

Ikram A, Aslam W, Aziz RHH, Noor F, Mallah GA, Ikram S et al (2022) Crop yield maximization using an IoT-based smart decision. J Sens 2022:1–15

ITPro (2024) One quarter of all data breaches due to employees swerving security policies. https://www.itpro.com/security/one-quarter-of-all-data-breaches-due-to-employees-swerving-security-policies. Accessed 02 Feb 2024

Jain S, Sethia D (2023) A review on applications of artificial intelligence for identifying soil nutrients. In: International conference on agriculture-centric computation. Cham, Springer Nature Switzerland, pp 71–86

Javaid M, Haleem A, Singh RP, Suman R (2022) Enhancing smart farming through the applications of Agriculture 4.0 technologies. Int J Intell Netw 3:150–164

Khan JI, Khan J, Ali F, Ullah F, Bacha J, Lee S (2022) Artificial intelligence and internet of things (AI-IoT) technologies in response to COVID-19 pandemic: a systematic review. IEEE Access 10:62613–62660

Khose SB, Mailapalli DR, Biswal S, Chatterjee C (2022) UAV-based multispectral image analytics for generating crop coefficient maps for rice. Arab J Geosci 15(22):1681

Khose SB, Dhokale KB, Shekhar S (2023) The role of precision farming in sustainable agriculture: advancements and impacts

Kokane SB, Kalamnurikar SS, Khose SB (2022) Robotics in food processing industries: a review. Trends Food Sci Technol 134:188–200

Kumar R, Singh UP, Bali A, Chouhan SS, Tiwari AK (2023) Adaptive control of unknown fuzzy disturbance-based uncertain nonlinear systems: application to hypersonic flight dynamics. J Anal 6:1–20

Li J, Tang Y, Zou X, Lin G, Wang H (2020) Detection of fruit-bearing branches and localization of litchi clusters for vision-based harvesting robots. IEEE Access 8:117746–117758

Liu T, Chen W, Wu W, Sun C, Guo W, Zhu X (2016) Detection of aphids in wheat fields using a computer vision technique. Biosyst Eng 141:82–93

Maddikunta PKR, Hakak S, Alazab M, Bhattacharya S, Gadekallu TR, Khan WZ, Pham QV (2021) Unmanned aerial vehicles in smart agriculture: Applications, requirements, and challenges. IEEE Sensors J 21(16):17608–17619

Medar R, Rajpurohit VS, Shweta S (2019) Crop yield prediction using machine learning techniques. In: 2019 IEEE 5th international conference for convergence in technology (I2CT). IEEE, pp 1–5

Mkrttchian V (2021) Artificial and natural intelligence techniques as IoP-and IoT-based technologies for sustainable farming and smart agriculture. In: Artificial intelligence and IoT-based technologies for sustainable farming and smart agriculture IGI Global, pp 40–53

Mohan P, Patil KK (2018) Deep learning based weighted SOM to forecast weather and crop prediction for agriculture application. Int J Intell Eng Syst 11:167–176

Mohanraj I, Gokul V, Ezhilarasie R, Umamakeswari A (2017) Intelligent drip irrigation and fertigation using wireless sensor networks. In: 2017 IEEE Technological Innovations in ICT for Agriculture and Rural Development (TIAR). IEEE, pp 36–41

Mojumdar MU, Chakraborty NR (2020) A computer vision technique to detect scab on malabar nightshade. In: 2020 11th international conference on computing, communication and networking technologies (ICCCNT). IEEE, pp 1–4

Nicolas C, Naila B, Amar RC (2022) TinyML smart sensor for energy saving in internet of things precision agriculture platform. In: 2022 Thirteenth International Conference on Ubiquitous and Future Networks (ICUFN). IEEE, pp 256–259

Nihalani R et al (2024) Long Short-Term Memory (LSTM) model for Indian sign language recognition. J Intell Fuzzy Syst:1–19. https://doi.org/10.3233/JIFS-233250

Nojavan M (2001) Principles of weed control. University of Urmia Press, Urmia

Oikonomidis A, Catal C, Kassahun A (2023) Deep learning for crop yield prediction: a systematic literature review. New Zealand J Crop Hortic Sci 51(1):1–26

Pant J, Pant RP, Singh MK, Singh DP, Pant H (2021) Analysis of agricultural crop yield prediction using statistical techniques of machine learning. Mater Today: Proc 46:10922–10926

Priyanka T, Soni P, Malathy C (2018) Agricultural crop yield prediction using artificial intelligence and satellite imagery. Eurasian J Anal Chem 13(7):6–12

Puranik V, Ranjan A, Kumari A (2019) Automation in agriculture and IoT. In: 2019 4th International Conference on Internet of Things: Smart Innovation and Usages (IoT-SIU). IEEE, pp 1–6

Ravi SS, Sai RGK (2022) Design of Deep Learning Model for Predicting Rainfall. In Proceedings of the 2022 8th International Conference on Advanced Computing and Communication Systems (ICACCS), Coimbatore, India, 25–26 March 1:1343–1347

Reynolds D, Ball J, Bauer A, Davey R, Griffiths S, Zhou J (2019) CropSight: a scalable and open-source information management system for distributed plant phenotyping and IoT-based crop management. Gigascience 8(3):giz009

Said S, Ihab R, Hesham O, Abou Tabl I, Maged N, Youssef S, Elagamy M (2021) AIOT-arch: furthering artificial intelligence in big data IoT applications. In: IOP conference series: materials science and engineering, vol 1051, No. 1. IOP Publishing, p 012008

Saxena A, Chouhan SS, Aziz RM, Agarwal V (2024) A comprehensive evaluation of Marine predator chaotic algorithm for feature selection of COVID-19. Evol Syst 4:1–4

Shaikh TA, Rasool T, Lone FR (2022) Towards leveraging the role of machine learning and artificial intelligence in precision agriculture and smart farming. Comput Electron Agric 198:107119

Sharma V, Tripathi AK, Mittal H (2022) Technological revolutions in smart farming: current trends, challenges & future directions. Comput Electron Agric:107217

Shekhar S (2022) Water and nutrient management in rice under alternate wetting and drying irrigation practice: field and modeling studies [dissertation]. IIT Kharagpur. Available from: http://www.idr.iitkgp.ac.in/jspui/bitstream/123456789/12050/1/NB17320_Abstract.pdf. (Accessed August 10, 2023)

Shekhar S, Dubey A, Pohshna C (2018) Estimation of irrigation scheduling for different cropping pattern at different growth stage of crop by using the CROPWAT model. Int J Curr Microbiol Appl Sci 7:3855–3862

Shekhar S, Mailapalli DR, Raghuwanshi NS, Das BS (2020) Hydrus-1D model for simulating water flow through paddy soils under alternate wetting and drying irrigation practice. Paddy Water Environ 18:73–85

Shekhar S, Mailapalli DR, Das BS, Mishra A, Raghuwanshi NS (2021a) Hydrus-1D for simulating potassium transport in flooded paddy soils. Commun Soil Sci Plant Anal 52:2803–2820

Shekhar S, Mailapalli DR, Raghuwanshi NS (2021b) Simulating nitrogen transport in paddy crop irrigated with alternate wetting and drying practice. Paddy Water Environ 19:499–513

Shekhar S, Tamilarasan R, Mailapalli DR, Raghuwanshi NS (2021c) Estimation of evapotranspiration for paddy under alternate wetting and drying irrigation practice*. Irrig Drain 70:195–206

Shekhar S, Mailapalli DR, Raghuwanshi NS (2022) Effect of alternate wetting and drying irrigation practice on rice crop growth and yield: a Lysimeter study. ACS Agric Sci Technol 2:919–931

Shekhar S, Mailapalli DR, Raghuwanshi NS, Pohshna C (2023a) Hydrus-1D model for simulating phosphorus transport in paddy crop irrigated with alternate wetting and drying practice. Commun Soil Sci Plant Anal 28:1–22

Shekhar S, Mailapalli DR, Raghuwanshi NS (2023b) Simulation and optimization of ponding water and nutrient management in rice irrigated with alternate wetting and drying practice under a humid subtropical region in India. Paddy Water Environ:1–19

Shirazi SN, Gouglidis A, Farshad A, Hutchison D (2017) The extended cloud: review and analysis of mobile edge computing and fog from a security and resilience perspective. IEEE J Sel Areas Commun 35(11):2586–2595

Singh KK (2018) An artificial intelligence and cloud-based collaborative platform for plant disease identification, tracking and forecasting for farmers. In: 2018 IEEE International Conference on Cloud Computing in Emerging Markets (CCEM), pp 49–56

Solanki S, Singh UP, Chouhan SS, Jain S (2023a) Brain tumour detection and classification by using deep learning classifier. Int J Intell Syst Appl Eng 11(2s):279–292

Solanki S, Singh UP, Chouhan SS (2023b) Brain tumor classification using ML and DL approaches. In: 2023 IEEE 5th International Conference on Cybernetics, Cognition and Machine Learning Applications (ICCCMLA), Hamburg, Germany, pp 204–208

Solanki S, Singh UP, Chouhan SS, Jain S (2024) A systematic analysis of magnetic resonance images and deep learning methods used for diagnosis of brain tumor. Multimedia Tools Appl 5:1–38

Somov A, Shadrin D, Fastovets I, Nikitin A, Matveev S, Hrinchuk O (2018) Pervasive agriculture: IoT-enabled greenhouse for plant growth control. IEEE Pervas Comput 17(4):65–75

Tejeda AI, Castro RC (2019) Algorithm of weed detection in crops by computational vision. In: 2019 International Conference on Electronics, Communications and Computers (CONIELECOMP), pp 124–128

Ukil A, Sen J, Koilakonda S (2011) Embedded security for Internet of Things. In: 2011 2nd National conference on emerging trends and applications in computer science, pp 1–6

Verma RC, Jitendrakumar PH, Ashoka P, Shekhar S, Pal A, Mondal K et al (2023a) A review of Millet grain phenolics, their health promotion and disease risk reduction. Int J Plant Soil Sci 35(20):863–874

Verma RC, Singh JP, Shekhar S, Pratap N, Rai AK, Singh S, Singh N (2023b) Enhancement of productivity through cluster frontline demonstrations and dissemination of production technology of pigeon pea (Cajanus cajan) in Ghazipur, Uttar Pradesh. Int J Stat Appl Math 8:558–561

Verma RC, Shekhar S, Singh JP, Pratap N, Rai AK, Singh S, Singh N (2023c) Enhance the lentil (Lens culinaris) production through cluster frontline demonstrations. Int J Stat Appl Math 8:553–557

Verma RC, Singh JP, Shekhar S, Pratap N, Rai AK, Singh S, Singh N (2023d) Increasing the productivity through cluster frontline demonstrations on mustard. Int J Stat Appl Math 8:587–590

Vivas E, de Guenni LB, Allende-Cid H, Salas R (2023) Deep lagged-wavelet for monthly rainfall forecasting in a tropical region. Stoch Environ Res Risk Assess 37(3):831–848

Waleed M, Um TW, Kamal T, Khan A, Iqbal A (2020) Determining the precise work area of agriculture machinery using internet of things and artificial intelligence. Appl Sci 10(10):3365

Xie BX, Chung SC, Chang CL (2020) Design and implementation of a modular AI-enabled shovel weeder. In: 2020 International Symposium on Computer, Consumer and Control (IS3C), pp 236–239

Xiong H, Dalhaus T, Wang P, Huang J (2020) Blockchain technology for agriculture: applications and rationale. Front Blockchain 3:7

Xu X, Gao P, Zhu X, Guo W, Ding J, Li C et al (2019) Design of an integrated climatic assessment indicator (ICAI) for wheat production: a case study in Jiangsu Province, China. Ecol Indic 101:943–953

Yang X, Shu L, Chen J, Ferrag MA, Wu J, Nurellari E, Huang K (2021) A survey on smart agriculture: Development modes, technologies, and security and privacy challenges. IEEE/CAA J Autom Sin 8(2):273–302

Yi S, Qin Z, Li Q (2015) Security and privacy issues of fog computing: a survey. In: Wireless algorithms, systems, and applications: proceedings of the 10th international conference, WASA 2015, Qufu, China, August 10–12, 2015, pp 685–695

Zhai Z, Martínez JF, Beltran V, Martínez NL (2020) Decision support systems for agriculture 4.0: survey and challenges. Comput Electron Agric 170:105256

Zhang X, Zhang J, Li L, Zhang Y, Yang G (2017) Monitoring citrus soil moisture and nutrients using an IoT based system. Sensors 17(3):447

Zhang T, Zhao Y, Jia W, Chen MY (2021) Collaborative algorithms that combine AI with IoT towards monitoring and control system. Future Gener Comput Syst 125:677–686

Zude-Sasse M, Fountas S, Gemtos TA, Abu-Khalaf N (2016) Applications of precision agriculture in horticultural crops. Eur J Hortic Sci 81(2):78–90

Enabling Digital Platforms: Toward Smart Agriculture

14

Dangi Pooja Arun and Ayush Mishra

Abstract

Smart, sustainable farming is a new way of doing things that uses cutting-edge technologies like the Internet of Things (IoT), cloud computing, and big data analytics to improve farming practices and care for the environment. Farmers can get real-time information about important things like soil moisture, temperature, and crop health by putting IoT monitors in their fields. This information is sent to cloud computing platforms, which have the computer power and storage needed to handle and analyze huge amounts of agricultural data. Algorithms for analyzing big data find patterns, trends, and connections. This lets farmers make smart choices about where to put resources, how to get rid of pests, how much water to use, and how to handle their livestock. Digital technologies like agricultural robots, drones, and mobile apps like farm calculators and cattle tracking can make farming more efficient, sustainable, and productive by giving farmers information on how to grow crops and reducing risks and uncertainties like diseases, pests, and drought. Real-time tracking and predictive analytics make it possible for farmers to find and fix potential problems quickly, which protects crops and increases production. By combining past and real-time data, farmers can learn important things about crop yields, market demand, and climate-related risks. This gives them the tools they need to plan their operations well and respond to changes in the market.

D. P. Arun (✉)
Department of Agricultural Economics and Extension, Lovely Professional University, Phagwara, Punjab, India

A. Mishra
Department of Agricultural Extension Education, CCS Haryana Agricultural University, Hisar, Haryana, India

S. S. Chouhan et al. (eds.), *Artificial Intelligence Techniques in Smart Agriculture*, https://doi.org/10.1007/978-981-97-5878-4_14

Keywords

Soil moisture · Smart agriculture · Weather forecasts · Cloud computing · Big Data analytics · Genomic sequencing · Feed and water management · Health monitoring · Disease surveillance

14.1 Introduction

Digital platforms have brought about a revolutionary change in the agriculture industry in recent years. These platforms, which utilize technological breakthroughs, data analytics, and networking, have transformed farming operations globally. Digital platforms in agriculture offer farmers the huge advantage of providing real-time information and tools. Farmers can obtain weather forecasts, market pricing, agricultural best practices, and pest management approaches through mobile applications, websites, and internet portals (Sam and Grobbelaar 2021).

Access to up-to-date and pertinent information enables farmers to make well-informed decisions, maximize resource efficiency, and reduce risks related to crop failure or market changes. Digital platforms enable precision agriculture practices, enabling farmers to monitor and manage their crops with exceptional accuracy (Elijah et al. 2018). Farmers can get data on soil moisture levels, crop health, and nutrient content using satellite photography, drones, and IoT sensors (Tzounis et al. 2017). Farmers can use this data to apply specific interventions like accurate irrigation and fertilizer application, resulting in increased yields, lower input costs, and less environmental harm (Ruiz-Real et al. 2020). Digital platforms are essential in reshaping the agricultural scene by providing market access. Online markets facilitate direct connections between producers and buyers, removing middlemen and allowing for more equitable pricing systems. This direct-to-consumer approach enhances farmers' profits and fosters transparency and traceability in the agricultural supply chain. Digital platforms provide financial services customized for farmers, particularly small-scale ones in developing nations (Kamal and Bablu 2023). Mobile banking, digital payments, and microloans empower farmers to obtain capital, handle their finances, and allocate funds for agriculture inputs (Ezeomah and Duncombe 2019) These financial services support financial inclusion by enabling farmers to overcome obstacles in obtaining financing and establishing strong livelihoods. Digital platforms not only empower individual farmers but also promote collaboration and knowledge sharing throughout the agricultural community. Various online platforms facilitate connections among farmers, researchers, policymakers, and agricultural extension workers. These platforms act as online centers for exchanging ideas, discussing best practices, and tackling common difficulties encountered by farmers globally. Furthermore, digital platforms also support capacity development by offering online training courses, tutorials, and extension services, which help farmers acquire the essential skills and information needed to implement new techniques. The four major types of digital platforms are the Internet of Things (IoT), Big Data Analytics, Artificial Intelligence, and Cloud Computing (Chouhan et al. 2024; Sharma et al. 2024; Panchal et al. 2024) (Fig. 14.1).

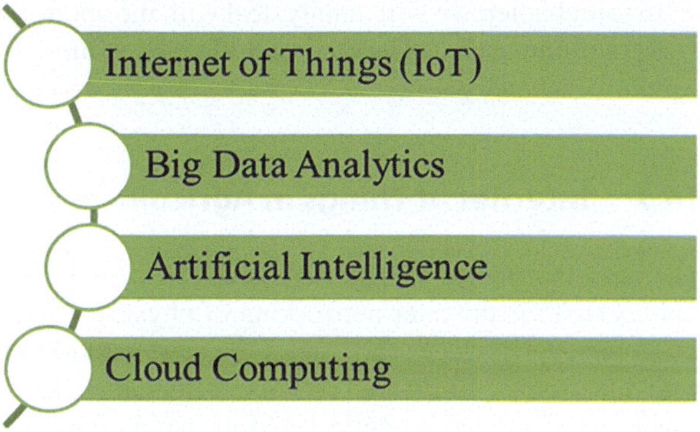

Fig. 14.1 Overview of digital platforms

Internet of Things (IoT): IoT is the process of linking devices, sensors, and machinery to the Internet to collect and share data. Agriculture benefits from IoT technology by allowing farmers to monitor environmental conditions, soil moisture levels, crop health, and equipment operation in real time (Elijah et al. 2018). This data assists farmers in making decisions based on data, optimizing resource utilization, and automating farm activities to enhance efficiency and output.

Big Data Analytics: Big Data Analytics is the process of analyzing extensive amounts of data to discover patterns, trends, and insights that may be used to make decisions. Big data analytics is utilized in agriculture to analyze data from IoT devices, weather forecasts, satellite photos, and historical crop yields (Joshi and Kaushik 2021). Farmers may use this data to adjust planting schedules, forecast pest outbreaks, pinpoint areas for yield enhancement, and make strategic management choices to boost farm production and profitability (Misra et al. 2020).

Artificial Intelligence (AI): Artificial Intelligence (AI) refers to the creation of computer systems that are capable of learning, solving problems, and making decisions—tasks that normally require human intelligence (Nihalani et al. 2024). AI is utilized in agriculture for tasks such as crop monitoring, disease detection, yield prediction, and autonomous farming machines (Misra et al. 2020). AI algorithms process data from IoT devices, drones, and satellite photography to offer practical insights and suggestions to farmers, enhancing crop productivity, resource use efficiency, and sustainability of farm management practices (Bannerjee et al. 2018).

Cloud Computing: Cloud computing enables the provision of computing services via the internet, granting users immediate access to a repository of collaborative computing resources, including storage, processing capabilities, and applications. Cloud computing in agriculture allows farmers to store, manage, and analyze extensive amounts of data gathered via IoT devices, satellite imaging, and other sources (Goraya and Kaur 2015). Farmers can utilize cloud-based platforms and services to access sophisticated analytics tools, work with stakeholders, and implement scalable solutions for precision agriculture, supply chain management, and farm management, without requiring a substantial initial investment in IT infrastructure (Patil et al. 2012).

In this chapter, we will mainly deal with the applications of IoT in different fields of Agriculture such as Genetics and Plant Breeding, Soil Health Management, and Animal Husbandry.

14.2 Internet of Things in Agriculture

The term Internet of Things was coined in 1999 by Kevin Ashton. The Internet of Things (IoT) is the inter-networking of physical devices, vehicles (also referred to as "connected devices" and "smart devices"), buildings, and other items embedded with electronics, software, sensors, actuators, and network connectivity, which enable these objects to collect and exchange data (Ogundokun et al. 2021). The implementation of the Internet of Things (IoT) in agriculture has introduced significant transformations in agricultural output. It not only enhances agricultural yield but also successfully enhances the quality of agricultural products, reduces labor expenses, boosts farmers' income, and genuinely achieves agricultural modernization and automation (Verdouw et al. 2016; Tzounis et al. 2017). IoT is used to monitor crops, detect diseases and pests, predict crop yields, for robotics-based harvesting of crops, and much more. The wireless sensor network (WSN) utilizes different wireless connections with specific range, bandwidth, and topology to identify mold and disease in agricultural products like apples. It can differentiate between fresh and moldy apples, as well as distinguish apples infected with various fungi. It can also detect the decay of oranges affected by *Penicillium Digitatum* and identify diseases in onions postharvesting (Kasera et al. 2024). Moreover, these gadgets are highly efficient tools for confirming the genuineness of food products and identifying any attempts at counterfeiting, especially in products such as milk, honey, argan oil, and saffron. The technology has been used to detect pesticides in agricultural products, differentiate between organic and inorganic products like table grapes, predict spraying days for pesticide residue detection on apples, evaluate pesticide residues in tea leaves, detect Chlorpyrifos (a common pesticide), and distinguish between different types of insecticide residues in vegetable fields. Furthermore, these devices are crucial in identifying the origins of crops and their surroundings, such as in the study of potato cultivars (Amkor and El Barbri 2023), and in determining the geographical source of table grapes, kiwi fruits, oranges, and so on (Violino et al. 2020).

IoT has also shown great potential for the dairy industry. IoT is increasingly been used to analyze the flavor components in milk and dairy products, and to classify milk and milk products from their corresponding flavor products (Brudzewski et al. 2004; Jia et al. 2019). Studying the volatile molecules in milk is a valuable method for obtaining information about milk and identifying various types of milk, particularly in relation to different heat treatments such as UHT, pasteurized, and sterilized milk (Akbar et al. 2020). Thermal processing of milk is a prevalent and crucial procedure in the dairy sector. It is done to enhance the microbiological safety of milk and extend its shelf life. The impact of heat treatment on the ultimate product's chemical, physical, and organoleptic characteristics is determined by the specific combination of temperature and duration of application (Neethirajan 2023).

14.3 Application of Digitized Platforms in Genetics and Plant Breeding

The Internet of Things (IoT) can play a significant role in revolutionizing genetics and plant breeding in agriculture. Here are some applications of IoT in this field:

Genomic Sequencing: IoT devices can be utilized for collecting genomic data from plants, facilitating high-throughput sequencing and analysis. Genomic sequencing in agriculture has immense potential to revolutionize crop breeding, disease resistance, and overall agricultural productivity (Joshi and Kaushik 2021). In India, genomic sequencing is increasingly being utilized to enhance crop varieties, improve yields, and address challenges such as climate change and pest resistance.

14.3.1 Genomic Sequencing in Rice Cultivation

Identification of Stress Tolerance Genes: Given the unpredictable climatic circumstances caused by shifting climate patterns, it is crucial to cultivate rice cultivars that possess the ability to endure several stressors, including drought, flooding, and salinity (Kilasi et al. 2018). Genomic sequencing facilitates the identification of genes linked to stress tolerance. Scientists have determined the specific genes that are accountable for drought tolerance in particular rice types through the process of sequencing their genomes (Wahab et al. 2020) (Fig. 14.2).

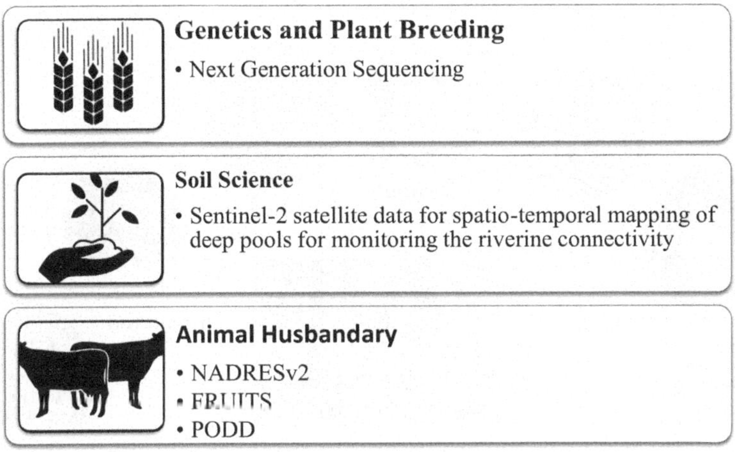

Fig. 14.2 Digitized platforms in agriculture

14.3.2 Isolation, Cloning, and Functional Validation of Pi54 Gene for Resistance to Rice Blast (Adopted from ICAR Gene Bank, 2022)

The Pi54 (Pi-kh) gene, which provides a significant level of resistance against various strains of the Magnaporthe oryzae fungus, was identified, isolated, and experimentally confirmed in 2005. Two orthologs of Pi54, namely, Pi54rh and Pi54, were identified from the blast-resistant wild species of rice, *Oryza rhizomatis* and *O. officinalis*, respectively. These orthologs were discovered utilizing an allele mining approach and their effectiveness was confirmed using gene complementation. The breeders in India and worldwide have successfully incorporated the very effective blast resistance gene Pi54 into numerous rice varieties through the utilization of marker-assisted selection (Figs. 14.3 and 14.4).

Accelerated Breeding Programs: Conventional breeding techniques are characterized by their time-consuming nature and frequent lack of efficiency. Genomic sequencing accelerates the process by enabling breeders to detect favorable characteristics at the molecular level. Breeders can enhance the features of rice varieties by analyzing the genetic makeup of parent plants with desired qualities and selecting the most favorable pairings for crossbreeding (Gudi et al. 2022).

Disease Susceptibility: Rice crops are vulnerable to a range of illnesses caused by pathogens including bacteria, fungi, and viruses. Genomic sequencing facilitates the identification of genes linked to disease resistance. Researchers have used genomic analysis to identify the specific genes that are responsible for resistance to bacterial blight, a highly destructive disease in rice. Marker-assisted selection and genetic engineering can be employed to generate resistant cultivars using this knowledge (Chen et al. 2019). The recent development of high-end computing devices and the adaptation of unmanned aerial vehicles has provided a mechanism to automate the agriculture industry's traditional practices. The on-field or aerial

Fig. 14.3 Plant genome sequences by National Institute for Plant Biotechnology (NIPB) (Source: ICAR Gene Bank, 2022)

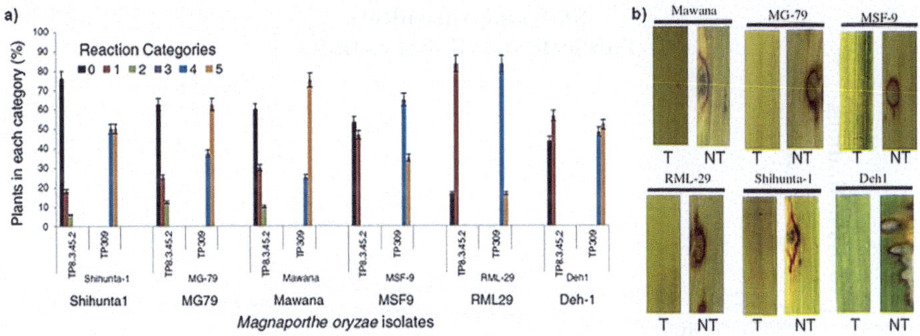

Fig. 14.4 Functional complementation analysis of Pi54-overexpression lines with highly virulent strains of M. oryzae, (**a**) MG-79 and (**b**) Mo-ni-25 (Source: ICAR Gene Bank, 2022)

images captured with the help of cameras are processed with the help of intelligence algorithms and an assessment can be drawn as an outcome (Solanki et al. 2023a, b). This practice efficiently provides a precise, timely, and economical decision-making system to overcome the problems of agricultural field experts and farmers, increasing the quality and quantity of crop yields.

14.3.3 Application of Internet of Things (IoT) in Next-Generation Sequencing (NGS) in Genetics and Plant Breeding

The integration of IoT with NGS technologies holds great promise for advancing genetics and plant breeding by providing insights into the complex interactions between genotype and phenotype and accelerating the development of improved crop varieties with desirable traits. The effectiveness of genome-based methods in enhancing agricultural output relies heavily on linking the observable characteristics of an organism with its genetic information through association studies, as well as on swiftly and precisely predicting breeding values (Fig. 14.5).

Genetic determinants for certain traits can be identified in plants and animals by employing thousands of SNP markers in Genome-Wide Association Studies (GWAS) and genetic mapping in family-based plant populations. This computer method identifies important genes, their favorable or unfavorable variations, and sets of genes inherited together, which has enabled the advancement of "Genome Editing" for creating tailored plant varieties. CRISPR and CRISPR-associated (Cas) proteins have been used for genome editing in plants and animals, as documented by Biswal et al. (2019) and Aglawe et al. (2018). The rapid and precise delivery of new breeds using gene-editing tools makes it the preferred choice for next-generation breeding programs. Bioinformatics is crucial for creating guide RNAs, improving nucleases, and detecting potential off-target effects (Balakrishnan et al. 2022).

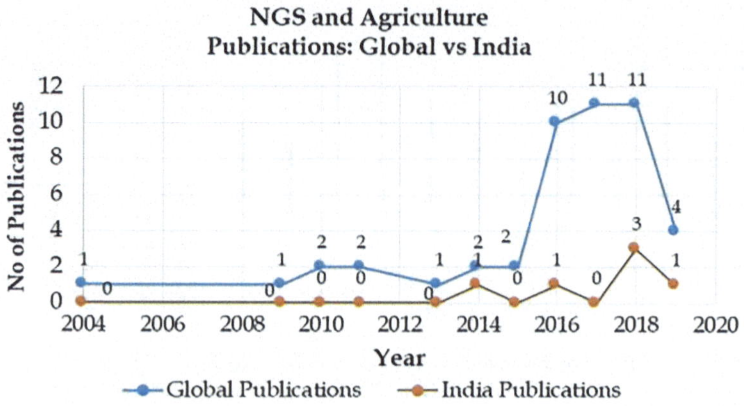

Fig. 14.5 Status of next-generation sequencing in agriculture

14.3.3.1 Advantages of Next-Generation Sequencing in Agriculture

1. Next-generation sequencing (NGS) enables laboratories to rapidly sequence the whole genome.
2. Enables the researchers to conduct a thorough sequencing of the specific regions of interest.
3. Facilitates the investigation of the human-associated microbiota.
4. It helps in discovering new disease causing microorganisms.
5. Analyze cancer samples using sequencing to investigate uncommon somatic mutations and tumor subclones.
6. Researchers can investigate epigenetic variables, such as genome-wide DNA methylation and DNA protein interactions, in a laboratory using next-generation sequencing.
7. The RNA sequencing approach enables the identification of novel RNA variations and splice sites.
8. It has facilitated researchers in amassing huge amounts of genomic sequencing data.
9. Comprehension and identification of intricate illnesses.
10. Comprehending the manifestation of modified genetic variations and their impact on organisms.
11. Approximately 90% of changes in the human genome are believed to be attributed to mutations in the exome, resulting in the development of illnesses. Therefore, this method facilitates the sequencing and comprehension of the exome patterns.
12. DNA next-generation sequencing is employed in gene therapy to locate isolated genes and supply an accurate replica of the faulty gene.
13. Additionally, studying the genes responsible for cancer and developing a gene that specifically targets and eliminates cancer cells, known as 'suicide gene therapy', can be beneficial.

14.3.4 Application of Digitized Platforms in Soil Health Management

The application of the Internet of Things (IoT) in soil science offers numerous opportunities to enhance soil monitoring, management, and sustainability practices. Here are several key ways in which IoT can be applied in soil science:

Soil Moisture Monitoring: Soil moisture monitoring involves the use of IoT sensors to track and measure the amounts of moisture in the soil, woods, or urban landscapes. This monitoring is done in real time, providing up-to-date information (Kodali and Sahu 2016). The utilization of this data can enhance the efficiency of irrigation schedules, mitigate the risks of excessive or insufficient watering, and maximize the utilization of water resources.

Soil Nutrient Monitoring: Soil nutrient monitoring involves the use of IoT sensors that are equipped with probes or spectrometers to accurately detect the quantities of various nutrients in the soil, such as nitrogen, phosphorus, potassium, and micronutrients. This data can guide fertilization strategies, maintaining the ideal nutrient levels for plant growth while reducing environmental consequences such as nutrient runoff or leaching (Raut et al. 2018).

Soil pH and Salinity Monitoring: Soil pH and salinity may be monitored using IoT sensors. These sensors are capable of measuring the levels of pH and salt in the soil. These elements are crucial in determining the growth of plants and the overall health of the soil. Real-time monitoring of these factors enables prompt adjustments to soil amendments or irrigation procedures to uphold optimal circumstances for plant growth.

Soil Temperature Monitoring: Soil temperature monitoring involves the use of IoT sensors to accurately measure and record changes in soil temperature over time, including fluctuations that occur within a day and between different seasons. This information is crucial for comprehending soil biological activity, seed germination, and crop development, as well as for enhancing soil management strategies.

Soil Erosion and Compaction Monitoring: Soil erosion and compaction can be monitored using IoT-enabled sensors, such as accelerometers or strain gauges, which are capable of detecting these events in real time. This data can provide valuable insights for implementing erosion control measures, adopting conservation practices, and devising land management strategies to effectively address soil degradation and enhance soil health.

14.3.4.1 Sentinel-2 Satellite Data for Spatiotemporal Mapping of Deep Pools for Monitoring the Riverine Connectivity (Adopted from Zhang et al. 2021)

Sentinel-2 satellite data can indeed be valuable for spatiotemporal mapping of deep pools for monitoring riverine connectivity. It can be utilized as:

High-Resolution Imagery: Sentinel-2 satellites offer high-quality multispectral imagery with a spatial resolution that varies from 10 to 60 meters, depending on the spectral band. This enables the recognition and demarcation of river characteristics, such as profound reservoirs, with a significant degree of precision.

Temporal Coverage: The temporal coverage of Sentinel-2 satellites entails revisiting the same point on the Earth every 5 days, thereby offering frequent and regular observations throughout time. The inclusion of temporal coverage is crucial for the surveillance of alterations in riverine characteristics, such as fluctuations in the dimensions, configuration, and interconnections of deep pools, which may arise from seasonal changes, hydrological incidents, or human-induced actions (Kantharajan et al. 2022).

Spectral Bands: The sensors of Sentinel-2 capture data in many spectral bands, encompassing the visible, near-infrared, and shortwave infrared wavelengths. The spectral bands possess sensitivity to changes in water depth, turbidity, and substrate composition, enabling the distinction between deep pools and adjacent riverine characteristics.

Water Index Algorithms: Water index techniques, such as the Normalized Difference Water Index (NDWI) or Modified Normalized Difference Water Index (MNDWI), can be used with Sentinel-2 data to emphasize water bodies, particularly deep pools, by analyzing their spectral properties. This enables the automated identification and delineation of profound reservoirs across extensive regions (Fig. 14.6).

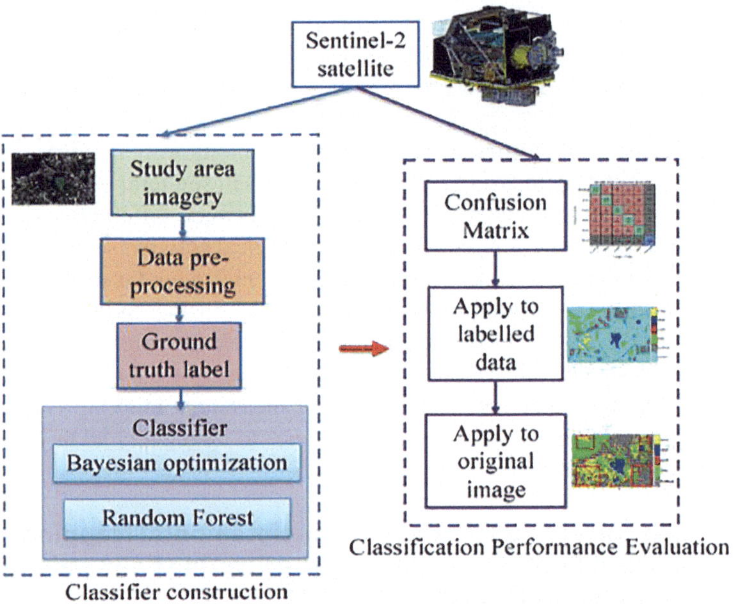

Fig. 14.6 The developed framework for land cover classification by using Sentinel-2 satellite and machine learning-based classifiers

14.3.5 Application of Digitized Platforms in Animal Husbandry

Livestock Monitoring: Animals can be equipped with IoT sensors to continuously monitor their health, behavior, and real-time location (Neethirajan 2023). Wearable gadgets, including smart collars or ear tags, are equipped with sensors that can monitor vital signs, activity levels, rumination patterns, and the GPS position of individual animals (Colditz 2023). This data enables farmers to promptly identify indications of sickness, estrus, or distress, facilitating immediate intervention and preventive healthcare.

Feed and Water Management: Animal feeders and waterers integrated with IoT technology and equipped with sensors can autonomously distribute accurate quantities of feed and water, considering the specific nutritional needs and consumption habits of cattle (Akbar et al. 2020). This optimizes feed utilization, decreases labor expenses, and guarantees continuous availability of fresh and uncontaminated water for the animals.

Heat Detection and Estrus Monitoring: IoT sensors can precisely identify indications of heat and estrus in dairy cows, beef cattle, and other livestock species by analyzing alterations in behavior, activity, and body temperature. This data aids in optimizing breeding operations, scheduling artificial insemination, and implementing reproductive management methods to promote fertility rates and genetic advancement.

Health Monitoring and Disease Surveillance: IoT-enabled health monitoring systems employ biosensors, cameras, or imaging technology to consistently monitor the health status of individual animals or large herds (Begum et al. 2014). By utilizing machine learning algorithms, this data can detect trends that are suggestive of disease outbreaks, injuries, or stress. This, in turn, allows for proactive disease management and the implementation of biosecurity measures.

Whole genomes of 12 bovine coronaviruses isolated from cow (3) and buffalo (9) nasal and fecal samples collected in 2020-2021 were sequenced. Phylogenetic analysis of the whole genomes with the global bovine coronavirus whole genome data reveals that nine of the viruses cluster with the classical bovine coronaviruses in subgroup GIa that has isolated from all over the world like China, the USA, and Germany. Three other viruses cluster with the GIb subgroup, which has majorly the isolates of France in 2017. This is indicative of the co-circulation of two different subgroups and also suggests multiple introductions.

14.3.5.1 Disease Informatics (Table 14.1; Fig. 14.7)

The PODD system's process for abnormal health surveillance is designed to promptly discover and respond to issues. Terdsak Yano developed the Participatory One Health Disease Detection (PODD) tool to identify emerging animal and environmental health threats, using Thailand as a model. The Faculty of Veterinary Medicine at Chiang Mai University, specializing in Food Animal Clinics (Fig. 14.8).

Table 14.1 Application of disease informatics

Device	Organization	Use
National Animal Disease Referral Expert System v2 (NADRESv2)	ICAR-NIVEDI	Dynamic geographic information and remote sensing-enabled expert system reports were communicated in the form of risk maps, bulletins, and post-prediction maps for necessary preparedness. The risk communication through SMSs to farmers
Farmer Registration and Unified Beneficiary Information System (FRUITS)	NIC, Govt of Karnataka	SMS alerts were sent to farmers in Karnataka for different animal diseases (anthrax, babesiosis, black quarter, bluetongue, FMD, theileriosis)
Sero-surveillance and sero-monitoring	DAHD, GoI	Monitoring of foot and mouth disease (FMD), brucellosis, Peste des Petits Ruminants (PPR), and classical swine fever (CSF) for each state/UTs of the country
Participatory One-health Digital Disease Detection (PODD)	Chiang Mai, Thailand	To detect abnormal deaths in backyard animals in order to elicit rapid investigation and response

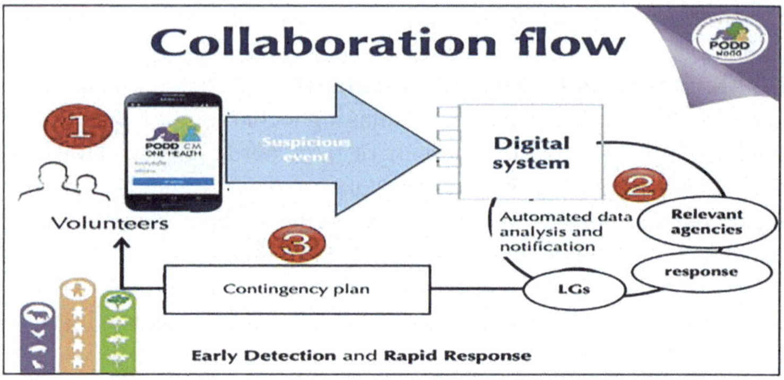

Fig. 14.7 The workflow of PODD system

14.4 Conclusion

Digital platforms have the potential to transform agriculture by enhancing its efficiency, sustainability, and resilience. Digital platforms enable farmers by providing access to information, markets, financial services, and collaborative networks, allowing them to improve their farming practices and livelihoods. Digital platforms like IoT are spurring innovation, and revolutionary change in agriculture by equipping researchers with novel tools, methodologies, and insights to tackle intricate difficulties and utilize on possibilities for sustainable agriculture and food security. Researchers can use IoT technologies to help create more efficient, robust, and environmentally sustainable agricultural systems to meet the increasing food demand in

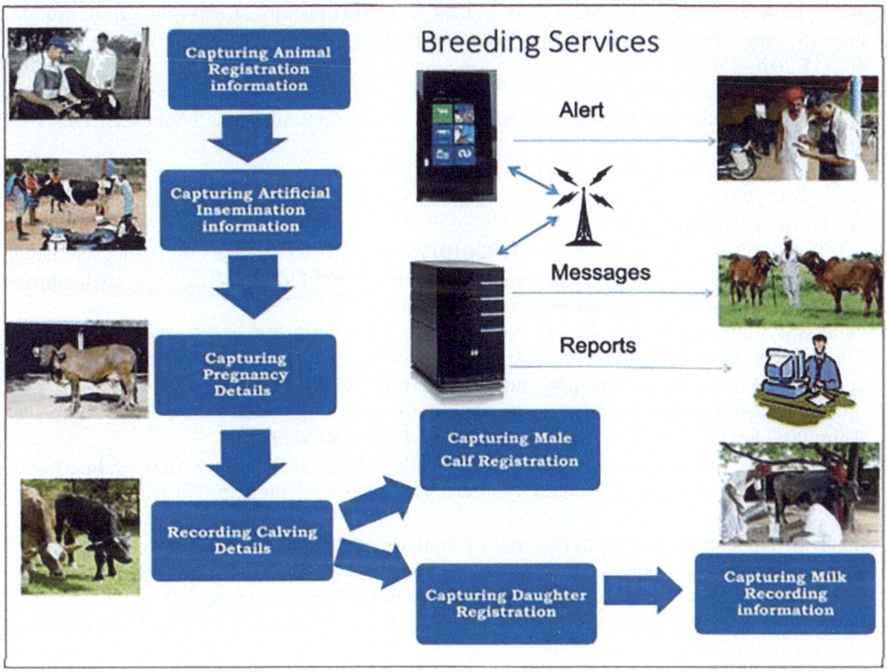

Fig. 14.8 Information network for animal productivity and health

a changing world. Farmers can utilize IoT technologies to leverage data and technology to address difficulties, streamline farm operations, and achieve sustainable growth in a complex and competitive agricultural environment. Continual innovation and investment in digital platforms have the potential to drive positive transformation and make the agriculture sector smart and resilient.

References

Aglawe SB, Barbadikar KM, Mangrauthia SK, Madhav MS (2018) New breeding technique "genome editing" for crop improvement: applications, potentials and challenges. Biotech 8:1–20

Akbar MO, Shahbaz Khan MS, Ali MJ, Hussain A, Qaiser G, Pasha M, Akhtar N (2020) IoT for development of smart dairy farming. J Food Qual 2020:1–8

Amkor A, El Barbri N (2023) Artificial intelligence methods for classification and prediction of potatoes harvested from fertilized soil based on a sensor array response. Sensors Actuators A Phys 349:114106. https://doi.org/10.1016/j.sna.2022.114106

Balakrishnan M, Thammi RD, Soam SK, Borah P, Mangrauthia SK, Pandey MK, Jaggi S, Srinivasa Rao C (2022) Bioinformatics in agriculture. Status paper. ICAR-National Academy of Agricultural Research Management, Hyderabad, India, p p24

Bannerjee G, Sarkar U, Das S, Ghosh I (2018) Artificial intelligence in agriculture: a literature survey. Int J Sci Res Comput Sci Appl Manage Stud 7(3):1–6

Begum MR, Anaruzzaman M, Khna MSI, Yousuf M (2014) Factors affecting the milk production of dairy cattle in northern rural areas of Bangladesh. Int J Agric Res Innov Technol 4(2355-2020-1569):41–45

Biswal AK, Mangrauthia SK, Reddy MR, Yugandhar P (2019) CRISPR mediated genome engineering to develop climate smart rice: Challenges and opportunities. Semin Cell Dev Biol 96:100–106

Brudzewski K, Osowski S, Markiewicz T (2004) Classification of milk by means of an electronic nose and SVM neural network. Sensors Actuators B Chem 98(2-3):291–298. https://doi.org/10.1016/j.snb.2003.10.028

Chen WL, Lin YB, Ng FL, Liu CY, Lin YW (2019) RiceTalk: rice blast detection using Internet of Things and artificial intelligence technologies. IEEE Internet Things J 7(2):1001–1010. https://doi.org/10.1109/JIOT.2019.2947624

Chouhan SS, Singh UP, Jain S (2024) Introduction to computer vision and drone technology. In: Chouhan SS, Singh UP, Jain S (eds) Applications of computer vision and drone technology in agriculture 4.0. Springer, Singapore. https://doi.org/10.1007/978-981-99-8684-2_1

Colditz IG (2023) A biological integrity framework for describing animal welfare and wellbeing. Anim Prod Sci 63(5):423–440. https://doi.org/10.1071/AN23029

Elijah O, Rahman TA, Orikumhi I, Leow CY, Hindia MN (2018) An overview of Internet of Things (IoT) and data analytics in agriculture: benefits and challenges. IEEE Internet Things J 5(5):3758–3773

Ezeomah B, Duncombe R (2019) The role of digital platforms in disrupting agricultural value chains in developing countries. In: International conference on social implications of computers in developing countries. Springer International Publishing, Cham, pp 231–247

Goraya MS, Kaur H (2015) Cloud computing in agriculture. HCTL Open Int J Technol Innov Res 16:2321–1814

Gudi S, Kumar P, Singh S, Tanin MJ, Sharma A (2022) Strategies for accelerating genetic gains in crop plants: special focus on speed breeding. Physiol Mol Biol Plants 28(10):1921–1938. https://doi.org/10.1007/s12298-022-01247-8

Jia W, Wang H, Shi L, Zhang F, Fan C, Chen X, Chu X (2019) High-throughput foodomics strategy for screening flavor components in dairy products using multiple mass spectrometry. Food Chem 279:1–11. https://doi.org/10.1016/j.foodchem.2018.12.005

Joshi A, Kaushik V (2021) Big Data and its analytics in agriculture. Bioinform Agric: High-Throughput Approach 1:71–83

Kamal M, Bablu TA (2023) Mobile applications empowering smallholder farmers: an analysis of the impact on agricultural development. Int J Soc Anal 8(6):36–52

Kantharajan G, Anand A, Krishnan P, Singh RK, Kumar K, Kumar Yadav A, Lal KK (2022) Applications of Sentinel-2 satellite data for spatio-temporal mapping of deep pools for monitoring the riverine connectivity and assessment of ecological dynamics: a case from Godavari, a tropical river in India (2016–2021). Environ Monit Assess 194(8):558. https://doi.org/10.1007/s10661-022-10089-6

Kasera RK, Gour S, Acharjee T (2024) A comprehensive survey on IoT and AI based applications in different pre-harvest, during-harvest and post-harvest activities of smart agriculture. Comput Electron Agric 216:108522. https://doi.org/10.1016/j.compag.2023.108522

Kilasi NL, Singh J, Vallejos CE, Ye C, Jagadish SK, Kusolwa P, Rathinasabapathi B (2018) Heat stress tolerance in rice (Oryza sativa L.): identification of quantitative trait loci and candidate genes for seedling growth under heat stress. Front Plant Sci 9:1578

Kodali RK, Sahu A (2016) An IoT based soil moisture monitoring on Losant platform. In: 2016 2nd International Conference on Contemporary Computing and Informatics (IC3I). IEEE, pp 764–768. https://doi.org/10.1109/IC3I.2016.7918063

Misra NN, Dixit Y, Al-Mallahi A, Bhullar MS, Upadhyay R, Martynenko A (2020) IoT, big data, and artificial intelligence in agriculture and food industry. IEEE Internet Things J 9(9):6305–6324

Neethirajan S (2023) Artificial intelligence and sensor innovations: enhancing livestock welfare with a human-centric approach. Human-Centric Intell Syst 1-16. https://doi.org/10.1007/s44230-023-00050-2

Nihalani R et al (2024) Long Short-Term Memory (LSTM) model for Indian sign language recognition. J Intell Fuzzy Syst:1–19. https://doi.org/10.3233/JIFS-233250

Ogundokun RO, Awotunde JB, Misra S, Abikoye OC, Folarin O (2021) Application of machine learning for ransomware detection in IoT devices. In: Artificial intelligence for cyber security: methods, issues and possible horizons or opportunities. Springer International Publishing, Cham, pp 393–420. https://doi.org/10.1007/978-3-030-72236-416

Panchal B, Pranjal P, Patel RK, Sharma A, Chouhan SS (2024) Assessing the quantity of a crop field using aerial images. In: Chouhan SS, Singh UP, Jain S (eds) Applications of computer vision and drone technology in agriculture 4.0. Springer, Singapore. https://doi.org/10.1007/978-981-99-8684-2_11

Patil VC, Al-Gaadi KA, Biradar DP, Rangaswamy M (2012) Internet of things (Iot) and cloud computing for agriculture: an overview. In: Proceedings of agro-informatics and precision agriculture (AIPA 2012), India, vol 292, p 296

Raut R, Varma H, Mulla C, Pawar VR (2018) Soil monitoring, fertigation, and irrigation system using IoT for agricultural application. In: Intelligent Communication and Computational Technologies: Proceedings of Internet of Things for Technological Development, IoT4TD 2017. Springer Singapore, pp 67–73. https://doi.org/10.1007/978-981-10-5523-2_7

Ruiz-Real JL, Uribe-Toril J, Torres Arriaza JA, de Pablo Valenciano J (2020) A look at the past, present and future research trends of artificial intelligence in agriculture. Agronomy 10(11):1839

Sam AK, Grobbelaar SS (2021) Research trends, theories and concepts on the utilization of digital platforms in agriculture: a scoping review. In: 20th Conference on e-Business, e-Services and e-Society (I3E). Springer International Publishing, pp 342–355

Solanki S, Singh UP, Chouhan SS (2023a) Brain tumor classification using ML and DL approaches. In: 2023 IEEE 5th International Conference on Cybernetics, Cognition and Machine Learning Applications (ICCCMLA), Hamburg, Germany, 2023, pp 204–208. https://doi.org/10.1109/ICCCMLA58983.2023.10346854

Solanki S, Singh UP, Chouhan SS, Jain S (2023b) Brain tumour detection and classification by using deep learning classifier. Int J Intell Syst Appl Eng 11(2s):279. Retrieved from https://ijisae.org/index.php/IJISAE/article/view/2624

Sharma A, Patel RK, Pranjal P, Panchal B, Chouhan SS (2024) Computer vision-based smart monitoring and control system for crop. In: Chouhan SS, Singh UP, Jain S (eds) Applications of computer vision and Drone technology in agriculture 4.0. Springer, Singapore. https://doi.org/10.1007/978-981-99-8684-2_5

Tzounis A, Katsoulas N, Bartzanas T, Kittas C (2017) Internet of Things in agriculture, recent advances and future challenges. Biosyst Eng 164:31–48

Verdouw C, Wolfert S, Tekinerdogan B (2016) Internet of Things in agriculture. CABI Rev 2016:1–12

Violino S, Ortenzi L, Antonucci F, Pallottino F, Benincasa C, Figorilli S, Costa C (2020) An artificial intelligence approach for Italian EVOO origin traceability through an open source IoT spectrometer. Food Secur 9(6):834

Wahab MMS, Akkareddy S, Shanthi P, Latha P (2020) Identification of differentially expressed genes under heat stress conditions in rice (Oryza sativa L.). Mol Biol Rep 47:1935–1948. https://doi.org/10.1007/s11033-020-05291-z

Zhang T, Su J, Xu Z, Luo Y, Li J (2021) Sentinel-2 satellite imagery for urban land cover classification by optimized random forest classifier. Appl Sci 11(2):543. https://doi.org/10.3390/app11020543

IoT and Drone-Based Field Monitoring and Surveillance System

15

Payel Kar and Sanhita Chowdhury

Abstract

In India, agriculture faces numerous challenges, including declining production, shifting climatic patterns, and environmental issues. Drone farming emerges as a promising approach to bolster sustainable agriculture across social, economic, and environmental dimensions. This chapter delves into the advantages of integrating drone technology into India's agricultural landscape. Over the past decade, advancements in information technology have propelled the widespread adoption of Internet-of-Things (IoT) devices, exemplified by the utilization of Unmanned Aerial Vehicles (UAVs) in environmental monitoring. The evolution of drone technology in agriculture is a focal point of analysis, highlighting its transformative potential. Precision Agriculture (PA) stands out as a key application domain for drones, enabling tasks such as pesticide spraying and crop monitoring with unparalleled efficiency and precision. Drones equipped with sensors gather real-time data on various agricultural parameters, including crop health, soil conditions, and environmental factors. This data empowers farmers to make informed decisions regarding fertilization, pesticide application, and irrigation practices. The integration of IoT sensors further enhances data collection capabilities, providing insights into soil moisture levels, meteorological conditions, and other critical parameters essential for comprehensive agricultural monitoring. Real-time transmission of data to cloud-based platforms facilitates analysis and visualization, enabling farmers to respond promptly to emerging challenges

P. Kar
Department of Entomology, Bidhan Chandra Krishi Viswavidyalaya, Mohanpur, Nadia, West Bengal, India

S. Chowdhury (✉)
Department of Entomology, Post Graduate College of Agriculture, Dr. Rajendra Prasad Central Agricultural University, Pusa, Samastipur, Bihar, India

S. S. Chouhan et al. (eds.), *Artificial Intelligence Techniques in Smart Agriculture*, https://doi.org/10.1007/978-981-97-5878-4_15

and optimize resource allocation. Despite the promising prospects of drone technology, persistent challenges such as cost constraints, limited battery life, and the need for enhanced image processing capabilities remain to be addressed. In conclusion, the convergence of drones and IoT holds tremendous potential for revolutionizing agriculture in India, bolstering crop yields, and ensuring food security. Addressing existing barriers is imperative to fully harness the transformative power of these technologies and unlock their benefits for farmers and the agricultural sector as a whole.

Keywords

Internet-of-Things (IoT) · Nutrient deficiency · Quadcopter · Hexacopter · Near Infrared (NIR) · Red Edge (RE) · Multispectral · Hyperspectral

15.1 Introduction

The word "drone" has found its way into the colloquial language in the past decade. The name brings forth a picture of a highly sophisticated piece of machinery dedicated to completing the subtlest of tasks. In reality, drones or Unmanned Aerial Vehicles (UAVs) are simply aircrafts that do not have any human on board, be it a pilot, passengers, or crew. Though they seem to be a recent technological advancement, records point the first attempt of constructing a drone operating on power (Taylor et al. 1977). Though used for the military in the initial stages, technological developments and related hardware miniaturization during the 1980s and 1990s have seen their prolific use in other sectors (such as agricultural, scientific, commercial, etc.) as well.

Another piece of technology that has been long time in development, as early as 1982, but has enjoyed the limelight in recent decades is the Internet of Things (IoT), more specifically, it is the IoT devices that have gained massive popularity and now seem to occupy every conceivable sphere of life. Any device that possesses sensors, can process information, has the software to achieve its tasks, and may share data by connecting mainly over the internet or some other networks of communication with other systems and devices can be called an IoT device. Examples include "smart home" devices, sensors, etc. With the mounting pressure of population and simultaneous reduction in farmland sizes, increasing agricultural production is the need of the hour. For instance, the population of the world is anticipated to hit 9.73 billion by 2050, showing an approximate increase of 25% from the current world population (Ullo and Sinha 2020). Rapid urbanization, changing land climate patterns, and increase in population density are contributing to the reduction in availability of arable land (Farooq et al. 2020). The total arable land has declined from 19.5 million square miles to 18.6 million square miles in about a decade (1991–2013) (Zhang et al. 2020). All of this points to an alarming gap between supply and demand of food, which is bound to increase as time progresses. Unless the production of food-grains is doubled, a dire situation will arise in which people will die of hunger (Kim et al. 2019; Panchal et al. 2024; Sharma et al. 2024).

Precision agriculture presents a promising prospect for increasing the crop yield to fulfill this ever-increasing need. Need-based application of agricultural inputs can go a long way in increasing the yield of crops. Integration of drones with IoT devices can help in gathering data about the condition (nutrient status, irrigation requirement, infestation of pests and diseases, weed infestation, etc.) of the standing crop in real time. This assists in deciding which cultivation practices (fertilizer application, pesticide spraying, irrigation, etc.) are to be undertaken. It also points out the areas in the field that need the most attention. These devices can gather field data in a fast, easy, and cost-effective manner as compared to traditional methods, which usually involve inspection by an expert, usually taking days. Also, human expert cannot provide the constant surveillance at all times irrespective of environmental conditions. In addition, drones can be used for pesticide spraying in the crops at precise doses and at required areas only. This has a few advantages, firstly it prevents humans from coming in contact with potentially dangerous chemicals and secondly, it reduces the wastage of these costly formulations and thirdly, by preventing unnecessary application, the problem of pesticide residues in the environment and pollution is curtailed.

15.2 A Brief Idea About Drones Used in Agriculture

According to the firm Goldman Sachs, in the coming years, the agriculture sector is going to be the second-largest user of drones in the world (Almalki 2018). Drones can be divided into two major types: fixed wing and multi-rotor. As the name suggests, the wings of fixed-wing drones are rigid in nature and they cut through the air at a specific angle. Multi-rotor drones use the speed and direction of the motors to move. They can further be divided into different types such as, single rotor, quadcopter, hexacopter, and octocopter (Ghazali et al. 2022). Compared to the fixed wing drone, the multi-rotor drone operates with a lower flight speed, smaller distance, and less duration because it needs a huge amount of power to generate lift and stay aloft (Tsouros et al. 2019). Some most popular type of drones used in precision agriculture are listed in Table 15.1. Along with a variety of applications in agriculture, several challenges become obstacles in the adoption of drone technology, which needs to be addressed (Fig. 15.1). Moreover, applications of drone technology are discussed below.

Typically, drones are equipped with brushless motors, propellers, flight controller, electric speed controller (ESC), and most importantly, a global positioning system (GPS). A radio channel transmitter and receiver control its operations (Mogili and Deepak 2018). One big advantage of using drones as surveillance devices is their flexibility; they may be altered to meet the operator's needs. This is achieved by installing some algorithms and associated hardware on it. In contrast with satellites or manned aircrafts, drones provide better functionality, are more budget-friendly, and a better spatial and spectral resolution.

Table 15.1 Commercial drones used in the market for different agricultural tasks

Drone	Type	Flight time (min)	Maximum distance (km)
1. Suitable for crop spraying			
DJI Agras T20, MS10 Max	Hexacopter	15	3
DJI MG-1P	Octocopter	20	5
Oryctes Dual	Hexacopter	15	2
Hylio AG-122	Octocopter	15	2
2. Suitable for mapping and crop monitoring			
DJI Phantom 4 RTK	Quadcopter	30	7
SenseFly eBee Ag	Fixed-Wing	55	36
DJI Matrice 200 V2	Quadcopter	38	8
Delair UX11	Fixed-Wing	52	47

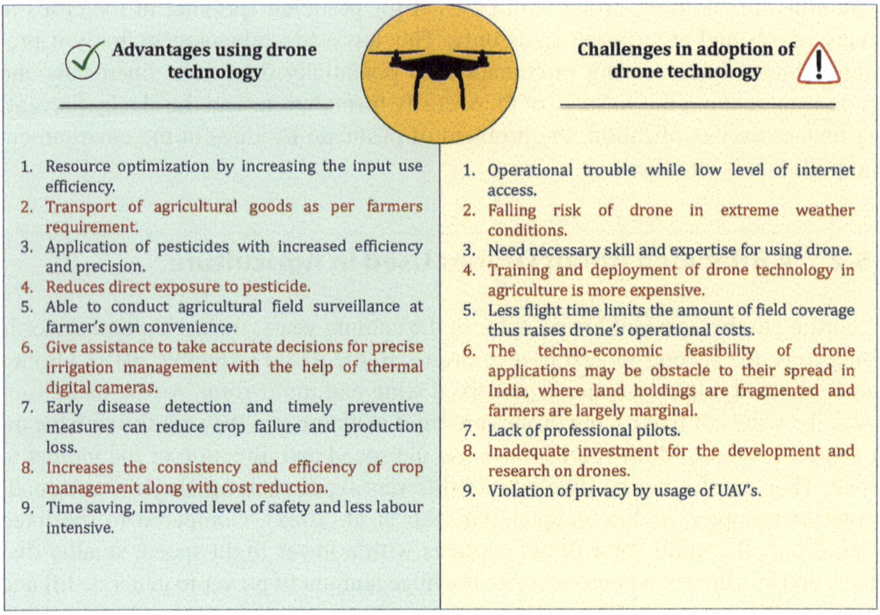

Fig. 15.1 Advantages and challenges of drone technology in agriculture

15.2.1 Sensors Used in Drones

There are specialized sensors mounted on drones for crop data acquisition. They have the role of capturing images with high spatial and temporal resolution. These images help to monitor the various characteristics of the vegetation. The sensors have to fulfill some conditions to do their job effectively. Lightweight, minimal energy consumption, and compact dimensions are essential prerequisites alongside the capacity to capture high-resolution images. The following types of sensors have the maximum utility in this respect:

1. Visible light sensors (RGB): Comparatively low cost, produce high-resolution images, lightweight, can operate independent of cloud cover, capture spectral information on the visible spectrum (380–700 nm).
2. Multispectral sensors: Higher cost, complex pre-processing methods required to get useful information, can capture spectral reflection and absorption on 5–12 channels.
3. Hyperspectral sensors: Same as multispectral sensors, but able to capture information in hundreds or thousands of bands, in a narrower bandwidth, data very specific and precise.
4. Thermal infrared sensors: Not frequently used, capture the information about temperature to produce images for very specific purposes (*e.g.*, irrigation scheduling).

It's fascinating how RGB sensors are sometimes altered to gather data from spectra beyond what's visible, mainly in Near Infrared (NIR) or Red Edge (RE) band. This is a clever way to save costs and involves swapping out one of the original optical filters with a near-infrared channel, creating a hybrid sensor like NIR-RGB. Another RGB sensor that is a part of the system now records the visible channel that the original optical filter was responsible for (Parraga et al. 2018; Gracia-Romero et al. 2018; Lambert et al. 2018; Zheng et al. 2018; Yonah et al. 2018; Näsi et al. 2018, Bali et al. 2023).

15.2.2 Method of Processing Data Acquired by Sensors

These sensors observe various aspects of vegetation, such as leaf color and texture, the shape of crops, and certain wavelengths of radiation, primarily in the red and near-infrared (NIR) regions. The collected data is then analyzed to provide insights into the condition of the crops, including their health, soil moisture levels, plant biomass, and other factors, at different stages of growth.

In UAV-based image acquisition for agricultural purposes, it is customary to obtain multiple overlapping images of crops. Typically, both frontal and lateral overlapping images are captured to facilitate the creation of three-dimensional models and/or orthophotos of the crops. The degree of overlap is contingent upon the specific application. Frontal overlap typically falls within the range of 60–95%, while lateral overlap spans 40–95% for 3D model generation and 25–40% for other applications.

Moreover, the altitude of UAV flights differs based on the application requirements and the desired spatial accuracy of the collected data. The gap between the UAV and the target crop significantly influences the level of detail captured, contingent upon the capabilities of the sensors and the resolution they provide. In most instances, the photographs have spatial resolution in the range 0.5 cm/pixel to 10 cm/pixel, depending on the intended application (Tsouros et al. 2019).

The following image-processing techniques are commonly used to analyze the images captured by the sensors:

1. **Photogrammetry techniques**: It refers to the accurate rendition of an area or object from overlapping images, producing 3D digital surfaces or terrain models as well as orthophotos. The 3D models and orthophotos contain details about the three-dimensional aspects of crops, such as their height, canopy structure, density, and so on, derived from the vegetation's arrangement. These outputs are particularly valuable for applications that rely solely on RGB imagery.
2. **Machine learning methods**: It is the use of gathered data to predict the trends and results. It can figure out certain factors related to how fast crops grow, spot diseases, and even recognize or tell apart objects in images. Lately, there's been a big surge in using machine learning, especially with all the progress that is happening in deep learning fields (Saxena et al. 2024; Solanki et al. 2023a, b, 2024; Kumar et al. 2023, Nihalani et al. 2024).
3. **Calculation of vegetation indices**: Vegetation Indices (VIs) stand out as key outcomes in remote sensing applications for precision agriculture (Table 15.2). These indices utilize different mathematical combinations or transformations of at least two spectral bands from the electromagnetic spectrum (Tsouros et al. 2019). The aim is to emphasize vegetation traits while reducing the influence of external factors. VI offers dependable spatial and temporal insights into monitored agricultural crops. Typically, multiple VIs are computed and analyzed to draw meaningful conclusions. They can be derived from individual photographs or from orthophotos covering the entire crop area. Calculating vegetation indices aids in pinpointing valuable crop attributes, including both physical and biological characteristics of the vegetation.

The data so acquired needs to be processed to get meaningful results. It can take up a lot of time, so there's been a bunch of software tools and techniques made to speed things up. Adobe photoshop is used to correct contortion, MATLAB is used to calculate vegetative indices, Agisoft Photoscan is used to construct 3D models and orthomosaics (Tsouros et al. 2019). Besides, other software tools like Erdas Imagine, eCognition, and PixelWrench 2 are also used.

Table 15.2 Mostly used vegetation indices

Vegetation Index (VI)	Formula
VI acquired from multispectral information	
Ratio Vegetation Index (RVI)	NIR/R
Normalized Difference Vegetation Index (NDVI)	(NIR – R)/(NIR + R)
Normalized Difference Red Edge Index (NDRE)	(NIR – RE)/(NIR + RE)
Green Normalized Difference Vegetation Index (GNDVI)	(NIR – G)/(NIR + G)
RGB-based VI	
Excess Greenness Index (ExG)	2*G-R-B
Normalized Difference Index (NDI)	(G – R)/(G+R)

[Abbreviations used: NIR-near infrared (720–1500 nm); R-Red (630–685 nm); G-Green (520–590 nm); B-Blue (440–510 nm); RE-Red Edge (670–720 nm)]

15.3 Components of Sensor Platform Based on IoT

Two main types of IoT sensors are employed to gather crop data for precision agriculture studies, *i.e.,* underground and aboveground. The underground sensors measure soil moisture while the aboveground ones measure everything else, such as meteorological conditions (temperature, humidity, wind direction, wind speed, and rainfall), light intensity in various spectral bands (visible, near-infrared, and ultraviolet), and even the air quality (Polinova et al. 2019; Yin et al. 2021).

The main job of the IoT sensor platform is to accumulate the data and transmit (preferably to a cloud device). Generally, it comprises an energy source, base stations and gateways (for data transmission), a cloud data center, and finally the end-user product—a mobile and/or web application accessible to only authorized people (Ali et al. 2021).

15.4 "Modus Operandi": Method of Data Acquisition and Transmission

Field data can be collected in two ways, one being an ad hoc communication type and the other being a server installation type. The former sends data to remote locations by direct communication between the sensors (Hussain et al. 2012). The latter accesses the server via the internet using a mobile network line or a wireless LAN; they can transfer the information collected to remote locations as well (Fukatsu and Hirafuji 2003; Saito et al. 2013; Nakayama et al. 2014).

Usually, the ad hoc type has a power-saving design, and the distance of communication is short. It is a point-to-point mode of communication with low power consumption and data transmission rate, but a long-range transmission. This makes the connection unstable due to environmental changes. In contrast, the main complaint with server-based systems is that they consume huge amounts of power and, as a result, require a stable power supply.

In regions where the field area is wide and the power supply is unstable, it is next to impossible for people to bring out the information-gathering terminals every day or maintain a server-based system. In this situation, UAVs become a reliable source for collecting environmental data (Kim et al. 2019).

The IoT platform consists of energy-supplying devices, IoT base stations, gateways, a cloud data center, and mobile or web application software (for end users, *i.e.,* farmers). Long Range (LoRa) sensor connection modules and TV white spaces (TVWS)) make up the IoT base station (Fig. 15.2). Data collected from the sensors is at first, transmitted to the base station and then forwarded to the Internet, generally to a cloud data center. LoRa modules having a long range of transmission are used to collect data from different sensors and transmit them to the gateway. TVWS refers to an unused TV signal band (radio video segment), originally allocated for broadcasting. The high bandwidth of TVWS allows for the transmission of videos or images, as they require a substantial amount of data to be transferred. The TVWS connects the gateway (commonly present in the farmers' houses) with the on-farm

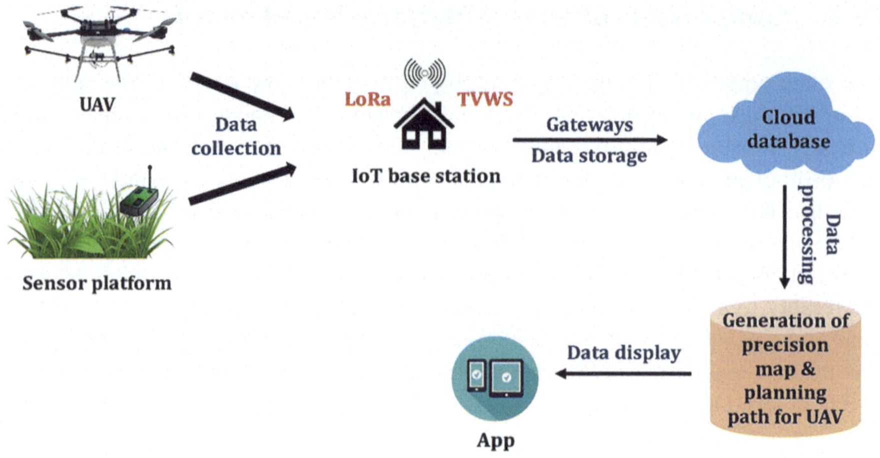

Fig. 15.2 Workflow diagram of the proposed IoT-based system

IoT base stations and the LoRa modules connect the sensors in the field with the base stations. In general, the data is forwarded from the gateway to the data cloud center via wired or wireless (preferably Wi-Fi) networks. Finally, the cloud data center processes the data received from the gateway (Gao et al. 2020).

The data, so collected, assists in the generation of a precision map and planning a path for the UAVs. The cloud center is the most important component of IoT architecture as it is responsible for data analysis. The results of such analysis show the conditions of crops for farmers in real time. The processed data is displayed in the app and helps the farmers visualize the problems in their fields so that management decisions can be taken at the proper time (Fujiwara et al. 2022).

The most effective way of collecting data is using a combination of in-field sensors for weather, soil health, plant developmental parameters, etc., and utilizing UAVs for collecting the data acquired by the sensors while using it as a sensor itself. UAVs have a limited amount of fuel, which puts a limit on the flight time. Hence, a flight path is to be preplanned so that the farm area can be covered effectively and the time can be used in an optimum manner. While planning the path, various environmental and geo-spatial parameters (wind direction, other weather conditions, nature of terrain, height, the field of view of the camera, required image quality, etc.) are to be kept in mind. It is better to design multiple flight paths to check the whole farmland, in which one is the preferred path and the rest are set aside as potential paths (Gao et al. 2020). Wind speed is a big factor and the flight angle and the speed of flight of the UAV is to be adjusted accordingly to achieve the longest possible flight time, utilizing the wind force in its favor. If proper adjustment is not done, valuable energy is wasted and flight time gets cut down. Specialized sensors for recording the wind direction and speed are present on the drones to adjust the flight path in real time. Therefore, a combination of preplanning and on-the-spot navigation is required to properly collect data from the sensors using UAVs.

15.5 Instances of Field Monitoring and Surveillance Based on IoT and Drones

Till date, IoT and drone technologies have found success in several field monitoring and surveillance operations, like weed mapping, crop growth monitoring, crop health monitoring, soil health status, water stress monitoring, etc.

15.5.1 Crop Health Monitoring

In the event of crops succumbing to pest and disease infestations, alterations in parameters such as coverage, biomass, Leaf Area Index (LAI), leaf cell structure, nitrogen levels, moisture content, pigment composition, and overall visual characteristics ensue. Consequently, these modifications contribute to variations in the reflectance spectrum across the visible to thermal infrared spectrum (Montero and Rueda 2018).

Multispectral indices can be calculated from the data gathered from reflection patterns at different wavelengths (visible, NIR, and thermal infrared rays) by sensors. These indices help in assessing the current condition of the crop and predict probable future scenarios, which help in decision-making. The drones can cover large areas of the Normalized Difference Vegetation Index (NDVI). It denotes the overall health of the vegetation (Simelli and Tsangaris 2015). It is calculated as follows-

$$\mathrm{NDVI} = \left(R_{\mathrm{NIR}} - R_{\mathrm{RED}}\right) / \left(R_{\mathrm{NIR}} + R_{\mathrm{RED}}\right)$$

where, R_{NIR} = Reflectance of the near infrared band; R_{RED} = Reflectance of the red band

The chlorophyll of green plant's leaves strongly reflects near-infrared light while absorbing red and blue ones. The index gives values −1 to +1, with values near zero indicating no vegetation and values near +1 (0.8–0.9) denoting highly dense green vegetation (Sato 2003). For instance, in a study conducted by Gao et al. (2020), the spectral reflectance of healthy and diseased wheat was compared. It was found that healthy vegetation results in a depression in the red region ("red valley") due to large amounts of radiation absorbance in that zone. At the same time, a peak is produced in the green region ("green peak") due to a high level of reflectance. As the disease progresses in the crop, both the peak and the valley disappear gradually, indicating that the vegetation is no longer robust.

15.5.2 Crop Growth Monitoring and Yield Estimation

Effective monitoring and estimation of crop yield are critical for enhancing agricultural productivity and quality. The lack of systematic monitoring tools poses a significant challenge in this regard, exacerbated by unpredictable weather patterns that affect crop micro-climates and jeopardize production. UAVs offer a promising solution by enabling regular data collection and visualization of crop growth dynamics. Recent research has focused on monitoring crop biomass, nitrogen levels, and yield prediction, crucial for optimizing fertilizer use and other interventions. UAVs facilitate the creation of detailed three-dimensional crop maps and the measurement of parameters like crop height and LAI (Tsouros et al. 2019).

15.5.3 Water Stress Monitoring

Characterizing water stress is a difficult task as the drought affects several factors of the crop physiology (Espinoza et al. 2017). Difference between air and canopy temperatures either by using thermal imagery or by measuring the temperatures directly is the most common method of detecting water stress in plants (Gonzalez-Dugo et al. 2013; Ludovisi et al. 2017).

Thresholds and regression equations worked out on thermal imagery data do not tend to hold when there is even a slight difference in conditions (environmental, physiological, genetic, etc.). Instead, using photochemical reflectance index (PRI) based on multispectral or hyperspectral images is preferable as it is sensitive to the changes in the carotenoids in the leaves (Berni et al. 2009; Zarco-Tejada et al. 2012, 2013).

15.5.4 Weed Mapping

Weeds, undesirable plants growing within crops, pose significant challenges by competing for essential resources (water and space), thereby reducing crop yields and growth potential. Excessive herbicide usage can lead to the development of herbicide-resistant weeds, further impacting crop productivity. Hyperspectral imaging enables discrimination between weed species with varying resistance to glyphosate, while RGB sensors aid in classifying different weed types. Researchers utilize drones equipped with hyperspectral sensors to monitor weed growth based on chlorophyll content and leaf density within the plant canopy. Site-specific weed management relies on precise weed cover maps derived from drone-collected images and data across entire fields. These maps facilitate targeted herbicide spraying, minimizing chemical usage and maximizing efficiency by pinpointing areas requiring treatment (Bah et al. 2018a, b).

15.5.5 Nutrient Status

To thrive and produce a good yield, appropriate levels of nutrients are required by plants (nitrogen for vegetative growth, potassium for disease resistance, etc.). Determining nutritional status by using plant color guides is the most common method, though they do not allow for quantitative assessments. Using vegetative indices and fitting them into regression models is the next best method for determining nutrient status (Dutta and Goswami 2020). Extensive studies need to be carried out to ascertain a proper method of assessing the nutrient status of the soil as well as the crop.

15.6 Conclusion

The necessity for transformative changes in the agriculture sector is emphasized to meet the increasing demands of a growing population, ensure farmers' welfare, and address emerging uncertainties on national and international levels, such as the COVID-19 pandemic. The rapidly emerging technology of drones is identified as a key player in this revolution due to its diverse applications. Drones not only enhance overall agricultural performance but also inspire farmers to embrace precision agricultural practices, transforming the profession from labor-intensive to technology-intensive. The reduction of human errors and inefficiencies in conventional agricultural practices is highlighted as a benefit of drone technology, acting as "an eye in the sky" to provide accurate and reliable information about the agricultural field's situation. The combination of software, sensors, cameras, and analytical tools autonomously captures and interprets data, reducing human errors significantly. Drones assist farmers in decision-making and can replace human labor by performing tasks with greater accuracy and efficiency. Despite the potential benefits, challenges hinder the widespread adoption of drone technology in agriculture. Addressing these challenges is deemed crucial for effectively leveraging the technology's potential and transforming the lives of millions of farmers. The analysis of large datasets obtained by drones and sensors provides insights into the relationship between pests, diseases, and the farm environment. The prevention and control of agricultural pests and diseases are highlighted as systematic projects requiring long-term observation and analysis. Despite advancements, challenges in drone technology adoption in agriculture persist, including cost, limited battery life, user literacy, and shortcomings in image processing and data analysis. Resolving these issues is deemed essential to raise the adoption rate of drones in the agricultural sector.

15.7 Future Prospect

Drone adoption rates need to be increased, but there are still many obstacles to overcome before drone technology can be widely used in agriculture. Factors such as the expense of the technology, the longevity of drone batteries, the risk of visual impairment, user proficiency, and deficiencies in image processing and data analysis constitute key concerns. Nevertheless, the incorporation of machine learning into drone systems has enabled the development of user-friendly solutions tailored to farmers' needs. In order to fully use drone technology's potential to revolutionize the agriculture industry and the lives of millions of farmers in India, it is imperative that these adoption hurdles be addressed.

Conflict of Interest None of the authors have conflict of interest to declare.

References

Ali H, Hang LY, Suan TY, Polaiah VR, Aluwi MIF, Zabidi AM, Elshaikh M (2021) Development of surveillance drone based internet of things (IoT) for industrial security applications. J Phys: Conf Ser 2107(1):012018. IOP Publishing. https://doi.org/10.1088/1742-6596/2107/1/012018

Almalki FA (2018) Optimisation of a propagation model for last mile connectivity with low altitude platforms using machine learning. Ph.D. Dissertation. Brunel University, London, UK

Bah MD, Hafiane A, Canals R (2018a) Deep learning with unsupervised data labeling for weed detection in line crops in UAV images. Remote Sens 10(11):1690. https://doi.org/10.3390/rs10111690

Bah MD, Hafiane A, Canals R (2018b) Weeds detection in UAV imagery using SLIC and the hough transform. In: Proceedings of the 2017 Seventh International Conference on Image Processing Theory, Tools and Applications (IPTA), Montreal, QC, Canada, 28 November–1 December, pp 1–6. https://doi.org/10.1109/IPTA.2017.8310102

Bali A, Chouhan SS, Kumar G et al (2023) Adaptive fault-tolerant control for pure-feedback stochastic nonlinear systems with sensor and actuator faults. Circ Syst Signal Process 42:5838–5867. https://doi.org/10.1007/s00034-023-02366-7

Berni JAJ, Zarco-Tejada PJ, Suarez L, Fereres E (2009) Thermal and narrowband multispectral remote sensing for vegetation monitoring from an unmanned aerial vehicle. IEEE Trans Geosci Remote Sens 47:722–738. https://doi.org/10.1109/TGRS.2008.2010457

Dutta G, Goswami P (2020) Application of drone in agriculture: a review. Int J Chem Stud 8(5):181–187. https://doi.org/10.22271/chemi.2020.v8.i5d.10529

Espinoza CZ, Khot LR, Sankaran S, Jacoby PW (2017) High Resolution multispectral and thermal remote sensing- based water stress assessment in subsurface irrigated grapevines. Remote Sens 9:961. https://doi.org/10.3390/rs9090961

Farooq MS, Riaz S, Abid A, Umer T, Bin Zikria Y (2020) Role of IoT technology in agriculture: a systematic literature review. Electronics 9(2):319. https://doi.org/10.3390/electronics9020319

Fujiwara R, Kikawada T, Sato H, Akiyama Y (2022) Comparison of remote sensing methods for plant heights in agricultural fields using unmanned aerial vehicle-based structure from motion. Front Plant Sci 13:886804. https://doi.org/10.3389/fpls.2022.886804

Fukatsu T, Hirafuji M (2003) Development of a field server for field monitoring. Agric Inf Res 12(1):1–12

Gao D, Sun Q, Hu B, Zhang S (2020) A framework for agricultural pest and disease monitoring based on Internet-of-Things and unmanned aerial vehicles. Sensors 20(5):1487. https://doi.org/10.3390/s20051487

Ghazali MHM, Azmin A, Rahiman W (2022) Drone implementation in precision agriculture–a survey. Int J Emerg Technol Adv Eng 12(4):67–77. https://doi.org/10.46338/ijetae0422_10

Gonzalez-Dugo V, Zarco-Tejada P, Nicolás E, Nortes PA, Alarcón JJ, Intrigliolo DS, Fereres EJPA (2013) Using high resolution UAV thermal imagery to assess the variability in the water status of five fruit tree species within a commercial orchard. Precis Agric 14:660–678. https://doi.org/10.1007/s11119-013-9322-9

Gracia-Romero A, Vergara-Díaz O, Thierfelder C, Cairns J, Kefauver S, Araus J (2018) Phenotyping conservation agriculture management effects on ground and aerial remote sensing assessments of maize hybrids performance in Zimbabwe. Remote Sens 10:349. https://doi.org/10.3390/rs10020349

Hussain R, Sahgal J, Mishra P, Sharma B (2012) Application of WSN in rural development, agriculture water management. Int J Soft Comput Eng 2(5):68–72

Kim J, Kim S, Ju C, Son HI (2019) Unmanned aerial vehicles in agriculture: a review of perspective of platform, control, and applications. IEEE Access 7:105100–105115. https://doi.org/10.1109/ACCESS.2019.2932119

Kumar R, Singh UP, Bali A et al (2023) Adaptive control of unknown fuzzy disturbance-based uncertain nonlinear systems: application to hypersonic flight dynamics. J Anal. https://doi.org/10.1007/s41478-023-00687-z

Lambert J, Hicks H, Childs D, Freckleton R (2018) Evaluating the potential of unmanned aerial systems for mapping weeds at field scales: a case study with Alopecurus myosuroides. Weed Res 58:35–45. https://doi.org/10.1111/wre.12275

Ludovisi R, Tauro F, Salvati R, Khoury S, Mugnozza Scarascia G, Harfouche A (2017) UAV-based thermal imaging for high-throughput field phenotyping of black poplar response to drought. Front Plant Sci 8:1681. https://doi.org/10.3389/fpls.2017.01681

Mogili UR, Deepak BBVL (2018) Review on application of Drone systems in precision agriculture. Procedia Comput Sci 133:502–509. https://doi.org/10.1016/j.procs.2018.07.063

Montero D, Rueda C (2018) Detection of palm oil bud rot employing artificial vision. In: IOP conference series: materials science and engineering, vol 437. IOP Publishing, Bristol, UK, p 012004. https://doi.org/10.1088/1757-899X/437/1/012004

Nakayama M, Ohno M, Hori H, Tanaka M, Toda N, Hirafuji M (2014) Research on the requirements for field server, and study of a design and manufacturing method to mass-produce field server. Agric Inf Res 23(1):29–37. https://doi.org/10.3173/air.23.29

Näsi R, Viljanen N, Kaivosoja J, Alhonoja K, Hakala T, Markelin L, Honkavaara E (2018) Estimating biomass and nitrogen amount of Barley and grass using UAV and aircraft based spectral and photogrammetric 3D features. Remote Sens 10(7):1082. https://doi.org/10.3390/rs10071082

Nihalani R et al (2024) Long Short-Term Memory (LSTM) model for Indian sign language recognition. J Intell Fuzzy Syst:1–19. https://doi.org/10.3233/JIFS-233250

Panchal B, Pranjal P, Patel RK, Sharma A, Chouhan SS (2024) Assessing the quantity of a crop field using aerial images. In: Chouhan SS, Singh UP, Jain S (eds) Applications of computer vision and Drone technology in agriculture 4.0. Springer, Singapore. https://doi.org/10.1007/978-981-99-8684-2_11

Parraga A, Doering D, Atkinson JG, Bertani T, de Oliveira Andrades Filho C, de Souza MRQ, Ruschel R, Susin AA (2018) Wheat plots segmentation for experimental agricultural field from visible and multispectral UAV imaging. In: Proceedings of the Sai intelligent systems conference, London, UK, 6–7 September 2018. Springer, Cham, Switzerland, pp 388–399

Polinova M, Salinas K, Bonfante A, Brook A (2019) Irrigation optimization under a limited water supply by the integration of modern approaches into traditional water management on the cotton fields. Remote Sens 11:2127. https://doi.org/10.3390/rs11182127

Solanki S, Singh UP, Chouhan SS (2023a) Brain tumor classification using ML and DL approaches. In: 2023 IEEE 5th International Conference on Cybernetics, Cognition and Machine Learning Applications (ICCCMLA), Hamburg, Germany, pp 204–208. https://doi.org/10.1109/ICCCMLA58983.2023.10346854

Saito Y, Kobayashi K, Suzuki T, Hirafuji M, Kiura T, Fukatsu T (2013) Agriserver: construction of a real-time field sensing network and utilization of acquired data. Agric Inf Res 22(1):1–11

Sato A (2003) The rmax helicopter uav: DTIC document

Saxena A, Chouhan SS, Aziz RM et al (2024) A comprehensive evaluation of marine predator chaotic algorithm for feature selection of COVID-19. Evolving Syst. https://doi.org/10.1007/s12530-023-09557-2

Sharma A, Patel RK, Pranjal P, Panchal B, Chouhan SS (2024) Computer vision-based smart monitoring and control system for crop. In: Chouhan SS, Singh UP, Jain S (eds) Applications of computer vision and drone technology in agriculture 4.0. Springer, Singapore. https://doi.org/10.1007/978-981-99-8684-2_5

Simelli I, Tsangaris A (2015) The Use of Unmanned Aerial Systems (UAS) in agriculture. In: HAICTA, pp 730–736

Solanki S, Singh UP, Chouhan SS, Jain S (2023b) Brain tumour detection and classification by using deep learning classifier. Int J Intell Syst Appl Eng 11(2s):279. Retrieved from https://ijisae.org/index.php/IJISAE/article/view/2624

Solanki S, Singh UP, Chouhan SS et al (2024) A systematic analysis of magnetic resonance images and deep learning methods used for diagnosis of brain tumor. Multimed Tools Appl 83:23929–23966. https://doi.org/10.1007/s11042-023-16430-6

Taylor J, William R, Munson K (1977) Jane's Pocket book of remotely piloted vehicles: robot aircraft today. Collier Books, New York, NY

Tsouros DC, Bibi S, Sarigiannidis PG (2019) A review on UAV-based applications for precision agriculture. Information 10(11):349. https://doi.org/10.3390/info10110349

Ullo SL, Sinha GR (2020) Advances in smart environment monitoring systems using IoT and sensors. Sensors 20(11):3113. https://doi.org/10.3390/s20113113

Yin H, Cao Y, Marelli B, Zeng X, Mason AJ, Cao C (2021) Soil sensors and plant wearables for smart and precision agriculture. Adv Mater 33:2007764. https://doi.org/10.1002/adma.202007764

Yonah IB, Mourice SK, Tumbo SD, Mbilinyi BP, Dempewolf J (2018) Unmanned aerial vehicle-based remote sensing in monitoring smallholder, heterogeneous crop fields in Tanzania. Int J Remote Sens 39:5453–5471. https://doi.org/10.1080/01431161.2018.1455241

Zarco-Tejada P, Gonzalez-Dugo V, Berni J (2012) Fluorescence, temperature and narrow-band indices acquired from a UAV platform for water stress detection using a micro-hyperspectral imager and a thermal camera. Remote Sens Environ 117:322–337. https://doi.org/10.1016/j.rse.2011.10.007

Zarco-Tejada PJ, González-Dugo V, Williams LE, Suarez L, Berni JA, Goldhamer D, Fereres E (2013) A PRI-based water stress index combining structural and chlorophyll effects: assessment using diurnal narrow-band airborne imagery and the CWSI thermal index. Remote Sens Environ 138:38–50. https://doi.org/10.1016/j.rse.2013.07.024

Zhang X, Cao Z, Dong W (2020) Overview of edge computing in the agricultural internet of things: key technologies, applications, challenges. IEEE Access 8:141748–141761. https://doi.org/10.1109/ACCESS.2020.3013005

Zheng H, Cheng T, Li D, Zhou X, Yao X, Tian Y, Cao W, Zhu Y (2018) Evaluation of RGB, color-infrared and multispectral images acquired from unmanned aerial systems for the estimation of nitrogen accumulation in rice. Remote Sens 10:824. https://doi.org/10.3390/rs10060824

IoT-Based Real-Time Farm Management System for Smart Agriculture

16

Jyotirmay Mahapatra, Ramesh K. Sahni,
Pradeepta Kumar Sahoo, and D. J. Shrinivasa

Abstract

There is an increasing global demand for food while the resources available are reducing day by day. Adopting proper management technology for agriculture helps in efficient utilization of human resource and time. Internet of Things (IoT) uses sensors to collect data, communication to transfer, edge devices or cloud to store, computers to analyze, smart devices to monitor, and actuators to implement. IoT has enabled farmers to automate many of the farm activities. Besides that, farmers can access and analyze farm data in real time resulting in advanced prediction of disease, pests, weather, and yield. The inputs like water, fertilizer, pesticides, weedicides, insecticides, fungicides, and seed can be optimized using gathered soil, crop, and environmental data. Thus, applying site-specific management helps in reducing input cost and environmental hazards. Autonomous

J. Mahapatra
ICAR-Central Institute of Agricultural Engineering, Bhopal, Madhya Pradesh, India

College of Agricultural Engineering and Technology, Odisha University of Agriculture and Technology, Bhubaneswar, Odisha, India

R. K. Sahni (✉)
ICAR-Central Institute of Agricultural Engineering, Bhopal, Madhya Pradesh, India

Center for Precision and Automated Agricultural Systems, Washington State University, Prosser, WA, USA
e-mail: ramesh.sahni@icar.gov.in

P. K. Sahoo
College of Agricultural Engineering and Technology, Odisha University of Agriculture and Technology, Bhubaneswar, Odisha, India

D. J. Shrinivasa
Department of Agricultural Engineering, Institute of Agricultural Sciences, Banaras Hindu University, Varanasi, Uttar Pradesh, India

S. S. Chouhan et al. (eds.), *Artificial Intelligence Techniques in Smart Agriculture*, https://doi.org/10.1007/978-981-97-5878-4_16

agricultural vehicles also apply IoT to share data. Artificial Intelligence (AI) techniques are often used for finding meaningful insights from collected data. IoT is utilized to maintain optimal soil moisture, lighting conditions, CO_2 concentration, humidity, temperature, and pH for maximizing crop growth under greenhouse like advanced agricultural structures. This chapter discusses the concept of smart agriculture, major sensors, and technologies used in IoT-based agricultural practices. Furthermore, data communication, storage and application of IoT and AI in agriculture have been discussed in detail.

Keywords

Actuators · RGB camera · Thermal camera · Leaf Area Index (LAI) · Extreme Learning Machine algorithm (ELM) · Genetic Algorithm (GA) · Wireless sensor networks (WSNs)

16.1 Introduction

Smart agriculture refers to the application of Internet of Things (IoT) technologies in agriculture. In agriculture, the IoT uses sensors, linked devices, and data analytics to collect and process data so that farmers may make more informed and effective decisions. This technology change has the power to completely transform the agriculture industry by increasing sustainability, productivity, and resource management.

Hamad et al. (2018) stressed the value of smart phones for accessing agronomy facts about various parameters, such as temperature, humidity, and soil moisture. In order to gather data in real time for precision agriculture, IoT/sensor nodes are essential. By gathering real-time data from the crop fields to improve precision in the agriculture system, these nodes have the potential to make the system more useful. Mahapatra and Sahoo (2023) discussed the role of different sensors in UAV to collect crop information. Data analytics and machine learning (ML) are incorporated to make the agriculture system more practical. These technologies are all incredibly useful in different domains. Precision agriculture has led to the development of numerous programs that provide farmers with timely information about crop status (Sahni et al. 2018a; Tiwari et al. 2019). Typically, the architecture for precision agriculture consists of three primary stages. In the first stage, a large number of sensors and IoT nodes are used to monitor various aspects of soil, environment, and plant life. For example, a soil moisture sensor measures the soil's wetness level, while a soil nutrient sensor determines the soil's fertility (Chouhan et al. 2024; Sharma et al. 2024; Panchal et al. 2024). We must gather this precise data in the second phase. Depending on the situation, we can either send the data to the cloud for further processing and remote monitoring, or we can store it locally at the closest fog node. The third stage of the design uses analytics techniques to determine the crop fields' current condition. The end users, who are farmers, are then given access to this information, which enables them to determine if the reading is above or below the threshold. As a result, they start a communication channel with the actuator that controls the watering system's on/off flow of water to the soil. Also, the farmer (or end user) may need to apply a potassium, nitrogen, and phosphorus fertilizer to the soil to restore its fertility. A response mechanism is triggered by the detection (sensing/predicting) of any crucial circumstance using analytics and actuators.

16.2 Major Sensors Used in IoT for Smart Agriculture

In IoT-based smart agriculture, various sensors are employed to gather real-time data from the field. These sensors help monitor and optimize factors such as soil conditions, weather, crop health, and overall environmental parameters. Some of the commonly used sensors in smart agriculture are discussed in Table 16.1.

Table 16.1 Major sensors used in IoT and their application in smart agriculture

Sensors	Working principle	Application in smart agriculture
Time Domain Reflectometry (TDR)	Velocity of pulse is slower in relatively moist soil than it is in drier soil	Measures the moisture content in the soil
Frequency Domain Reflectometry (FDR)	Difference between the frequency of the output wave and the return wave	
Neutron probe	Density and hydrogen concentration of slow neutron clouds rise with soil moisture content	
Infrared remote sensing	Absorption or reflection of the light depends on soil moisture content	
NIR spectroscopy	Different wavelengths of light can be absorbed to reveal details about the chemical makeup of the soil, including its nutrient level	Measurement of nutrient availability in soil
Magnetic sensor	Estimate iron content, indirectly by using magnetic characteristics	
Ion-selective electrode	Electrodes have a specific ion selectivity, and the concentration of that ion in the soil affects how potential changes	
NPK sensor	Generates a detectable electrical signal by means of an electrode that selectively reacts with nitrate, phosphate, or potassium ions	
GPS	Travel time of signal from a satellite to GPS receiver is used to determine its distance from each satellite	Track movement and pinpoint exact location.
PAR (Photosynthetically Active Radiation) meter	Counts the number of photons falling on a square meter within the PAR range every second	Measure the amount of light available is sufficient for photosynthesis.

(continued)

Table 16.1 (continued)

Sensors	Working principle	Application in smart agriculture
LiDAR (Light Detection and Ranging)	By sending out pulses on a constant basis and timing the duration of each pulse's return, it produces a dense collection of distance measurements	Measures distances and produces accurate, three-dimensional data about the features of nearby objects
Incremental encoder	Count pulses with respect to reference point	Precise measurement of degree of rotation or displacement
Absolute encoder	Distinct binary or gray code that indicates the shaft's absolute position	
RGB Camera	Gathers visible light (400–700 nm), then organizes that information to render images and video streams	Computer vision-based object detection
Multispectral camera	Record data in a few distinct bands (often less than 10). Each substance has an own spectral signature	Material identification, crop stress (disease, pest infection, water scarcity, nutrition requirement, etc.) detection
Hyperspectral camera	Record data in hundreds or even thousands of tiny, contiguous bands	
Thermal camera	Infrared radiation rises with temperature, and warmer objects emit more of it	Crop stress like water scarcity, nutrient shortages, or infections can be identified
Leaf Area Index (LAI) sensor	Energy emitted and reflected by plants, as well as incoming solar radiation, are used to calculate LAI	Crop stress such as disease or nutritional deficiencies and yield can be predicted
SPAD (Single Photon Avalanche Diode)/ Chlorophyll Sensor	The sensor measures how much light (wavelength: 660 and 730 nm) is reflected or passes through the leaves	Used in harvesting, phenotyping, crop stress, nutrient status analysis, and fertilizer application optimization
NDIR CO_2 sensor	Counts the quantity of infrared light that enters an air sample after the sensor produces light at a predetermined wavelength	Helps in maintaining optimal conditions for photosynthesis inside greenhouse
MOS CO_2 sensor	Variations in gas concentration cause changes in the metal oxide's conductivity	
Conductive	Change in conductivity	Useful for gas identification, pH, salinity, and soil moisture measurements
Capacitive	Change in capacitance	Sense touch and measure proximity, liquid level, displacement, motion, humidity, angular location, etc.
Inductive	Change in inductance	Position, proximity, metal detection
Resistive	Change in resistance	Temperature, light intensity, pressure, force, humidity, and gas measurement

16.2.1 Soil Moisture Sensor

It measures the moisture content in the soil to optimize irrigation and prevent overwatering or underwatering. Several types of soil sensors are used in agriculture such as capacitive, Time Domain Reflectometry (TDR) based, Neutron probe, and infrared (Fig. 16.1). Capacitive sensors assess soil moisture by identifying changes in the dielectric characteristics of the soil between two electrodes. Sensors that use TDR use parallel rods that function as transmission lines. The rods receive an applied voltage, which is then reflected back to the sensor for processing. The apparent permittivity of the substrate has an impact on the velocity or speed of the voltage pulse as it moves along the rod. The pulse's velocity is slower in relatively moist soil than it is in drier soil. Salt concentration and salinity can have an impact on the dielectric characteristics of soil. Nevertheless, a lot of probes detect EC separately and use that information to compensate the impact of salinity on moisture readings. An oscillator is used in frequency domain reflectometry (FDR) to send an electromagnetic signal via a metal tin or other wave guide. Soil moisture is assessed by taking the difference between the frequency of the output wave and the return wave. The change in frequency of a reflected radio wave or resonance frequency can be used to calculate capacitance.

The neutron source, which is buried in the soil to be tested, is used in another technique. Continuous fast neutron emissions cause different atomic ions to collide with the soil's medium. The atoms are then slowed down by the rapid neutrons as they release energy. Fast neutrons lose the greatest energy and slow down the fastest when they collide with hydrogen atoms. The density and hydrogen concentration of slow neutron clouds both rise as soil moisture content does. The functional link between the density of the slow neutron cloud and the molecule of water is how the neutron meter calculates the moisture content of the soil.

An efficient way to get information on soil moisture at the regional level is to employ the infrared remote sensing approach, which involves transmitting or

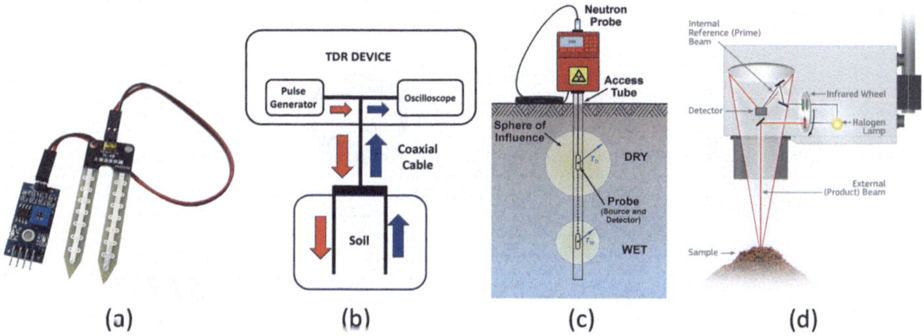

Fig. 16.1 Different types of soil moisture sensors: (**a**) capacitive, (**b**) TDR (**c**) neutron probe, and (**d**) infrared

reflecting electromagnetic waves through measurements of the soil surface properties (Yang et al. 2010).

16.2.2 Soil Nutrient Sensor

To determine nutrient contents and examine the optical characteristics of the soil, fluorescence or absorption spectroscopy is frequently used. In NIR spectroscopy, near-infrared light is shone into the soil, and the reflected light is analyzed. Different wavelengths of light can be absorbed to reveal details about the chemical makeup of the soil, including its nutrient level. Magnetic sensors estimate soil nutrient concentrations, particularly iron content, indirectly by using magnetic characteristics. The purpose of ion-selective electrode sensors is to quantify particular ions present in the soil. These ion-selective electrodes have a specific ion selectivity, and the concentration of that ion in the soil affects how potential changes. By evaluating the dielectric properties of soil, capacitive sensors determine the amount of nutrients and moisture in the soil. Soil NPK sensor generates a detectable electrical signal by means of an electrode that selectively reacts with nitrate, phosphate, or potassium ions. The amount of nitrogen present in the soil is then correlated with this signal.

16.2.3 Actuators

Actuators are vital parts of the IoT ecosystem because they translate digital signals or orders into physical actions, allowing devices to interact with and affect the real environment. Actuators that are electric, hydraulic, or pneumatic are widely used. Nonetheless, a variety of electric actuator types are frequently employed such as DC motor, stepper, servo, BLDC, and solenoid valves (Fig. 16.2). The selection of these actuators is based on the necessary speed, torque, and precision.

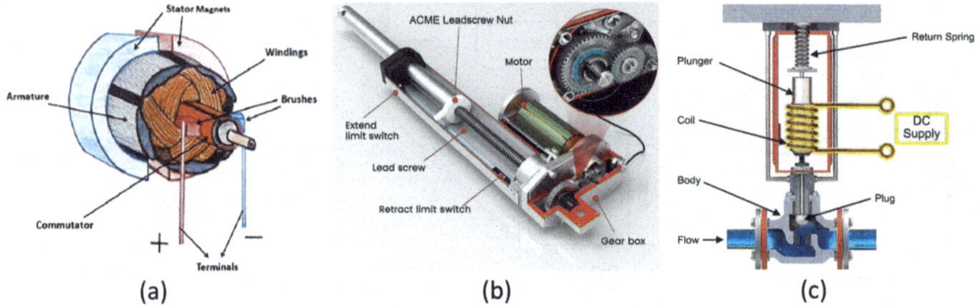

Fig. 16.2 Different types of actuators used for automatic control of agricultural activities: (**a**) DC motor, (**b**) linear actuator, and (**c**) solenoid valve

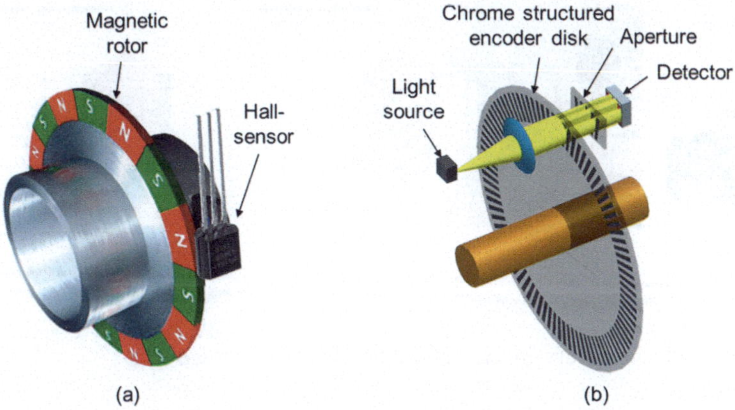

Fig. 16.3 Encoders used in smart agriculture: (**a**) magnetic encoder and (**b**) optical encoder

16.2.4 Encoders

Encoders transform mechanical motion into digital data or electrical signals. When the shaft or object they are attached to rotates, incremental encoders produce a series of pulses. The object's position or velocity is then determined by counting these pulses. Encoders are broadly divided into two categories: absolute encoders and incremental encoders (Fig. 16.3). Absolute encoders give each point on the encoder disc a distinct binary or gray code that indicates the shaft's absolute position. Magnetic sensors and a magnetized disc with unique magnetic patterns for every position are used in magnetic absolute encoders. The number of pulses per revolution determines the resolution of incremental encoders, whereas the number of distinct positions that absolute encoders may represent determines the resolution of those encoders.

16.2.5 RGB, Multispectral, and Hyperspectral Camera

Hyperspectral cameras can record a far greater number of color bands, frequently covering the ultraviolet (UV), visible, and infrared (IR) sections of the spectrum, than regular RGB camera, which can only record three color bands (red, green, and blue). Hyperspectral cameras record data in hundreds or even thousands of tiny, contiguous bands, whereas multispectral cameras record data in a few distinct bands (often less than ten) (Fig. 16.4). This makes it possible to analyze the spectral properties of the scene under observation in more detail (Chowdhury et al. 2024). Each substance has its own spectral signature, or how it reacts to various light wavelengths. Material identification and distinction based on spectral characteristics is made possible by hyperspectral imaging.

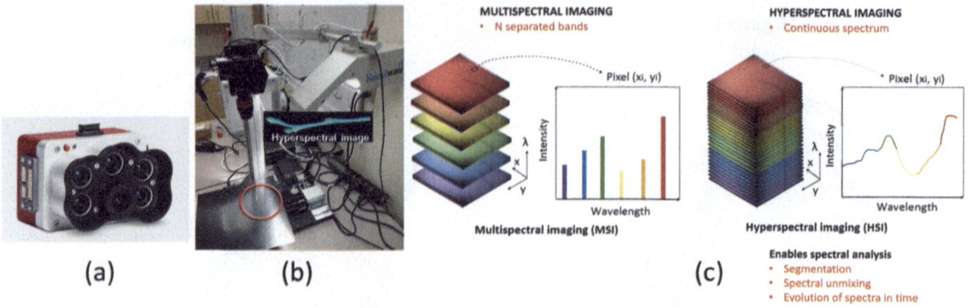

Fig. 16.4 (**a**) Multispectral camera (**b**) imaging with hyperspectral camera, and (**c**) spectral signature showing different wavelength indicating different crop parameters

Fig. 16.5 (**a**) mobile-based thermal camera, (**b**) thermal Imaging in the night for frost prediction, and (c) thermal image of a grapes bunch

16.2.6 Thermal Camera

Crop stress factors like water scarcity, nutrient shortages, or infections can be identified with the aid of thermal cameras (Fig. 16.5). The thermal signatures of healthy plants are usually consistent, and any deviation may point to possible problems. Temperature variations are sometimes the first sign of plant illnesses before any outward symptoms show. Early detection of these temperature fluctuations with thermal cameras allows farmers to take preventive action, such as isolating problematic regions or applying targeted pesticides. Based on temperature variations, thermal cameras are able to distinguish between weeds and crops. Precision weeding can make use of this data. Livestock health can also be observed with thermal cameras. These cameras can also be used to control temperature changes within greenhouses. These sensors can also be used to identify early frost, so that protective measures can be taken.

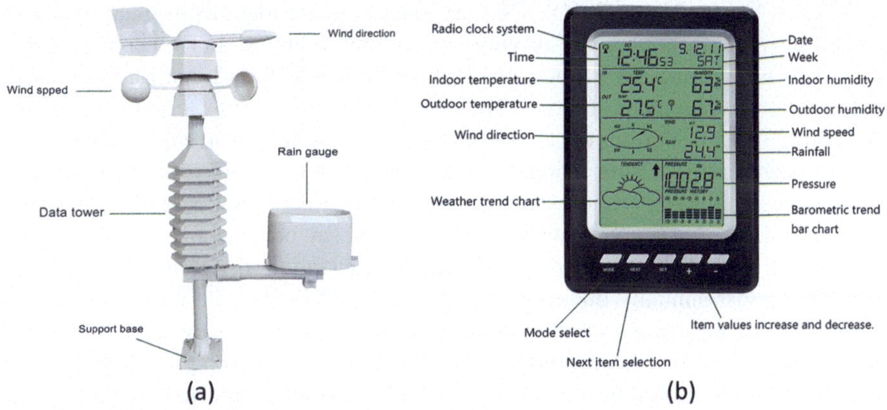

Fig. 16.6 (**a**) weather station and (**b**) GUI for visualization and retrieval/storage of data

16.2.7 Leaf Area Index (LAI) Sensors

By determining the quantity and density of foliage, LAI sensors assist in tracking the growth of crops. A farmer can forecast yield by utilizing the acquired data. Using LAI, crop stress factors such as disease or nutritional deficiencies can be examined. In crop phenotyping research, these sensors are employed to describe and measure the physical characteristics of plants. This ratio may be associated with processes of gas-vegetation exchange, including transpiration and evaporation, carbon flux, rainfall interception, and photosynthesis.

16.2.8 Weather Station

Typically, these stations are made up of a variety of sensors that detect several aspects of the surrounding environment, such as air pressure, CO_2 concentration, temperature, daylight hours, rainfall, humidity, and solar radiation. Farmers can make well-informed decisions about irrigation, planting, harvesting, and general crop management with the use of the gathered data. These weather stations could have an integrated data logger, communication system for remote access, and solar panel for energy. A weather station and a GUI (Graphical User Interface) are shown in Fig. 16.6.

16.3 IoT-Based Technologies

IoT offers new and workable solutions in various areas including smart agriculture. All farming tools and equipment can be connected to one another through the use of technology to decide on the optimal irrigation and fertilizer strategies (Kumar et al. 2018a). Intelligent systems improve the precision and performance efficiency of equipment used for raising animals and even tracking the growth of plants.

Wireless Sensor Networks (WSNs) are used to gather data from many types of sensing devices. Robotics, self-driving agricultural vehicles, and other automated equipment require the application of ground sensors, ICT, and control systems to enable smart management of farms. The growth of smart systems requires fast communicating internet, cutting-edge mobile devices, and satellites to supply data (visuals and positioning). Using many satellite images and sensors placed throughout fields, James et al. (2019) successfully used IoT real-time to find and recognize diseases of leaves that impede agricultural growth (in banana and paddy crops). According to a study by Food and Agriculture Organization (FAO) in 2017, 20–40% of crops are lost annually due to pests, illnesses, and inadequate crop monitoring. Thus, by using sensors and smart systems, it is feasible to keep an eye on the weather, fertility levels, and the correct amount of fertilizer needed for crop growth. Real-time benefits of smart agriculture technology, which is based on the IoT, include plant protection, disease prediction, irrigation, and improved product quality. The fundamental framework for the development and implementation of an agricultural IoT system is shown in Fig. 16.7. In agricultural IoT, data transmission is crucial, and

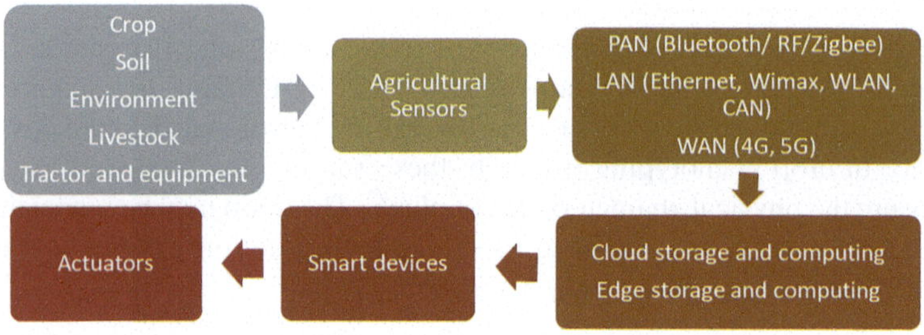

Fig. 16.7 Architecture of agricultural IoT

Table 16.2 IoT-based agricultural applications

Activity	Parameters and sensors	Microcontroller	Data transmission	Reference
Green house air conditioning	Temperature and humidity	PIC18F4520 microcontroller	Wireless sensor network (WSN), ZigBee, and fuzzy controller	Yang et al. (2013)
Irrigation in green house	Soil humidity, light, air humidity, and temperature	ARM processor	ZigBee and CC2530 protocol, WSN and transceiver hardware	Xu et al. (2012)
Monitoring pest, insect trapper	Camera	CC1110F32 SoC	3G/4G internet communication	López et al. (2012)
Irrigation	Soil moisture, temperature, and humidity	Arduino UNO	Global System for Mobile Communication (GSM)	Krishnan et al. (2020)

(continued)

Table 16.2 (continued)

Activity	Parameters and sensors	Microcontroller	Data transmission	Reference
Indoor lighting conditions	CO_2, humidity, temperature, spectrums, photoperiod, and intensity	Artmel-ATMega	Zigbee, Dash-7 and WiFi	Harun et al. (2019)
Hydroponic System	Ambient temperature, lighting, humidity, water level, pH	Raspberry Pi, Arduino UNO		Mehra et al. (2018)
Aeroponics	Temperature and humidity		USART	Idris et al. (2012)
Aeroponics	Humidity, temperature, light intensity, pH level, nutrition concentration, and water level			Bedjoirawan et al. (2023)
Irrigation	Rain, soil moisture, water flow	Raspberry pi	WiFi	Kumar et al. (2017)
Irrigation, humidity of mushroom farm	Humidity, temperature, and soil moisture	nodeMCU	Internet	Muangprathub et al. (2019)
Soil fertilizer recommendation system	Soil nutrient	Raspberry pi	Internet	Bodake et al. (2018)

WSNs are a key method of information sharing. Communication technology should not be selected at random for transferring agricultural IoT data; rather, it should be selected in accordance with the project needs. Table 16.2 presents some of the IoT-based research for the regulation of particular agricultural operations.

16.4 Data Communication, Storage, and Mining

16.4.1 Data Communication

The rise of sensing systems for data collection, processing, and analysis has coincided with the use of large volumes of data in agricultural management. This lowers the 4G network's ability to join each of the widely distributed parts of the smart network. With the introduction of the ultrafast 5G transition, data processing and transfer have recently become more straightforward procedures. The 5G network offers incredibly quick data transformation time. Since the 5G network's upload and download rates are many times faster than those of the 4G network, data transfer over it is quicker than over other networks.

Table 16.3 A comparison of agricultural communication technology parameters used in IoT (Xu et al. 2022)

Parameters	Wireless WAN (4G/5G/GPRS)	WiFi	Bluetooth	ZigBee	Lora
Area of application	Voice, data	Media	Media, cable alterative	Monitoring, sensos	Data transparent transmission
Power consumption	High	High	Low	Low	Low
Transmission distance	Long distance	< 100 m	< 10 m	10–100 m	Up to 2–5 km in towns and 15 km in suburbs
Advantage	Large coverage good service quality	High flexibility	Low price and simple configuration	Low power consumption, flexible network topology	Strong immunity and stable operation

A range of protocols for communication are frequently employed in farming with the goal of smart agriculture. By utilizing these protocols, growers may communicate effectively and decide on smart farming strategies that will monitor and enhance crop growth. The most popular wireless technologies are Mobile Communications Standards, WiFi, LoraWan, Low Rate Wireless Personal Area Networks, WiMax, RFID, Bluetooth, and ZigBee. Table 16.3 compares each of these wireless protocols used in the agricultural field.

16.4.2 Cloud Storage

When it comes to cloud computing in agriculture, sensors are employed for data collection and interpretation, especially in irrigation-based systems. When using cloud computing for irrigation, the monitored data is often stored and then retrieved as needed. Cloud computing users can store, analyze, and show data on a variety of platforms and devices using both free and paid solutions. The cloud's capabilities have made it possible for many agriculture-related companies to store and retrieve data that enhances productivity and effectiveness.

Furthermore, by developing irrigation algorithms, cloud computing has generated alerts. Many risks and hazards that would have otherwise happened have been decreased by employing these alerts. Using these signals makes it simpler to adjust work procedures and implement the necessary safety precautions to reduce the possibility of any issues. A number of cloud-related software has been developed to support job productivity. Every program has a distinct goal and significance, but how it is implemented depends on a variety of elements, such as cost, services provided, adaptability, and others. The maintenance of irrigation system looks challenging when involved risks, expenses, and complexity are considered. According to Fernández-Ahumada et al. (2019), cloud technology is utilized by irrigation workers to reduce risks and improve labor outputs, thereby achieving predefined targets.

16.4.3 Data Mining for IoT

In order to find the necessary knowledge, information analysis is the procedure of gathering, transforming, cleansing, and modeling data. The obtained data and findings are conveyed through recommendations for conclusions and assistance in making decisions. In order to uncover hidden correlations and other valuable information, data analytics involves analyzing big data sets with a range of data kinds.

Analytical tools' processes and algorithms must find hidden connections and patterns in the data. Multimedia field data gathered by the sensors that were deployed aids in the detection of diseases in leaves, stems, roots, and fruits as well as fruit quality and health (Suksawat and Komkum, 2015). Additionally, it helps with smart irrigation and weed identification. Drones and Internet of Things (IoT) sensor devices working together seems to be a good way to produce additional food of greater quality; the same gadgets also regularly monitor the situation to ensure the best possible results (Anand et al. 2023; Sahni et al. 2024). Huge volumes of data in a variety of formats are produced by devices like environmental sensors, drones, plant sensors, soil sensors, etc. This generated data can be cloud-stored using standard hardware. Crop rotations, crop patterns, weather parameters, soil types, environmental conditions, soil nutrients, GPS data, Geographic Information System (GIS) data, data from agricultural machinery and farmer records, such as Variable Rate Fertilizers (VRF), and yield monitoring are the main subjects of precision agriculture datasets (Bendre et al. 2015). Large data files in distributed and shared fault-tolerant databases hold these massive volumes of data. IoT data sets that have been stored can be analyzed by analytics tools like Map Reduce, Spark, and Skytree to extract value. This calls for the integration of ML and IoT data analytics. It is also necessary to construct forecasting models for managing crops, which include yield and disease prediction.

16.5 Application of AI in Agricultural IoT

The application of ML in farming is prevalent in fields such as weather prediction, crop quality, soil condition, disease and weed detection, irrigation planning, and yield detection. The use of AI in agricultural IoT has the capability to completely transform farming practices, improving sustainability, efficiency, and productivity. Below are several applications of AI in agricultural IoT

16.5.1 Crop and Yield Management

In agriculture, yield prediction and monitoring are essential because they provide users with the knowledge they need to take prompt action and minimize loss. SVM-based system with the data collected from the weather stations produces precise forecast of the crop development process. Automatic fruit counting helps in agricultural process planning and loss reduction for the farmer. Agricultural yield prediction and pest control analysis systems that are dependent on the surrounding conditions are mainly based on neural network.

A hybrid ML model incorporating IoT was presented by Gupta and Nahar (2023) for yield prediction. Preprocessing, feature selection (FS), and classification are the three stages of the job. The dataset is first preprocessed, and Correlation and the Variance Inflation Factor (VIF)-Based FS (CBFS) algorithms are used for FS. Lastly, a two-tier ML model for an Internet of Things-based smart farming system is suggested. The Adaptive k-Nearest Centroid Neighbor Classifier (aKNCN) model is suggested in the first tier in order to classify the soil samples according to the input attributes of soil and estimate the soil quality. The Extreme Learning Machine algorithm (ELM) is utilized in the second tier to forecast the crop output. A modified Butterfly Optimization Algorithm (mBOA) is used in the optimized strategy to update the weights in order to increase the ELM's performance accuracy while minimizing error values. Aggelopoulou et al. (2011) have predicted the site-specific yield using precision agricultural techniques such as image processing and data analysis algorithms. With an accuracy of over 80% in forecasting the apple orchard yield, they demonstrated the strong association between bloom density and fruit yield. By analyzing crop photos taken in the field at different levels of illuminations, a workable and meaningful solution is offered for assessing the maturity age of the crop groups using the combination of ANN + GA technique (Sharma et al. 2022).

16.5.2 Disease/Pest Management

The creation of early warning models is one of the latest developments in pest management. These big data-driven algorithms first gather several kinds of data that may lead to the expansion of pests. In order to anticipate the level of pest presence in the field, a neural network is trained using real-time data gathering from the fields pertaining to soil, weather, and other environmental factors. Pest identification in large farms and greenhouses is also found to be greatly aided by object detection utilizing Faster R-CNN (faster region-based convolutional neural network).

Geetha (2015) predicted the occurrence of late blight disease in potatoes applying IoT technology and ML techniques. The prediction was 94% accurate using a back-propagation network. Five on-site weather stations were used for data collection in orchards by Mazilu and Trandafir (2002). Models utilizing the gathered data can be applied to forecast the periods of apple scab infection. These models can assist farmers in determining if fungicide applications are necessary to control apple scab, or not. Cetişli and Büyükçingir (2013) have suggested a novel prediction model for apple scab warning. The suggested model is predicated on time series prediction and artificial intelligence. Rather than using the sum of the wetness duration, the apple scab infection period was assessed as the time series prediction model. Using feature selection techniques such as Fisher's Linear Discriminant Analysis, Pearson's Correlation Coefficient, and Adaptive Neuro fuzzy Classifier with Linguistic hedges, important hours were identified. The model used for prediction was an adaptable neural network. It takes 24 h to collect the measurements in order to determine the apple scab infection. Five meteorological parameters: temperature, rainfall, daylight, relative humidity, and leaf wetness were measured. Time was an additional component for time series prediction because data was gathered every 12 min. They noted a relationship between climatic data and apple scab to identify the time for infection.

Foughali et al. (2018) described a unique decision support system (DSS) that integrates WSN, cloud, and IoT to prevent late blight. One of the most important components of implementing greenhouses efficiently is integrated pest management, or IPM. Image processing techniques can further automate manual inspection. For analysis, traditional ML techniques such as Support Vector Machines (SVM) with a suitable kernel (radial basis kernel) can be applied. Entomologists label the insects in order to properly model and train the data for the identification of unfamiliar pests. It is also discovered that object identification models work well for spotting microscopic pests in greenhouses (Li et al. 2021).

16.5.3 Autonomous Tractor and Field Robots

The tractor performance monitors assist in remotely visualizing the complete operation by measuring and recording it. The criteria that are typically considered are wheel slip, draft, power, and fuel consumption. Enhancing these variables can significantly boost the tractor's efficiency. To measure the draft, a strain gauge was affixed to a ring transducer located at the front of the drawbar. A positive displacement flow meter was used to measure the fuel consumption, while toothed gears and magnetic pick-ups were used to detect wheel speed (Grogan et al. 1987).

Deep learning algorithms, sensors, and stereo cameras are major components of modern autonomous agricultural systems (Rathnayake et al. 2022). By using stereo matching, an RGB camera equipped with several infrared cameras is able to obtain depth images. For object detection, the well-liked CNN-based object detection algorithm YOLO (Redmon et al. 2016) is employed. This identifies the items within the pixel bounding boxes. Tractor manufacturers such as Case IH and John Deere have begun to provide autonomous tractors to farmers. Cameras and LiDAR, which can precisely identify obstacles, were added to Case IH's concept tractor.

Researchers have built autonomous tractors, harvesters, and transplanters using deep learning-based computer vision algorithms, with almost the same human efficiency (Nihalani et al. 2024). According to Blok et al. (2016), the autonomous fruit harvester prototype was primarily composed of an image acquisition module with an image manipulator module installed on a self-propelled carrier. To identify the fruits and vegetables that needed to be harvested, the input was entered into an object detection algorithm based on computer vision. Agriculture robots must plan their routes carefully before entering a farm to carry out a certain activity. ANN brings solution to the path planning problem of agricultural autonomous vehicles. Path detection techniques for agricultural mobile robots operating in a shadowed environment have been developed by Bo et al. (2010). An essential role that the ANN technique can play is reducing the amount of human labor in the greenhouse.

16.5.4 Soil Analysis

There are several models available for predicting the physical characteristics of soil. Understanding soil is crucial for farmers, as improper crop selection can cause significant harm if the physical characteristics and moisture content of the soil is not known (Sahni et al. 2018b). Numerous studies are conducted to identify soil texture utilizing methods such as Naïve Bayes, SVM, and decision trees in an effort to address these issues. To begin researching soil, a database with every physical characteristic of each type of soil must be established. The system is trained with a variety of learning algorithms to anticipate the properties of the soil. A review of the literature reveals that supervised learning algorithms are used in the majority of soil prediction-related research. The primary use of SVM is the classification of labeled linear data sets. Decision trees and Naïve Bayes are employed in addition to SVM. The SVM, often referred to as support vector networks, is a technique for supervised learning. Soil pattern classification is another application for SVM. The Naive Bayes algorithm is predicated on likelihood. It indicates the highest probability of discovering the answer. The test cases are located on several nodes in the decision tree. It is shown as a tree structure, where one condition is contained in each node. These are a few of the algorithms that are employed in agricultural research. In agricultural research, artificial neural network (ANN) approaches are also utilized to forecast the soil moisture content of a given region. ANN algorithms are used to train the collected data, yielding optimal outcomes.

16.5.5 Irrigation Management

For predicting soil moisture content, data-driven ML models are preferred over physical/mechanistic models. This is because data-driven models need comparatively less data for calibration than mechanistic model calibration. Furthermore, ML models have superior predictive accuracy compared to physical models and can process spatial and temporal data with ease and speed (Solanki et al. 2023, 2024). The irrigation systems based on ML models contain a closed-loop controller that uses feedback from the system's comparison of preprocessed data with measured real-time data to learn from calibration (Kumar et al. 2018b).

Using soil and climate data, Navarro et al. (2016) have suggested a smart irrigation decision support system (SIDSS) for weekly irrigation. The SIDSS forecasts the amount of water for a week using the Adaptive Neuro-Fuzzy Inference System (ANFIS). A linear time series model has been suggested in Delgoda et al. (2016) to forecast the root zone's deficiency in soil moisture. Though these models perform better in irrigation control, their calibration is more time consuming and restricted to the particular location in which it is performed.

According to Hinnell et al. (2010), ANNs were created to forecast the geographical distribution of water in the subsurface for use in neuro drip irrigation systems. Water distribution in the lowest soil layer is critical to the drip irrigation method's proper operation. Here, ANNs produce the forecast that the user finds useful, which facilitates quick decision-making. After the water from the emitter, which is on the surface of the land, has permeated the soil, ANN models provide the first and second wetting patterns. The user is thus given continuous patterns by the ANN model.

16.5.6 Frost Management

An intelligent weather station was used to provide smart frost management using neural networks in greenhouses (Castañeda-Miranda and Castaño-Meneses 2020). Furthermore, the knowledge and learning rules derived from the data are adjusted based on the ANN. As a result, a fuzzy system is attached to turn on an irrigation system on the greenhouse's plastic surface in order to estimate the uncertainty that represents the subjective perception of temperature. In this instance, a diffuse map enables the classification of frost and can be used to turn on the non-freeze control, which keeps the inside temperature warmer during the frost period. The temperature of the frost intelligent control was compared to the temperature measured in the croplands using the fuzzy system that was used (Wang et al. 2015). After that, the traditional Proportional Integrative Derivative (PID) control systems were compared with the Fuzzy Logic barrier control systems and Neural Network. The PID system needs an adaptive system that uses generic algorithms or control in the frequency domain to synchronize, as well as the ability to tune the values of three constants: proportional, integrative, and derivative. Additionally, the use of fuzzy logic systems enables the management of subjective and imprecise data, such as the description of low, medium, and high temperatures using fuzzy sets. The barrier control system that uses a neural network and fuzzy logic is trained based on the inputs and outputs of the system that are realized during adjustment or learning. Since this type of inference is made inductively, no particular physical modeling is needed in this case. Furthermore, the fuzzy system allows for subjective evaluations, sensor physical limitations, the aggregation of contradicting information, redundant data, and contradictory information. It also allows for the possibility of incomplete or inaccurate modeling. These configurations, referred to as neuro-fuzzy systems, combine the neural network's learning capability with the fuzzy system rules' ease of use when handling ambiguous data.

16.6 Conclusion

As the resources for agriculture are continuously shrinking and production demand is growing, automation is essentially required. IoT integrates advanced sensors, microcontrollers, communication technologies, data analysis techniques, GUI, and actuators for automatic or remote monitoring. This innovative approach harnesses the power of connected devices, sensors, and data analytics to deliver a dynamic and responsive framework for farmers. The system's ability to provide real-time insights into crop health, environmental conditions, and livestock well-being empowers farmers with the knowledge needed to make informed decisions promptly. The advanced autonomous tractor or agricultural robots are connected to the internet and can be operated based on analyzed sensor data. UAV may be integrated to provide sensed data to VRT equipment for IoT-based automated precision agriculture. The precision and efficiency achieved through automated processes, resource optimization, and predictive analytics contribute not only to increased agricultural productivity but also to environmentally conscious and sustainable farming practices. As the synergy between IoT and agriculture continues to evolve, this transformative technology is poised to reshape the landscape of farming, offering a more resilient, adaptive, and technologically advanced future for the agricultural sector. The IoT-based Real-Time Farm Management System emerges as a cornerstone in the journey toward smarter, more efficient, and sustainable agriculture.

References

Aggelopoulou AD, Bochtis D, Fountas S, Swain KC, Gemtos TA, Nanos GD (2011) Yield prediction in apple orchards based on image processing. Precis Agric 12:448–456. https://doi.org/10.1007/s11119-010-9187-0

Anand R, Sahni RK, Kumar SP, Thorat DS, Kumar AK (2023) Advancement in agricultural practices with use of drones in the context of precision farming. Global J Eng Sci 11(2):1–7. https://irispublishers.com/gjes/pdf/GJES.MS.ID.000760.pdf

Bedjoirawan NA, Izdaharra AM, Sintia S, Widianto MH (2023) Aeroponics cultivation of Bok Choy with IoT-based monitoring and automation system. In: International Conference on Informatics, Multimedia, Cyber and Informations System (ICIMCIS). IEEE, pp 155–160. https://doi.org/10.1109/ICIMCIS60089.2023.10349015

Bendre MR, Thool RC, Thool VR (2015) Big data in precision agriculture: Weather forecasting for future farming. In: International conference on next generation computing technologies (NGCT). IEEE, pp 744–750. https://doi.org/10.1109/NGCT.2015.7375220

Blok PM, Barth R, van den Berg W (2016) Machine vision for a selective broccoli harvesting robot. IFAC-PapersOnLine 49:66–71. https://doi.org/10.1016/j.ifacol.2016.10.013

Bo Z, Hua MW, He SZ, Rong ME (2010) Path recognition method of agricultural wheeled-mobile robot in shadow environment. In: International Conference on E-Health Networking Digital Ecosystems and Technologies (EDT), vol 1. IEEE, pp 284–287. https://doi.org/10.1109/EDT.2010.5496583

Bodake K, Ghate R, Doshi H, Jadhav P, Tarle B (2018) Soil based fertilizer recommendation system using Internet of Things. MVP J Eng Sci 1(1):13–19

Castañeda-Miranda A, Castaño-Meneses VM (2020) Internet of things for smart farming and frost intelligent control in greenhouses. Comput Electron Agric 176:105614. https://doi.org/10.1016/j.compag.2020.105614

Cetişli B, Büyükçingir E (2013) Time series prediction of apple scab using meteorological measurements. Afr J Biotech 12:5444–5451. https://doi.org/10.5897/ajb12.394

Chouhan SS, Singh UP, Jain S (2024) Introduction to computer vision and Drone technology. In: Chouhan SS, Singh UP, Jain S (eds) Applications of computer vision and drone technology in agriculture 4.0. Springer, Singapore. https://doi.org/10.1007/978-981-99-8684-2_1

Chowdhury M, Anand R, Dhar T, Kurmi R, Sahni RK, Kushwah A (2024) Digital insights into plant health: exploring vegetation indices through computer vision. In: Chouhan SS, Singh UP, Jain S (eds) Applications of computer vision and Drone technology in agriculture 4.0. Springer, Singapore, pp 7–30. https://doi.org/10.1007/978-981-99-8684-2_2

Delgoda D, Saleem SK, Malano H, Halgamuge MN (2016) Root zone soil moisture prediction models based on system identification: Formulation of the theory and validation using field and AQUACROP data. Agric Water Manage 163:344–353. https://doi.org/10.1016/j.agwat.2015.08.011

Fernández-Ahumada LM, Ramírez-Faz J, Torres-Romero M, López-Luque R (2019) Proposal for the design of monitoring and operating irrigation networks based on IoT, cloud computing and free hardware technologies. Sensors 19:2318. https://doi.org/10.3390/s19102318

Foughali K, Fathallah K, Frihida A (2018) Using Cloud IOT for disease prevention in precision agriculture. Procedia Comput Sci 130:575–582. https://doi.org/10.1016/j.procs.2018.04.106

Geetha MCS (2015) Application of classification technique in data mining for agricultural land. IJARCCE:352–355. https://doi.org/10.17148/ijarcce.2015.4384

Grogan J, Morris DA, Searcy SW, Stout BA (1987) Microcomputer-based tractor performance monitoring and optimization system. J Agric Eng Res 38:227–243. https://doi.org/10.1016/0021-8634(87)90091-6

Gupta A, Nahar P (2023) Classification and yield prediction in smart agriculture system using IoT. J Amb Intel Human Comput 14:10235–10244. https://doi.org/10.1007/s12652-021-03685-w

Hamad MAA, Eltahir MES, Ali AEM, Hamdan AM (2018) Efficiency of using smart-mobile phones in accessing agricultural information by smallholder farmers in North Kordofan—Sudan. SSRN Electr J. https://doi.org/10.2139/ssrn.3240758

Harun AN, Mohamed N, Ahmad R, Rahim ARA, Ani NN (2019) Improved Internet of Things (IoT) monitoring system for growth optimization of Brassica chinensis. Comput Electron Agric 164:104836. https://doi.org/10.1016/j.compag.2019.05.045

Hinnell AC, Lazarovitch N, Furman A, Poulton M, Warrick AW (2010) Neuro-Drip: estimation of subsurface wetting patterns for drip irrigation using neural networks. Irrig Sci 28:535–544. https://doi.org/10.1007/s00271-010-0214-8

Idris I, Sani MI (2012) Monitoring and control of aeroponic growing system for potato production. In: IEEE conference on control, systems & industrial informatics. IEEE, pp 120–125. https://doi.org/10.1109/CCSII.2012.6470485

James A, Saji A, Nair A, Joseph D (2019) Cropsense—a smart agricultural system using iot. J Electr Design Eng 5(3). https://doi.org/10.5281/zenodo.3566563

Krishnan RS, Julie EG, Robinson YH, Raja S, Kumar R, Thong PH, Son LH (2020) Fuzzy logic based smart irrigation system using Internet of Things. J Cleaner Prod 252:119902. https://doi.org/10.1016/j.jclepro.2019.119902

Kumar A, Surendra A, Mohan H, Valliappan KM, Kirthika N (2017) Internet of things based smart irrigation using regression algorithm. In: International Conference on Intelligent Computing, Instrumentation and Control Technologies (ICICICT). https://doi.org/10.1109/icicict1.2017.8342819

Kumar M, Sahni RK, Waghaye AM, Nayak AK (2018a) Precision irrigation technologies: a present need to save for future. www.biotecharticles.com. https://biotecharticles.com/Agriculture-Article/Precision-Irrigation-Technologies-A-Present-Need-to-Save-for-Future-4382.html

Kumar M, Sahni RK, Waghaye AM, Nayak AK, Kumar D (2018b) Automated irrigation system for rice: a review. Andhra Agric J 65:324–329

Li W, Wang D, Li M, Gao Y, Wu J, Yang X (2021) Field detection of tiny pests from sticky trap images using deep learning in agricultural greenhouse. Comput Electron Agric 183:106048. https://doi.org/10.1016/j.compag.2021.106048

López O, Rach M, Migallon H, Malumbres M, Bonastre A, Serrano J (2012) Monitoring pest insect traps by means of low-power image sensor technologies. Sensors 12:15801–15819. https://doi.org/10.3390/s121115801

Mahapatra J, Sahoo R (2023) Application of Drones in farming. Agric Mag 3(3):332–335. ISSN: 2583-1755. https://theagricultureonline.com/volume-03-issue-03-december-2023

Mazilu IM, Trandafir R (2002) POMOSAT—'An orchard system for monitoring and modeling apple scab (Venturia Inaequalis)'. Technical University of Civil Engineering, Bucharest, Romania

Mehra M, Saxena S, Sankaranarayanan S, Tom RJ, Veeramanikandan M (2018) IoT based hydroponics system using Deep Neural Networks. Comput Electron Agric 155:473–486. https://doi.org/10.1016/j.compag.2018.10.015

Muangprathub J, Boonnam N, Kajornkasirat S, Lekbangpong N, Wanichsombat A, Nillaor P (2019) IoT and agriculture data analysis for smart farm. Comput Electron Agric 156:467–474. https://doi.org/10.1016/j.compag.2018.12.011

Navarro-Hellín H, Martínez-del-Rincon J, Domingo-Miguel R, Soto-Valles F, Torres-Sánchez R (2016) A decision support system for managing irrigation in agriculture. Comput Electron Agric 124:121–131. https://doi.org/10.1016/j.compag.2016.04.003

Nihalani R et al (2024) Long Short-Term Memory (LSTM) model for indian sign language recognition. J Intel Fuzzy Syst 1-19. https://doi.org/10.3233/JIFS-233250

Panchal B, Pranjal P, Patel RK, Sharma A, Chouhan SS (2024) Assessing the quantity of a crop field using aerial images. In: Chouhan SS, Singh UP, Jain S (eds) Applications of computer vision and drone technology in agriculture 4.0. Springer, Singapore. https://doi.org/10.1007/978-981-99-8684-2_11

Rathnayake AP, Sahni RK, Khot LR, Hoheisel GA, Zhu H (2022) Intelligent sprayer spray rates optimization to efficiently apply chemicals in modern apple orchards. J ASABE 65(6):1411–1420. https://doi.org/10.13031/ja.14654

Redmon J, Divvala S, Girshick R, Farhadi A (2016) You only look once: Unified, real-time object detection. In: Proceedings of the IEEE conference on computer vision and pattern recognition, pp 779–788. https://doi.org/10.1109/CVPR.2016.91

Sahni RK, Kumar V, Kumar SP, Chandel NS, Tiwari PS (2018a) Precision agriculture technologies. www.biotecharticles.com.https://biotecharticles.com/Agriculture-Article/Precision-Agriculture-Technologies-4383.html.

Sahni RK, Kumar D, Tiwari PS, Kumar V, Kumar SP, Chandel NS (2018b) DGPS based on-the-go soil nutrient mapping system: a review. Andhra Agric J 65:1–6

Sahni RK, Thorat D, Rajwade Y, Jyoti B, Ranjan J, Anand R (2024) Drone spraying system for efficient agrochemical application in precision agriculture. In: Chouhan SS, Singh UP, Jain S (eds) Applications of computer vision and drone technology in agriculture 4.0. Springer, Singapore, pp 225–244. https://doi.org/10.1007/978-981-99-8684-2_13

Sharma A, Georgi M, Tregubenko M, Tselykh A, Tselykh A (2022) Enabling smart agriculture by implementing artificial intelligence and embedded sensing. Comput Electron Agric 165:107936. https://doi.org/10.1016/j.cie.2022.107936

Sharma A, Patel RK, Pranjal P, Panchal B, Chouhan SS (2024) Computer vision-based smart monitoring and control system for crop. In: Chouhan SS, Singh UP, Jain S (eds) Applications of computer vision and drone technology in agriculture 4.0. Springer, Singapore. https://doi.org/10.1007/978-981-99-8684-2_5

Solanki S, Singh UP, Chouhan SS, Jain S (2023) Brain tumour detection and classification by using deep learning classifier. Int J Intel Syst Appl Eng 11(2):279–292. https://ijisae.org/index.php/IJISAE/article/view/2624

Solanki S, Singh UP, Chouhan SS, Jain S (2024) A systematic analysis of magnetic resonance images and deep learning methods used for diagnosis of brain tumor. Multimed Tools Appl 83:23929–23966. https://doi.org/10.1007/s11042-023-16430-6

Suksawat B, Komkum P (2015) Pineapple quality grading using image processing and fuzzy logic based on Thai Agriculture Standards. In: International conference on control, automation and robotics. IEEE, pp 218–222. https://doi.org/10.1109/ICCAR.2015.7166035

Tiwari PS, Sahni RK, Kumar SP, Kumar V, Chandel NS (2019) Precision agriculture applications in horticulture. Pantnagar J Res 17(1):1–10. https://www.gbpuat.res.in/uploads/archive/Vol.17(1).1.pdf

Wang J, Damevski K, Chen H (2015) Sensor data modeling and validating for wireless soil sensor network. Comput Electron Agric 112:75–82. https://doi.org/10.1016/j.compag.2014.12.016

Xu Z, Lou B, Shao G (2012) An intelligent irrigation system for Greenhouse Jonquil based on ZigBee Wireless Sensor Networks. In: Wang Y, Zhang X (eds) Internet of Things, communications in computer and information science, vol 312. Springer, Berlin, Heidelberg, pp 200–207. https://doi.org/10.1007/978-3-642-32427-7_27

Xu J, Gu B, Tian G (2022) Review of agricultural IoT technology. Artifi Intel Agric 6:10–22. https://doi.org/10.1016/j.aiia.2022.01.001

Yang T, Gong HL, Li XJ, Zhao WG, Meng D (2010) Progress of soil moisture monitoring by remote sensing. Acta Ecolog Sin 30(22):6264–6277

Yang M-T, Chen C-C, Kuo Y-L (2013) Implementation of intelligent air conditioner for fine agriculture. Energy Build 60:364–371. https://doi.org/10.1016/j.enbuild.2013.01.034

SPRINGER NATURE

GPSR Compliance

The European Union's (EU) General Product Safety Regulation (GPSR) is a set of rules that requires consumer products to be safe and our obligations to ensure this.

If you have any concerns about our products, you can contact us on ProductSafety@springernature.com

In case Publisher is established outside the EU, the EU authorized representative is:

Springer Nature Customer Service Center GmbH
Europaplatz 3
69115 Heidelberg, Germany

The manufacturer's authorised representative in the EU is Springer Nature Customer Service Centre GmbH, Europaplatz 3, 69115 Heidelberg, Germany. If you have any concerns regarding our products, please contact ProductSafety@springernature.com

Printed and bound by CPI Group (UK) Ltd, Croydon, CR0 4YY
01/10/2026
02225956-0009